PIC Microcontrollers

D0082541

PIC Microcontrollers
An Introduction to Microelectronics

Second Edition

Martin Bates

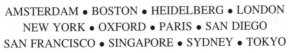

AMSTERDAM • BOSTON • HEIDELBERG • LONDON
NEW YORK • OXFORD • PARIS • SAN DIEGO
SAN FRANCISCO • SINGAPORE • SYDNEY • TOKYO

Newnes is an imprint of Elsevier

Newnes
An imprint of Elsevier
Linacre House, Jordan Hill, Oxford OX2 8DP
200 Wheeler Road, Burlington, MA 01803

First published 2000 by Arnold
Second edition 2004

British Library Cataloguing in Publication Data
A catalogue record for this book is available from the British Library

Library of Congress Cataloguing in Publication Data
A catalogue record for this book is available from the Library of Congress

ISBN   0 7506 6267 0

> For information on all Newnes publications
> visit our website at http:// books.elsevier.com

Typeset in 10/12 pt Times by Integra Software Services Pvt. Ltd, Pondicherry, India
www.integra-india.com

Printed and bound in Meppel, The Netherlands by Krips bv.

Contents

Preface to the First Edition

The Microchip™ PIC 16F84 microcontroller is an unremarkable looking 18-pin chip – so why write a whole book on it? The answer is that it contains within its ordinary looking plastic case most of the technology that students of microelectronics need to know about in order to understand microprocessor and computer systems. It also represents a significant new development in microelectronics and, importantly, it offers an easier introduction to the world of digital processing and control than conventional microprocessors. The microcontroller is a self-contained, programmable device, and the student, hobbyist or engineer can put it to use without knowing in too much detail how it works. On the other hand, we can learn a great deal about microelectronics by looking inside.

Studying the PIC chip will give the user a valuable insight into the technology behind the explosion in microprocessor-controlled applications which has occurred in recent years, which has been based on cheap, mass-produced digital circuits. Mobile phones, video cameras, digital television, satellite broadcasting and microwave cookers – there are not many current electronic products which do not contain some kind of microprocessor. Industrial control systems have seen similar developments, where complex computer control systems have steadily increased productivity, quality and reliability. The key, of course, is the increase in power of microprocessors and related technology, while the cost of these clever little chips continues to fall.

The microcontroller is essentially a computer on one chip, which can carry out a complex programmed sequence of actions, with the minimum of additional components. As an example, in this book a motor control circuit will be described which allows the motion of a small dc motor to be programmed and controlled by the PIC chip. The only additional major components required are power transistors to provide the current drive to the motor. In the past, equivalent control and interface circuits for such an application would have required many more components, and been much more complicated and expensive to design and produce. The small microcontroller also makes it easier for a device such as a motor to be individually controlled as part of a larger system.

When I first came across the PIC chip a few years ago, it was immediately obvious that this would be an ideal device for teaching and learning microprocessor software techniques, especially for students with minimal prior knowledge and skills. It is relatively cheap, and, even better, it has non-volatile program memory that is electrically reprogrammable (Flash ROM). In addition, the manufacturers, Arizona Microchip, had the foresight to make development system software required to develop programs for the chip widely available. Packages are available for DOS and Windows, and the support hardware and software are being added to all the time, by the manufacturers, independent suppliers and enthusiasts. On the other hand, a complete set of more powerful development tools is also available for the professional user.

Both DOS and Windows versions of the PIC development system have been used to prepare the sample applications in this book, and the programs downloaded using the PICSTART-16B programming unit. However, there are many designs for inexpensive programmers available in magazines and on the Internet, usually with their own software. The current Windows version of the program development package, MPLAB, can be downloaded free of charge from the Internet at 'http://www.microchip.com', along with data sheets and all the latest product development information. The data sheet for the PIC 16F84 is reprinted in full, because it is an excellent document which contains the definitive information on the chip, presented in a clear and concise manner.

The objective of this book is to ensure that any beginner, student or engineer, will quickly be able to start using this chip for their own projects and designs. When I started using it in my teaching, I put together a teaching pack and was expecting a range of suitable reference books to quickly appear. Indeed, the chip soon started to feature in numerous electronics magazine projects and was clearly popular, but all the books that I obtained seemed to assume quite a lot of prior knowledge of microprocessors. I wanted to use the PIC with students who were new to the subject, and eventually I realised that if I wanted a suitable book, I would have to do it myself! I hope that the reader finds the result useful.

Martin P. Bates
Lecturer in Microelectronics
Hastings College of Arts & Technology
July 1999

Preface to the Second Edition

The revisions required in the second edition of this book are mainly due to the rapid development of microcontroller technology. As the PIC family of devices has grown, more features have been incorporated at lower cost. So, while the focus of the first edition was the popular 16F84 chip, and this remains a valuable reference point for the beginner, the scope has been expanded so that a broader understanding of the range microcontroller types and applications can be gained.

One of the reasons the 16F84 was originally selected was its flash memory, which allows easy reprogramming, making it a good choice for education and training. Flash memory is now available in a wider range of devices, making the choice of chip less obvious. On the one hand we now have more small 8-pin chips which can be used in simple systems requiring fewer inputs and outputs, as well as a proliferation of more powerful devices incorporating a variety of serial data interfaces, as well as analogue inputs and many other advanced features.

For this reason the focus has been shifted away from the 16F84. A wider selection of devices and I/O methods is now discussed, and a more general treatment attempted. Application development software has also moved on, and new methods of programming and debugging introduced. I hope I have been able to reflect these developments adequately without introducing too many complications for the beginner, to whom this text is still firmly addressed.

Part A is a general introduction to microelectronics system technology, and can be skipped if appropriate. In Part B, the PIC microcontroller is described in detail from first principles. Part C contains practical advice on implementing PIC projects, with examples. Part D contains new material on the more advanced features of other PIC MCUs (Microcontroller Units) as well as a review of a range of other control system technologies.

I have tried to incorporate a systematic approach to project development, making the design process as explicit as possible. The book will thus support the delivery of the microelectronic systems and project modules of, for example, UK BTEC electronics programmes which incorporate an Integrated Vocational Assignment, which requires the student to develop a specific project and document the process in detail. The PIC is a good choice for producing interesting, but achievable, projects which incorporate a good balance between hardware and software design, and allow the design process to be clearly documented through every stage.

Acknowledgement is due to Microchip Technology Inc. for their kind permission to reproduce the PIC 16F84A data sheet, to Microsoft Corporation and Labcenter Electronics for the application software used to produce documents, drawings, circuit schematics and layouts for this book, and to all for use of their trademarks.

Finally, thanks to the following for their help, advice and tolerance: Melvyn Ball (Hastings College), Jason Guest (General Dynamics, Hastings), Chris Garrett (University of Brighton) and, of course, Julie at home; also, to all colleagues who commented on the first edition, and students who bought it!

Martin Bates
December 2003
mbates@hastings.ac.uk

Introduction

Let's admit one thing straight away – microprocessor systems are quite complicated! However, they are now found in so many different products that all students of engineering need to know something about how they work.

In this book we are going to look specifically at the PIC family of microcontrollers. Microcontrollers have all the essential features of a full-size computer, but all on a single chip. By contrast, conventional microprocessor systems, such as the PC (personal computer), are built with a separate processor, memory, input and output chips. The extra hardware and software required to make these chips work together makes the system more difficult to understand than our single chip microcontroller unit (MCU).

As well as being easier to understand, microcontrollers are important because they make electronic circuits cheaper and easier to build. 'Hard-wired' circuits can be replaced with a microcontroller and its software, reducing the number of components required. Importantly, the software element (control program) can be reproduced at minimal cost, once it has been created. So the development costs may be higher, but the production costs will be lower in the long run. It is also easier to change software if the product is to be modified. In general, software is increasingly replacing hardware in electronic designs. For example, to design a system like a video recorder without microprocessors or microcontrollers would be very complicated and expensive, if not impossible.

Using the PIC, we will find that we can quite quickly work out some simple, but useful, applications. These will illustrate the universal principles of microprocessor systems that apply to more complex computer and control systems. At first, however, we do not have to worry too much about exactly how the chip works – we will go back to that later. The big problem with microprocessors and microcontrollers is that in order to fully understand how the system works, we have to understand both the hardware and the software at the same time. Therefore we have to circle round the subject, looking at the system from different angles, until a reasonable level of understanding is built up.

We will approach microcontroller and microprocessor systems (microsystems) step by step, assuming very little prior knowledge. The operation of the PC will be outlined first, because most students will be familiar with how it works from the user's point of view. We will look at how the hardware and software interact, and the function of the Pentium microprocessor in controlling the input (keyboard, mouse), output (screen) and memory and disks.

Some basic microelectronic system principles will then be covered. One objective is to understand the hardware diagrams in the PIC data sheets, so that external circuits connected to the PIC input/output pins can be designed correctly. Also, it is necessary to understand the internal hardware configuration of a microcontroller to fully understand the programming of the chip. The clarity and completeness of these data sheets is an important reason for choosing the PIC as our typical microcontroller. We can then start to look specifically at the PIC microcontroller and develop simple applications which will illustrate the essential hardware features and basic programming ideas. More details will then be added using further application examples.

In the final section, the complete application design process will be described, including use of the PIC development system and hardware design methods. The range of PIC microcontrollers and the more advanced features of some of them will then be described, plus some other types of control technologies which can carry out similar functions to microcontrollers, such as programmable logic controllers.

All reference material can be downloaded from www.microchip.com and other manufacturers' websites.

Part A
Microelectronic Systems

Chapter 1
Computer Systems

We will begin our study of microsystems with something familiar, by looking at how the PC (personal computer) works when running a wordprocessor. Most readers will be familiar with using a wordprocessor and will know more or less how it functions from the user's point of view. Some basic microsystem concepts will be introduced by analysing how the software operates with the computer hardware, to allow the user to enter, store and process documents. For example, we will see why different kinds of memory are needed to support the system operation.

It is also useful to get some idea of how a PC works because it is used as the hardware platform for the PIC program development system. The programs for the PIC are written using a text editor, and the machine code program created and downloaded to the PIC chip using the PC. The PIC development system hardware can be seen connected to the PC in Fig. 1.1(a). A simplified diagram, Fig. 1.1(b), allows us to see the main parts of the system more clearly.

We will then have a quick look at a microcontroller system, set up to operate as a simple equivalent of the microprocessor-based PC system, so we can see how it compares. The microcontroller has a keypad with only 12 keys instead of a keyboard, and a seven-segment display instead of a screen. Its memory is much smaller than the PC, yet it can carry out the same basic tasks. In fact, it is far more versatile; the Pentium™processor used in the PC is designed specifically for that system. The microcontroller can be used in a great variety of circuits. Also, it is much cheaper!

1.1   The PC System

The PC hardware is based on the Intel™ series of microprocessors with Microsoft Windows™ operating system software. The standard PC hardware comprises a main unit, separate keyboard and mouse, VDU (visual display unit) and possibly a printer and connection to a network. The circuit board (motherboard) in the main unit carries a group of chips which work together

(a)

(b)

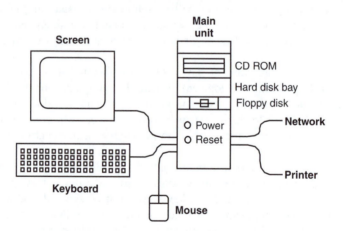

Figure 1.1 (a) The PC system (with PIC development system); (b) Diagram of PC system.

to provide digital processing of information and control of input and output devices. A power supply for the motherboard and the peripheral devices is included in the main unit.

The processor must have access to software (programs) to allow useful work to be done by the hardware. These are usually stored on a hard disk inside the main unit; this can hold large amounts of data which is retained when the power is off. There are two main types of software required – the operating system (Windows™) and the application (Word™). As well as the operating system and application software, the hard disk stores the data created by the user (document files). Documents can also be stored on floppy disk for backup or portability.

The keyboard is used for data input, and the VDU displays the resulting document. The mouse provides an additional input device, allowing operations to be selected from menus or by clicking on icons and buttons. This is called the graphical user interface (GUI). There may be a network card fitted in the PC to exchange information with other users, download data or applications, or share resources such as printers over a local area network (LAN). In addition, a modem can give direct access to a wide area network (WAN), usually the Internet. A CD ROM drive allows large volumes of reference information stored on optical disk to be accessed, and is also used to load application software.

If we remove the cover from the main unit, the main components can be identified fairly easily. In the photograph, Figure 1.2(a), the power supply is top left, with the hard disk drive below and the motherboard vertical at the back of the tower case. The disk and video interface cards are visible at the bottom, slotted via edge connectors into the motherboard, with a modem in the middle in the dark casing. The connections to the video board and modem are available at the rear (left) of the case, with the floppy disk at the front (top right). In current PC designs, some of these interfaces are built into the motherboard, so the whole package is more compact.

Block diagrams are useful for showing the main parts of a complex system, and how they connect together, in a simplified form. Figure 1.2(b) shows the components of the PC system and the direction of the information flow between them. In the case of the disk drives and network it is bidirectional (flowing in both directions), representing the process of saving data to, and retrieving data from, the hard disk or floppy disk.

1.1.1 PC Hardware

Inside the PC main unit, the motherboard has slots for expansion boards and memory modules to be added to the system. The power supply and disk drives are fitted separately into the main unit frame. The keyboard and mouse interfaces are usually on the motherboard. In older designs, the expansion boards carried interface circuits for the disk drives and external peripherals such as the display and printer, but these functions now increasingly incorporated into the motherboard itself. Note that the functional block diagram does not show any difference between internally and externally fitted peripherals, because it is not relevant to the overall system operation.

The PC is a modular system, which allows the hardware to be put together to meet the individual user's requirements, and allows subsystems, such disk drives and keyboard to be easily replaced if faulty. The modular design also allows upgrading (for instance, fitting extra memory chips) and also makes the PC architecture well suited to industrial applications. In this case, the PC can be 'ruggedised' (put into a more robust casing) for use on the factory floor. This modular architecture is one of the reasons for the success of the PC as a universal hardware platform.

1.1.2 PC Motherboard

The main features of a typical motherboard are shown in Fig. 1.3. The heart of the system is the microprocessor, a single chip, which is also called the central processing unit (CPU). This name refers back to the days when the CPU was built from discrete components and could be the size of a washing machine! In Fig. 1.3(a), the CPU is under the cooling fan at the lower right. The CPU controls all the other system components, but must have access to a suitable program in memory before it can do anything useful. The blocks of program required are provided by the operating system software and the application software which are downloaded to memory from the hard disk on startup.

(a)

(b)

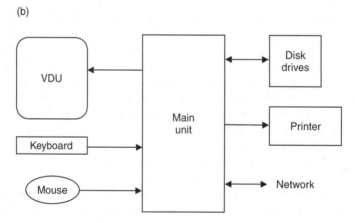

Figure 1.2 The PC system main unit. (a) View of PC main unit; (b) Block diagram of PC system.

The Intel CPU has undergone continuous development since the introduction of the PC in 1981, with the Pentium processor being the current standard. Intel processors are classified as CISC (complex instruction set computer) devices, which means they have a relatively large number of instructions which can be used in a number of different ways. This makes them powerful, but relatively slow compared with more streamlined processors which have fewer instructions. These are classified as RISC chips (reduced instruction set computer), of which the PIC microcontroller is an example.

As stated above, CPU cannot work on its own; it needs some memory and input/output devices for getting data in, storing it and sending it out again. The main memory block is

(a)

(b)

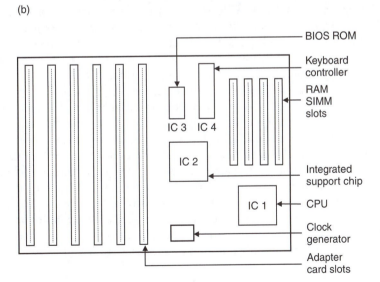

Figure 1.3 PC motherboard. (a) PC motherboard in the main unit; (b) Layout of PC motherboard.

made up of RAM (read and write memory) chips, which are mounted in SIMMs (single in-line memory modules). Higher capacity DIMMs (dual in-line memory modules) are used currently. These can be seen at the top of the photograph in Fig. 1.3(a). Additional peripheral interfacing boards are fitted in the expansion card slots to connect the main board to the disk drives, VDU, printer and network. Spare slots allow additional peripheral interfaces and more memory to be added if required. Each peripheral interface is a sub-circuit which is built around a specific input/output chip (or set of chips) which handles the data transfer.

The integrated support device (ISD) is a chip which provides various system control and memory management functions in one chip, and is designed for that particular motherboard. The motherboard itself can be represented as a block diagram (Fig. 1.4) to show how the components are interconnected.

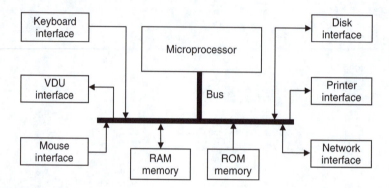

Figure 1.4 Block diagram of PC motherboard.

The block diagram shows that the CPU is connected to the peripheral interfaces by a set of bus lines. These are groups of connections on the motherboard which work together to transfer the data from the inputs, such as keyboard, to the processor, and from the processor to memory. When the data has been processed and stored, it can be sent to an output peripheral, such as the screen. We will look at how this is achieved in more detail later.

Busses connect all the main chips in the system together, but, because they operate as shared connections, can only pass data to or from one peripheral interface or memory location at a time. This arrangement is used because separate connections to all the main chips would need an impossible number of tracks on the motherboard. The disadvantage of bus connection is that it slows down the program execution speed, because all data transfers use the same set of lines, and only one data word can be present on the bus at any one time. To help compensate for this, the bus connections are typically 16, 32 or more bits wide, that is, there are 16 or 32 connections working together, each carrying one bit of a data word simultaneously. This parallel data connection is faster than a serial connection, such as the keyboard input or network connection, which can only carry one bit at a time. In the microcontroller, these system bus connections are hidden inside the chip, making circuit design easier.

1.1.3   PC Memory

There are two types of memory in the PC system. The main memory block is RAM, where input data is stored before and after processing in the CPU. The operating system and application program are also copied to RAM from disk for execution, because access to data in RAM is faster. Unfortunately, RAM storage is 'volatile', which means that the data and application software disappear when the PC is switched off, and these have to be reloaded each time the computer is switched back on.

This means that some ROM (read only memory), which is non-volatile, is needed to get the system started at switch on. The BIOS (basic input/output system) ROM chip, seen at the left of Fig. 1.3(a), contains enough code to check the system hardware and load the main operating system (OS) software from disk. It also contains some basic hardware control routines so that the keyboard and screen can be used before the main system has been loaded.

The hard disk is a non-volatile, read/write storage device, consisting of a set of metal disks with a magnetic recording surface, read/write heads, motors and control hardware. It provides a large volume of data storage for the operating system, application and user files. A number of applications can be stored on disk and then selected as required for loading into memory; because the disk is read and write device, user files can be stored, applications added and software updates easily installed.

1.2 Wordprocessor Operation

In order to understand the operation of the PC microprocessor system, we will look at how the wordprocessor application uses the hardware and software resources.

1.2.1 *Starting the Computer*

When the PC is switched on, the BIOS ROM program starts automatically. It checks that the system hardware is working properly and displays messages to report the results. If there is a problem, the BIOS program attempts to diagnose the fault, and will display an error message. If all is well, it loads (copies) the main operating system software (Windows) from hard disk into RAM. As you will probably have noticed, this all takes some time; this is an indication of the amount of data transfer required, and the relatively slow access to the hard drive.

1.2.2 *Starting the Application*

Windows displays an initial screen with icons and menus which allows the application to be selected using the mouse and on-screen pointer. Word is started by clicking on its icon; Windows converts this action to a command which runs the executable file (WINWORD.EXE) stored on disk. In older machines the operating system, MSDOS (Microsoft disk operating system), required this command to be typed in to start the application.

The application program is transferred from disk to RAM, or as much of it as will fit in the available memory. If necessary, application program blocks can be swapped into memory when needed. The wordprocessor screen is displayed and a new document file can be created or an existing one loaded by the user from disk for updating.

1.2.3 *Data Input*

The main data input is obviously from the keyboard, which consists of a grid of switches which are scanned by a dedicated microcontroller within the keyboard unit. This chip detects when a key has been pressed, and sends a corresponding code to the CPU via a serial data line in the keyboard cable. The serial data is a sequence of high and low voltages on a single wire, which represent a binary code, each key generating a different code. The keyboard interface converts this serial code to parallel form for transfer to the CPU via the system data bus. It also signals separately to the CPU that a keycode is ready to be read into the CPU, by generating an 'interrupt' signal. This serial-to-parallel (or parallel-to-serial) data conversion process is required in all the interfaces that use serial data transfer, namely, the keyboard, VDU, network and modem. Binary coding, interrupts and other such processes will be explained in more detail later.

In Windows, and other GUIs, the mouse can be used to select commands for managing the application and its data. It controls a pointer on the screen; when the mouse is moved, the ball turns two rollers, which have perforated wheels attached. The holes are detected using an opto-detector, which sends pulses representing movement in two directions. These pulse sequences are passed to the CPU via the mouse interface and used to modify the position of the pointer on the screen. The buttons, used to select an action, must also be input to the CPU.

1.2.4   Data Storage

Each character of the text being typed into the wordprocessor is stored as an 8-bit (one byte) binary code, which occupies one location in RAM. Each bit of data must be stored as a charge on small capacitor in the RAM chip. The parallel data is received by the CPU, then sent back via the same data bus lines from the CPU to the RAM. The RAM stores the data bytes at numbered locations; these address numbers are identified by the CPU using the system address bus. The data is transferred on the data bus to the address in RAM selected by the CPU via the ISD, which provides the additional logic required to handle the data transfers.

1.2.5   Data Processing

In the past, programs running on the DOS operating system required less processing power, partly because the screen was simpler, being divided up into one space for each character. The video interface would convert the stored character code into the pattern for the character, and output it to the correct position on the screen.

The Windows screen is more complicated, because the text is displayed in graphics (drawing) mode, at a higher resolution, so that the text size, style and layout appears on screen as it will be printed. Graphics, tables and special characters can be embedded in the text. This means the CPU has far more work to do in displaying the page, and this is one reason why Windows needs more memory and a more powerful CPU than former DOS-based wordprocessors. The processor must also manage the WIMP (Windows, Icons, Mouse, Pointer) interface, which allows actions to be selected on screen. Word now has many more features than earlier wordprocessors, and there is now little difference between a typical wordprocessor and so called desk-top publishing (DTP) programs, which provide comprehensive page layout control.

1.2.6   Data Output

The characters must be displayed on the screen as they are typed in, so the character codes stored in memory are also sent to the VDU via the system data bus and video interface. The display is made up of single coloured dots (pixels) organised in lines across the screen, which are output in sequence from a video amplifier. This is known as a scanned display. The shape of the character on screen must be generated from its code in memory, and sent out on the correct set of lines at the right time on the video signal. The display is therefore formed as a two-dimensional image made up from a serial data stream which sets the colour of each pixel on the screen in turn, line by line.

If a file is transferred on a network, it must also be sent in serial form. The characters (letters) in a text file would typically be sent as ASCII code, along with formatting information and network control codes. ASCII code represents one character as one byte (8 bits) of binary code, and is therefore a very compact form of the data. The code for 'A' for example is 01000001.

The printer works in a similar way to the screen, except that the output is generated as lines of dots of ink on a page. If you watch an inkjet printer working, you can see the scanning operation take place. In older printers, the data is sent in 8-bit parallel form, along with control codes, via the printer port. If the printer itself is capable of formatting the final output, only the character code and any formatting codes are needed. For cheaper printers, the computer itself must generate the page layout, and send a 'bit-map' of the page, where one bit (or group of bits) is the code for one coloured dot on the page; this will take longer.

The operation of the wordprocessor can be illustrated using a flowchart, which is a graphical method of describing a sequential process. Figure 1.5 describes only the basic process of text input and word wrapping at the end of each line. Flowcharts will be used later to represent microcontroller program operation.

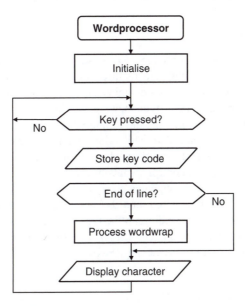

Figure 1.5 Wordprocessor flowchart.

1.3 PC Microprocessor System

As we have seen, the PC working as a wordprocessor carries out the following functions:

- Data input
- Data storage
- Data processing
- Data output

All microprocessor systems perform these same basic functions. To carry them out, the microprocessor system needs a set of supporting chips with suitable interconnections. The system will therefore typically consist of:

- CPU
- RAM
- ROM
- I/O (Input/Output) ports
- ISD
- XTAL (crystal) clock generator

These devices must be interconnected by:

- address bus
- data bus
- various control lines

These busses and control lines originate from the CPU, which is in overall charge of the system.

1.3.1 System Operation

The PC motherboard components are connected as shown in Fig. 1.6. The address and data busses, control lines and support chip are required to handle the data transfer between the CPU, memory and ports. The clock circuit contains a crystal oscillator as found in watches and clocks, which produces a precise fixed frequency signal which drives the microprocessor. The CPU operations are triggered on the rising and falling edges of the clock signal, allowing their relative timing to be precisely controlled. This allows events in the CPU to be completed in the correct sequence, with sufficient time allowed for each step.

The CPU generates all the main control signals based on this timing reference. This is why the CPU should not be operated at a frequency above its rated clock speed – correct completion of each step can no longer be guaranteed, and the system could crash. A given CPU can be used in different system designs, depending on the type of application, the amount of memory needed, the I/O requirements and so on. The ISD is designed to assist the processor to handle memory and I/O operations within a particular design.

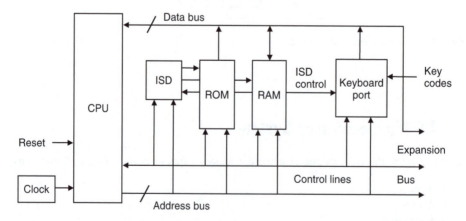

Figure 1.6 Block diagram of PC microprocessor system.

For simplicity, only the keyboard port is shown in the block diagram, as this was sometimes (in older designs) the only I/O device on the main board. However, other ports, such as the printer, modem and so on are connected in the same way, whether they are part of the motherboard or fitted as expansion cards. The signal connections to the plug-in peripheral interfaces will be made to the system busses and the relevant control lines via the expansion bus, which appears on the motherboard as edge connectors. This allows the system to be upgraded by replacing or adding to these cards. In current designs, where upgrading is less likely to be required, the VDU, disk and network tend to be integrated onto the main board. Additional RAM memory may be fitted in a similar way if spare slots are available.

1.3.2 Program Execution

The ROM and RAM memory contain program information and data in numbered locations. The ISD contains address decoding logic which allocates a particular memory chip to a range of addresses. The I/O port registers, which are set up to handle the data transfer in and out of the system, are also allocated particular addresses by the system designer, and accessed by the CPU in the same way as memory locations.

A register is a temporary store for a data word within a port chip or the CPU. In the port chip it can hold data, or a control code which sets up how the port will operate. For example, the bits in the data direction register control whether each port pin operates as an input or an output. The data being sent in or out is then stored temporarily in the port data register. More of this later!

The wordprocessor program consists of a list of instructions in binary code stored in memory, with each instruction and any associated data (operands) being stored in sequential locations. The program instruction codes are fetched into the CPU and decoded. The CPU sets up the internal and external control lines as necessary and carries out the operation specified in the program, such as read a character code from the keyboard port into the CPU. The instructions are executed in order of their addresses, unless the instruction itself causes a jump to another point in the program, or an interrupt is received.

1.3.3 Execution Cycle

Program execution is illustrated in Fig. 1.7. Assuming that the application program code is in RAM, the program execution cycle proceeds as follows:

1. The CPU outputs (1) the address of the location (memory slot) containing the required instruction. This address is kept in the program counter. The sample address is shown in decimal (3724) in Fig. 1.7, but it is output in binary form on the address lines from the processor. The ISD uses the address to select the RAM chip which has been allocated to this address. The address bus also connects directly to the RAM chip to select the individual location.

2. The instruction code is returned to the CPU from the RAM chip via the data bus (2). The CPU reads the instruction from the data bus into an instruction register. The CPU then decodes and executes the instruction (3). The operands (data to be processed) are fetched (4) from the following locations in RAM via the data bus, in the same way as the instruction.

3. The instruction execution continues by feeding the operand(s) to the data processing logic (5). Additional data can be fetched from memory (6) (this would be the text data in our

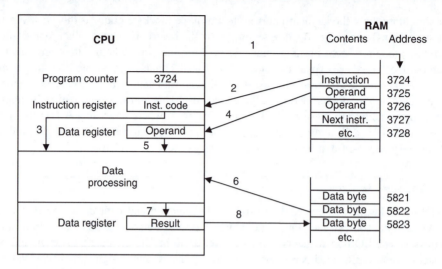

Figure 1.7 Program execution sequence.

wordprocessor). The result of the operation is stored in a data register (7), and then, if necessary, in memory (8) for later use. In the meantime, the program counter has been incremented (increased) to the address of the next instruction code. The address of the next instruction is then output and the sequence repeats from step 2.

The operating system, the wordprocessor program and the text data are stored in different parts of RAM during program execution, and the wordproccessing application program calls up operating system routines as required to read in, process and store the text data. Current CISC processors such as the Pentium series have instructions which are more than 8 bits in size which are stored in multiple locations, and use complex memory management techniques, to speed up program execution. These long instructions and data words are normally multiples of 8 bits, as this is how the memory is organised.

1.4   PC Engineering Applications

The PC can be used as a standard hardware platform in a variety of engineering systems by fitting special interfacing hardware in the expansion slots and programming the PC to control the target system through this I/O hardware (Fig. 1.8). This type of arrangement is increasingly used in manufacturing systems where the PC might control a machine tool, robot or assembly system, or be used to run an instrumentation or data logging application. The PC provides a

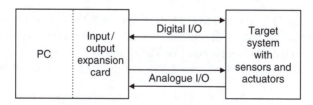

Figure 1.8 PC engineering application.

standard network interface so that commands or design data can be sent to the PC and status information and other measurement data can be returned to a supervisory computer.

The PC has the advantage of using a standard operating system and programming languages which allow control programs to be written in high level languages such as 'C' or Visual Basic. Graphical programming tools are also available for designing control and instrumentation applications more quickly and easily. An example of this type of system is given in Chapter 16.

1.5   The Microcontroller

We have now looked at some of the main ideas to be used later in explaining microcontroller operation: hardware, software, how they interact and how the function of complex systems can be represented in a simplified form such as block diagrams and flowcharts. We can now compare the PC system with an equivalent microcontroller system.

The microcontroller can provide, in a simplified form, all the main elements of the conventional microprocessor system on a single chip. As a result, less complex applications can be designed and built quickly and cheaply. A working system can consist of a microcontroller chip and just a few external components for feeding data and control signals in and out.

1.5.1   A Microcontroller Application

A simple equivalent of the word processing application described above could be built as shown in Fig. 1.9, around an MCU (microcontroller unit).

The basic function of the system shown is to store and display numbers which are input on the keypad. The microcontroller chip can be programmed to scan the keypad and identify any key which has been pressed. The keys are connected in a 3×4 grid of rows and columns, so that a row and a column are connected together when the key is pressed. The microcontroller can identify the key by selecting a row and checking the columns for a connection. Thus, four input lines and three outputs are required for connection to the microcontroller. In order to simplify the drawing, these parallel connections are represented by the block arrows.

Seven-segment displays show the input numbers as they are stored in the microcontroller. Each display digit consists of seven light emitting diodes (LEDs) which show as a line segment of the number when lit. Each number from 0 to 9 is displayed as a suitable pattern of lit segments.

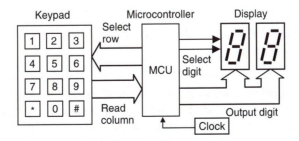

Figure 1.9   Microcontroller keypad display system.

The basic display program could work as follows: when a key is pressed, the digit is displayed on the right (least significant) digit, and subsequent keystrokes will cause the previously entered digit to shift to the left, to allow decimal numbers up to 99 to be stored and displayed. Calculations could then be performed on the data, and the result displayed.

The starting point for writing the program for the microcontroller is to convert the general description given above into a description of the operations which can be programmed into the chip using the set of instructions which are available for that microcontroller. The instruction set is defined by the manufacturer of the device. The process whereby the required function is implemented is called the program algorithm, which can be graphically represented by a flowchart (Fig. 1.10).

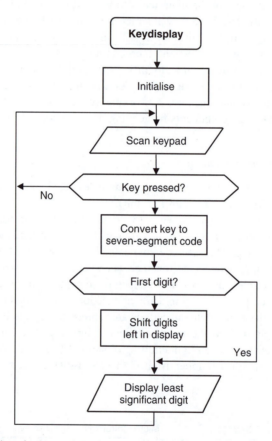

Figure 1.10   Flowchart for keypad display program.

With suitable development of the software and/or hardware, the system could be modified to work as a calculator, message display, electronic lock or similar application. Additional digits could be added to the display as required. Keyboard scanning and display driving are standard operations for microcontrollers, and the techniques required to create the working application will be explained in later chapters.

1.5.2 Programming a Microcontroller

Some microcontrollers have ROM program memory, which is programmed before the chip is fitted into the application circuit, and cannot be changed. One-time programmable (OTP) devices are generally used for longer production runs where the program is known to be correct. We will be using PIC chips which have flash program memory, which can be erased and re-programmed many times, which is invaluable when learning. A PIC device is programmed by placing it in a special programming unit which is attached to a host computer (Fig. 1.11). Note the zero insertion force (ZIF) socket which will accept different-sized chips for programming.

The program is written and converted to machine code in the host computer using suitable development system software and downloaded via a serial data link to the chip in the programmer unit. The microcontroller is then taken out of the programmer, and placed in the application circuit. The circuit can then be checked for correct operation.

Having introduced some basic ideas concerning microprocessors and microcontrollers, in the next chapter we will review some principles of digital circuits and microprocessor systems. The process of creating microcontroller applications such as the example outlined above can then be tackled.

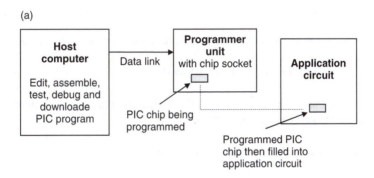

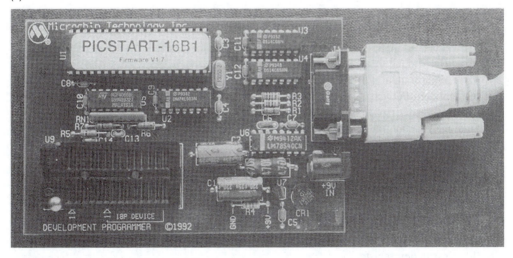

Figure 1.11 Programming a PIC microcontroller. (a) PIC program downloading; (b) PIC programming unit.

Summary

- The PC consists of data input, storage, processing and output devices.

- The main unit is a modular system, consisting of the motherboard, power supply, disk drives and expansion cards containing interfacing circuits plugged into the motherboard.

- The motherboard carries the microprocessor (CPU) chip, RAM memory modules, a BIOS ROM, ISD and keyboard interface.

- The CPU communicates with the main system chips via a shared set of address and data bus lines. The address lines select the device and location for the data to be transferred on the data bus.

- The microcontroller provides, in simplified from, most of the features of a conventional microprocessor system on one chip.

Questions

1. Name at least two PC input devices, two output devices and two storage devices.

2. Why is the BIOS ROM needed when starting the PC?

3. Why are shared bus connections used in the typical microprocessor system, even though it slows down the program execution?

4. State two advantages of the modular PC hardware design.

5. Why does the PC take so long to start up?

6. Sort these data paths into serial and parallel:
   (a) internal data bus
   (b) keyboard input
   (c) VDU output
   (d) printer output
   (e) modem I/O

7. State the function, in ten words or less, of the:
   (a) CPU
   (b) ROM
   (c) RAM
   (d) ISD
   (e) address bus
   (f) data bus
   (g) program counter
   (h) instruction register

8. Explain the difference between a typical microprocessor and microcontroller.

1. Study the messages which appear on the screen when PC is switched on, and explain their significance with reference to the system operation.

2. Under supervision if necessary, and with reference to relevant manuals, carry out the following investigation: Disconnect the power supply and remove the cover of the main unit of a PC and identify the main hardware subsystems – power supply, motherboard and disk units. On the motherboard, identify the CPU, RAM modules, expansion slots, keyboard interface, VDU interface, disk interface, printer interface. Is there an internal modem or network card? Are there any other interfaces fitted?

3. Run the wordprocessor and study the process of word-wrapping which occurs at the end of each line. Describe the algorithm that determines the word placement, and the significance of the space character in this process. Draw a flowchart to represent this process.

Chapter 2
Information Coding

This chapter introduces some methods for representing information within microprocessor systems. Binary and hexadecimal number systems will be outlined, so that data storage and program coding methods can be explained.

Much of modern technology is based on the use of mathematical models to represent information and processes in the real world. These mathematical models are used in engineering to help design new systems and products. For instance, the three-dimensional drawing of a suspension arm for a car created on a CAD (computer aided design) system screen is generated from a digital representation of the shape of the part in the memory of the computer. The advantages of the computer model are fairly obvious – it can be stored on disk, transferred electronically and modified much more easily than the equivalent information on paper. The component design can also be mathematically analysed in the computer prior to construction. For example, the stresses and strains to which the component will be subject at its final position in the suspension assembly can be studied. Further, when a component design is finished, the design data can be converted directly to a program for a machine tool which will automatically manufacture the part.

The programs for our microcontroller applications will be stored in the same way; we therefore need to know something about how such data is represented in the computer.

2.1   Number Systems

Mathematics is based on number systems, which use a set of characters to represent numerical values. The characters used are simply symbols, just squiggles on a page, but the number systems they are part of have been developed over thousands of years – because they are so useful.

In microprocessor, microcontroller and digital electronic systems, numerical processing is carried out using binary codes, a number system which has only come into common use with the development of digital computers. We therefore have to understand binary numbering in order to use a microcontroller. Another number system, hexadecimal, is also useful here because it provides a more compact way of representing binary code.

2.1.1 Decimal: Base 10

The name of each number system refers to the 'base' of the number system, which corresponds to the number of symbols used in representing values. In decimal, ten symbols are used, with which, hopefully, you are familiar:

0   1   2   3   4   5   6   7   8   9

Why use a particular base number? The reason for using ten is simple – we humans have ten fingers which can be used for counting, so the decimal system was developed as a way of writing this down and doing calculations on paper (or stone!) instead of on our fingers. We use the term 'digit' to refer to fingers and numbers and 'digital' to describe binary electrical circuits. The use of written numbers was essential for the original development of industry and commerce.

Assuming that we know how to count and write down numbers in decimal, let's analyse what a typical number means. Take the number 274; in words, it is two hundred and seventy four. This means: take two hundreds, seven tens and four units and add them together. The position of each digit in the number is literally significant; each column has a weighting which applies to the digit in that column. As you know, the least significant digit is conventionally placed at the right, and the most significant at the left. More digits are added at the left hand end as the number size increases. In decimal, the columns have a weight 1, 10, 100, etc. Note that these correspond to a power series of 10, the number system base. Another example is detailed in Table 2.1.

A number system can be used with any base you like, but some are more useful than others. For instance, relics of the base 12 system are still in use – think of clocks, boxes of eggs and measurement of angles. Base 12 is useful because 12 is divisible by 2, 3, 4 and 6, giving lots of useful fractions – a half, a third, a quarter and one-sixth. However, the decimal system is our standard system, so the analysis of other systems will still be based on decimal for comparison of number values.

2.1.2 Binary: Base 2

Binary is used in digital computer systems because it represents the way that values are stored and processed. The binary digits, 0 and 1, represent two voltage levels used in digital circuits, typically 5 V and 0 V. We can understand the binary system by comparing it with decimal – the basic rules are the same for any number system.

In binary, the base is 2, so the column weighting is a power series of 2, as shown in Table 2.2 (note that any number to the power zero has the value 1). With a base of 2, only the digits

Table 2.1 Structure of a decimal number

Column weight	1000	100	10	1
Power of base	10^3	10^2	10^1	10^0
Digits	**3**	**6**	**5**	**2**
Total value $= (3 \times 1000) + (6 \times 100) + (5 \times 10) + (2 \times 1)$				

Table 2.2   Structure of a binary number

Most significant bit (MSB) ⌐→   Least significant bit (LSB) ⌐→

Column weight	2^7	2^6	2^5	2^4	2^3	2^2	2^1	2^0
Decimal weight	128	64	32	16	8	4	2	1
Example number	1	0	1	0	0	0	1	1
Decimal equivalent	$128 + 0 + 32 + 0 + 0 + 0 + 2 + 1 = 163$							

0 and 1 are available, so the numbers tend to have lots of digits. For instance, a 32-bit computer uses 32-digit binary numbers. An example with 8 digits is given showing what the digits represent and how to convert the value back to decimal.

The decimal equivalent in all number systems can be calculated by multiplying the digit value by its weighting in decimal, and then adding the resulting column products. In binary, because the digit value is 1 or 0, the result can be obtained by simply adding the digit weight where the digit value is a '1', because any number multiplied by zero is zero. When decimal data is entered into a computer, the values are converted to binary. The program instructions which process input and output data are also stored as binary codes.

2.1.3   Hexadecimal: Base 16

Binary numbers have lots of digits, so they are not very easy to understand when written down or printed out. Conversion to decimal is not particularly straightforward, so hexadecimal is used as a way to represent binary numbers in a compact way, while allowing easy conversion back to the original binary.

Hexadecimal (base 16), or 'hex' for short, uses the same digits as the decimal system from 0 to 9, then uses letters A to F, as a single character representation for numbers 10–15. Thus, characters which are normally used to make words, are here used as numbers, because the symbols are already available. A binary number can then be easily converted to hex by writing it down in groups of 4 bits, and then converting each group to its equivalent hex digit, as in Table 2.3.

The base of the number can be shown as a subscript where necessary to avoid confusion. All number systems use the same set of characters, so if the base of the number given is not obvious from the context, it can be specified. For example, the number 100 (one, zero, zero) could have the decimal value 4 in binary, 100 (one hundred) in decimal or 256 in hexadecimal. A letter following a number can also indicate its base, such as A9h for hexadecimal. Later, we will see other ways of indicating numerical type when programming.

Some examples of equivalent values are given in Table 2.4. The numbers are printed in 'Courier' type, as used on old-fashioned typewriters, because each character occupies the same space, so all the digits line up neatly in columns.

2.1.4   Counting

A list of equivalent numbers, counting from zero, is given in Table 2.5, with some comments on important values. This table also defines memory capacity in microprocessor systems; for example, '1k' of memory is 1024 locations. Notice that $1024 = 2^{10}$. This is worth remembering as a starting point in calculating memory capacity.

Table 2.3 Hexadecimal digits

Decimal	Binary	Hexadecimal
0	0000	0
1	0001	1
2	0010	2
3	0011	3
4	0100	4
5	0101	5
6	0110	6
7	0111	7
8	1000	8
9	1001	9
10	1010	A
11	1011	B
12	1100	C
13	1101	D
14	1110	E
15	1111	F

Table 2.4 Examples of equivalent values

Decimal	Binary	Hexadecimal
16_{10}	$1\ 0000_2$	10_{16}
31_{10}	$1\ 1111_2$	$1F_{16}$
100_{10}	$110\ 0100_2$	64_{16}
169_{10}	$1010\ 1001_2$	$A9_{16}$
255_{10}	$1111\ 1111_2$	FF_{16}
1024_{10}	$100\ 0000\ 0000_2$	400_{16}

Table 2.5 Significant equivalent numbers

Decimal (Base 10)	Binary (Base 2)	Hex (Base 16)	Comment
0	0	0	All the same
1	1	1	All the same
2	10	2	[2^1] Use 2nd column in binary
3	11	3	Maximum 2-bit count
4	100	4	[2^2] Use 3rd column in binary
5	101	5	
6	110	6	
7	111	7	Maximum 3-bit count
8	1000	8	[2^3] Use 4th column in binary
9	1001	9	Decimal and hex same until 9
10	1010	A	Use letters in hex
11	1011	B	
12	1100	C	
13	1101	D	

continued...

Table 2.5 continued

Decimal (Base 10)	Binary (Base 2)	Hex (Base 16)	Comment
14	1110	E	
15	1111	F	Maximum 4-bit count
16	1 0000	10	$[2^4]$ Use 2nd column in hex
17	1 0001	11	Use space to clarify binary
18	1 0010	12	
19	1 0011	13	
20	1 0100	14	
21	1 0101	15	
22	1 0110	16	
23	1 0111	17	
24	1 1000	18	
25	1 1001	19	
26	1 1010	1A	
27	1 1011	1B	
28	1 1100	1C	
29	1 1101	1D	
30	1 1110	1E	
31	1 1111	1F	Maximum 5-bit count
32	10 0000	20	$[2^5]$
33	10 0001	21	
34	10 0010	22	
. .	. . . . . .	. .	
62	11 1110	38	
63	11 1111	39	Maximum 6-bit count
64	100 0000	40	$[2^6]$
65	100 0001	41	
. .	. . . .	. .	
127	111 1111	79	Maximum 7-bit count
128	1000 0000	80	$[2^7]$
129	1000 0001	81	
. .	. . . .	. .	
254	1111 1110	FE	
255	1111 1111	FF	Maximum 8-bit count
256	1 0000 0000	100	$[2^8]$
. .	. . . .	. .	
511	1 1111 1111	1FF	Maximum 9-bit count
512	10 0000 0000	200	$[2^9]$
. .	. . . .	. . .	
1023	11 1111 1111	3FF	Maximum 10-bit count
1024	100 0000 0000	400	$[2^{10}]$ = 1k
. .	. . . .	. . .	
2047	111 1111 1111	7FF	Maximum 11-bit count
2048	1000 0000 0000	800	$[2^{11}]$ = 2k
. .	. . . .	. . .	
4095	1111 1111 1111	FFF	Maximum 12-bit count
4096	1 0000 0000 0000	1000	$[2^{12}]$ = 4k
. .	. . . .	. .	
65535	1111 1111 1111 1111	FFFF	Maximum 16-bit count

The rules for counting in any number system are given below.

1. Start with all digits set to zero.

2. In the right digit position (LSB), count up from zero to the maximum digit available (1 in binary, 9 in decimal, F in hexadecimal).

3. If a column value is at its maximum, reset it to zero, and increment (add 1 to) the next column to the left.

In microprocessors, there is a fixed number of digits in the registers which store binary numbers (8, 16, 32 bits or more). If the number storage space has a fixed number of digits, leading zeros must be used to fill the empty positions, because each register bit must be either 1 or 0, and leading zeros do not alter the value.

2.1.5 Bits, Bytes and Words

One binary digit represents a 'bit' of information. A group of 8 bits is called a 'byte', and larger binary codes are called 'words'. This last term is used fairly loosely, but it sometimes refers to a 16-bit code, with a 32-bit code called a 'long word', specifically in the Motorola 68000 CPU, which was widely used in the past. As we now know, in hexadecimal four bits are represented by one hex digit, so a byte is 2 hex digits, and so on. Thus, register and memory values are typically displayed as hexadecimal numbers with 2, 4, 8, 16... digits.

2.2 Machine Code Programs

Microcontrollers store their program code and data in binary form, typically using voltage levels of $+5$ V and 0 V to represent binary 1 and 0. The program is normally stored in non-volatile ROM, and is executed by passing each code in turn to a decoding circuit which sets up the processor to carry out that particular instruction. The processor then operates on input or stored data, and switches the outputs as required.

2.2.1 Data Words

Conventional microprocessors handle the code in 8-bit binary words, or multiples of 8 bits. The data word size has increased with the complexity of the integrated circuits available; some examples are given in Table 2.6.

The first generation of popular UK home computers, such as the Commodore, Apple, BBC and Spectrum used 8-bit microprocessors; that is, the program and data words were all 8-bit numbers. Second generation home games machines such as the Atari and Amiga used the 16-bit 68000 chip, which was also the processor used in the Apple Mac, the first mass-produced computer to use a WIMP interface.

The original IBM PC was a business-oriented personal computer using the Intel 8088, which handled 16 bits inside the CPU, but only 8 bits externally. The Intel processor then went through a progressive development, leading to the 32-bit Pentium processor, and on to the current generation. At the same time clock speeds increased, and the processor complexity developed, so that the data processing capability of the current Pentium PC is massive compared with the original 8-bit machine.

Table 2.6 Comparison of microprocessors and microcontrollers

Microprocessor/ microcontroller	Computer/ application	Address bus (bits)	Data bus (bits)	Instruction (bits)	Internal CPU data (bits)
Zilog Z80	Spectrum	16	8	8/16/24	8
Rockwell 6502	Commodore/BBC	16	8	8/16/24	8
Motorola 68000	Atari/Amiga/Mac	24	16	16/32/48	16
Intel 8086/8	PC XT	16 + 4	8/16	16/32/48	16
Intel Pentium	Pentium PC	32	32	16/32/64	32
Intel 8051	Industrial/Control	Internal 16	Internal 8	16	8
PIC 16F84	Industrial/Control	Internal 13	Internal 8	14	8

The 8051 was one of the first widely used microcontrollers and is well established in the industrial control market. The PIC family is a more recent challenger for the position of leading microcontroller type. Its manufacturer, Microchip, has succeeded by initially specialising in small, cheap, re-programmable devices which were good for beginners, and then expanding the range, providing free development tools along the way.

2.2.2 Machine Code

Microprocessor machine code is a list of binary codes which are often shown in hexadecimal. An example of 6502 code is listed in Table 2.7.

The program code is a list of 8-bit binary numbers, stored in numbered memory locations, here starting at 0200_{16}, forming a list of instructions for the microprocessor to execute. The function of this particular program is to load a number given in the program (55_{16}) into the main data register (called A), and then store it in a memory location (0300_{16}). The program shows two instructions, each of which starts with the instruction (operation) code itself ($A9_{16}$, $8D_{16}$), which are followed by data required by the instruction (a number to load, and a memory address to store it in). These are called the operands. Note that in 6502 programs, the complete instruction may consist of 1, 2 or 3 bytes.

Table 2.7 6502 machine code

	Memory address	Hex code	Meaning
First instruction	0200	A9	Load the main data register A
	0201	55	with the number 55
Second instruction	0202	8D	Store the contents of register A
	0203	00	in the memory location
	0204	03	whose address is 0300
Next instruction	0205	XX	Next instruction code . . .
	0206	XX	Next operand . . .

2.2.3 8086 Machine Code

The Intel 8086 was the CPU developed for in the original IBM PC. It is useful to know something about 8086 machine code (see Table 2.8) because this is the native language of the PC, and it can be studied without access to any other hardware system. As with other processor families, the same basic instruction set has been expanded for later processors, but the basic syntax is the same, so 8086 code should run on a Pentium processor. Backward code compatibility has always been a major feature of the Intel/Microsoft product line.

Table 2.8 PC machine code

Address segment: Offset	0	1	2	3	4	5	6	7	8	9	A	B	C	D	E	F
1B85:0100	0F	00	B9	8A	FF	F3	AE	47–61		03	1F	8B	C3	48	12	B1
1B85:0110	04	8B	C6	F7	0A	0A	D0	D3–48		DA	2B	D0	34	00	74	1B
1B85:0120	00	DB	D2	D3	E0	03	F0	8E–DA		8B	C7	16	C2	B6	01	16

8086 code can be viewed on a PC by selecting the 'MS-DOS prompt' from the Windows Start button menu. An MS-DOS window should open with a '>' symbol and flashing cursor. Text commands can then be entered (before Windows, all operating system actions had to be entered this way). Type 'debug', which is a text command to the operating system. A '-' prompt appears, to indicate that Debug commands will be accepted. If 'd' (dump) is entered, the contents of a block of PC program memory will be dumped to the screen as 2-digit hex codes.

The addressing system in the Intel processor was more complicated than that in most other processors, with the address derived from the combination of a 16-bit segment address and an offset. This system was originally devised when the 16-bit 8086 was introduced, to maintain compatibility with older 8-bit systems. The memory was at this stage divided into $10 \times 64k$ segments (64k is the maximum memory space addressable with a 16-bit address). Thus each address is shown in the form 'SSSS:OOOO', where SSSS is the 4-hex-digit segment address and OOOO is the 4-bit offset. For example, if $SSSS = 1B85_{16}$ and $OOOO = 0100_{16}$, then the actual address will be $1B850 + 0100 = 1B950_{16}$. In theory, this system could address up to 4 gigabytes of memory.

If the Debug command 'u' is entered, the source assembly language 'mnemonics' are displayed for the current memory range. It can be seen that each instruction can contain 1, 2, 3 or 4 bytes. Assembly language is the programming method used for writing machine code programs, because the assembler mnemonics are easier to remember than the corresponding binary codes. An assembler utility is required to convert the source code mnemonics into executable machine code. This idea will be explained in more detail later, using the PIC instruction set as an example.

2.2.4 PIC Machine Code

The PIC machine code program is easier to interpret than the 8086 code, because it has instructions which are of fixed length (14 bits in the 16F84). In hexadecimal, a 14-bit instruction must be represented with four digits, with the most significant two bits unused. The default program start address (which is used if the programmer does not specify another), is 0000 (zero). A simple PIC machine code program is shown in Table 2.9.

The machine code for the first PIC program, BIN1, that we will be studying later is listed. It consists of five instructions, stored at addresses 0000–0004 in the program memory.

Table 2.9 Simple PIC machine code program BIN1

Program memory address	Program machine code	Meaning of Machine Code
0000	3000	Move the number 00 into the working register
0001	0066	Copy this code into port B data direction register
0002	0186	Clear port B data to register to zero
0003	0A86	Increase the value in port B data register by one
0004	2803	Jump back to Address 0003

The meaning of each instruction is given, but a fuller explanation will have to wait for now. Each instruction is 14-bits long, but the actual operation code and operand length varies within the fixed total, as shown in Table 2.10.

Table 2.10 PIC machine code in binary form

Memory address	Binary machine code	Meaning
000	**11 00**00 *0000 0000*	Load W with *00000000* (0)
001	**00 0000 0101** *0110*	Copy W to direction register of port *110* (6)
002	**00 0001 1000** *0110*	Clear data register of port *110* (6)
003	**00 1100 1000** *0110*	Increment register of port *110* (6)
004	**10 1***000 0000 0011*	Jump back to address *0000000011* (3)

The operation code part of the 14-bit instruction is shown in bold, while the operand is shown in italics. The operands refer to numbered registers and addresses within the PIC chip. For example, the last instruction operand contains the address of the third instruction, because the program jumps back and repeats from this point.

The PIC machine code can be seen in the programmer software (MPLAB) window prior to downloading, or printed in the source program list file. When the PIC chip is placed in the programmer unit, the binary codes for the program can be sent to its program memory, in serial form, one bit at a time. Each 14-bit code is stored at the address location (0000–0004) specified. When the program is later executed, the codes are interpreted by the processor block in the chip and the action carried out. The meaning and use of the registers will be explained later, when this program will be analysed in more detail.

2.3   ASCII Code

ASCII (American standard code for information interchange) is a type of binary code for representing alphanumeric characters, as found on your computer keyboard. The basic code consists of seven bits. For example, capital (or 'upper case') 'A' is represented by binary code 100 0001 (65), 'B' by 66, and so on to 'Z' $= 65 + 25 = 90 = 101\ 1010_2$. Lower case letters and other common keyboard characters such as punctuation, brackets and arithmetic signs, plus some special control characters also have a code in the range 0–127. The numerical characters also have a code, for example '9' $= 011\,1001_2$, so you sometimes need to make it clear if the code is the binary equivalent (1001_2) or the ASCII code ($011\,1001_2$).

We will not be using ASCII codes a great deal in this book, but we need to know of them, as they are the standard coding method for text files. When a program is typed into the computer to create a 'source code file', this is how the text is stored. Later, the ASCII codes must be converted into corresponding binary machine code instructions. If this is confusing, come back to this point when we have looked at programming in more detail!

Summary

- Programs and data in a microprocessor system are stored in binary form, typically as '0' $= 0$ V and '1' $= 5$ V.

- The binary codes can be displayed and printed in hexadecimal form, where 1 hex digit $= 4$ binary bits.

- A microprocessor program consists of a sequence of binary codes representing instructions and data which are decoded and executed by the CPU.

- The microprocessor memory contains a set of locations, numbered from zero, where the program is stored.

- Each program instruction consists of an operation code and (often) an operand.

- Each complete instruction may occupy a fixed number of bits or a variable number of bytes.

Questions

A calculator which converts between number systems is required for this exercise. Attempt the following calculations manually, and then check the answer on a calculator.

1. Refer to Table 2.5.
   (a) Predict the binary equivalent of 35_{10}, 61_{10} and 1025_{10}.
   (b) Convert the numbers in (a) from binary to hexadecimal.
   (c) Work out the 8-bit binary code for the 6502 program code in Table 2.7.
   (d) Write down the 16-bit binary code for the hex address 0203_{16}.

2. Write down the hex code, and work out the decimal equivalent number for the binary numbers:
   (a) 101_2
   (b) 1100_2
   (c) $1001\,1110_2$
   (d) $0011\,1010\,1111\,0000_2$

3. Light emitting diodes are often used to display output codes in simple test systems, where a binary '1' lights the LED. For an 8-bit output, work out the binary and hex code required:
   (a) to light all the LEDs,
   (b) to switch them all off,

   (c)   to light alternate LEDs, with the LSB = 1.

        Now work out the hex data sequence of eight 2-digit hex numbers which will produce:

   (d)   a bar graph effect on a set of eight LEDs (all off, then LSB on, two on, three on and so on until all eight are on),

   (e)   a scanning effect (switch on one LED at a time, in order, LSB first).

Answers

1. (a) $35_{10} = 100011_2$
         $61_{10} = 111101_2$
         $1025_{10} = 10000000001_2$

    (b) $35_{10} = 23_{16}$
         $61_{10} = 3D_{16}$
         $1025_{10} = 401_{16}$

    (c) $A9_{16} = 10101001_2$
         $55_{16} = 01010101_2$
         $8D_{16} = 10001101_2$
         $00_{16} = 00000000_2$
         $03_{16} = 00000011_2$

    (d) $0203_{16} = 0000\,0010\,0000\,0011_2$

2. (a) $101_2 = 5_{16} = 5_{10}$
    (b) $1100_2 = C_{16} = 12_{10}$
    (c) $1001\,1110_2 = 9E_{16} = 158_{10}$
    (d) $0011\,1010\,1111\,0000_2 = 3AF0_{16} = 15088_{10}$

3. (a) $1111\,1111_2 = FF_{16}$
    (b) $0000\,0000_2 = 00_{16}$
    (c) $0101\,0101_2 = 55_{16}$
    (d) $00, 01, 03, 07, 0F, 1F, 3F, 7F, FF_{16}$
    (e) $00, 01, 02, 04, 08, 10, 20, 40, 80_{16}$

Activities

1.   The seven-segment display is a device which we will use later as an output device for the PIC chip. Digits are displayed by illuminating selected segments. A diagram showing the connections to the LED segments is given in Fig. 2.1(a). The segments are identified by letter: 'a' for the top segment, 'b' is the next clockwise round the outside, and so on up to 'f' for the top left segment, with the middle segment called 'g'.

     These are connected as shown to a port data register bits 1–7, with the LSB not connected. Work out the binary and hex codes required to obtain the displayed characters 0–F shown in Fig. 2.1(b), if the display operates 'active high', that is, a '1' in the register switches the corresponding segment on. Assume that bit 0 = '0'.

2.   Debug is a DOS utility which allows you to operate at machine code level in the PC system. At the DOS prompt on a PC, enter the command 'debug'; a prompt '-' is obtained.

   (a)   Enter '?' and the debug commands are displayed.

   (b)   Enter 'd' (dump) and the contents of the current memory range are displayed in hex bytes (2 digits), with the ASCII character equivalent at the right. The 4-digit codes

(a)

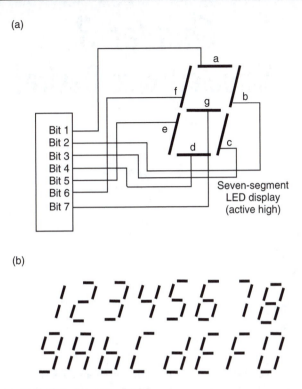

(b)

Figure 2.1 Seven-segment display of hex digits.

on the left are the segment address and offset, separated by a colon. The addresses are displayed at intervals of 16 (=10H) locations, since each row shows 16 bytes.

(c) Enter 'u' (un-assemble), and the assembly code is displayed, one instruction per line. Note the presence of instructions such as MOV (Move data), ADD (add data), INC (increase value by 1) and so on. Note also the variable instruction length.

(d) Enter 'r' and the processor registers are displayed. Note that at least one of the segment registers (CS, DS, SS, ES) contains the segment address, and the instruction pointer (IP) contains the offset.

(e) Enter 't' to trace the machine code execution. The code is executed one instruction at a time so that you can track the changes in the registers. Note that the system continues to run normally in background during debugging, so that the screen remains visible and further debug operations can be called.

(f) Enter 'q' to quit.

Chapter 3
Microelectronic Devices

We have seen in Chapter 2 that a microcontroller program consists of a list of binary codes, stored in non-volatile memory. The instructions are executed in sequence, processing data obtained from the chip registers or inputs. The results are stored back in the registers, in RAM locations or sent to an output device. We will now look briefly at the basic circuit elements needed to provide these functions. The intention is to explain the operation of basic elements of logic devices, which include microcontrollers, in enough detail to allow the reader to understand PIC data sheets.

3.1   Digital Devices

The binary codes which make up the program and data in the microcontroller are stored and processed as electronic signals. The binary numbers are conventionally represented as follows:

$$\text{Binary } 0 = 0\,\text{V}$$
$$\text{Binary } 1 = +5\,\text{V}$$

A $+5$ V supply, usually derived from the mains, is therefore required to power the circuits. It must be able to provide sufficient current for the processor circuits, at a voltage which must be between 4.75 V and 5.25 V for standard TTL (Transistor–Transistor Logic).

The power consumed in operating a digital circuit simply appears as waste heat, which must be removed from the chip. This is why a large complex device such as the Pentium processor typically has a heatsink and fan attached. The power consumption is the product of the supply voltage and current drawn at the power supply pins of the chip:

$$P = VI \qquad \text{Watts} \qquad (I = \text{chip current})$$
$$ = V^2/R \qquad \text{Watts} \qquad (R = \text{input resistance of chip})$$

If the same logic function can be implemented with less power consumed, this problem of power dissipation can be reduced and system efficiency increased. There are two ways to do this: to use a low power transistor type in the circuits, or to reduce the voltage, or both. A supply of 3.3 V is now commonly used to reduce power consumption in large chips. The heating effect is proportional to the supply voltage squared (see above), so the voltage reduction from 5 V to 3.3 V will reduce heating and power consumption by 66%.

In the original small-scale chips, bipolar transistors were used to form TTL gates. However, these have relatively large power dissipation, and run at correspondingly high temperatures. This limits the number of gates that can be operated on one chip, so VLSI (very large-scale integrated) circuits normally use FET-based (field effect transistor) logic gates, because of their lower power consumption. Also, these chips can run from a wider range of supply voltages, so are more suitable for battery-powered applications, such as laptop computers. There is continuing development of logic technologies, to obtain higher speed, lower cost and lower power dissipation in increasingly complex chips.

The PIC chip is a CMOS (complementary metal-oxide semiconductor) device, using FETs as digital switches. These, when combined together in various ways, create logic gates that can process binary data. For example, we will see how logic gates can be combined to create a binary adder, which is an essential feature of any microprocessor, allowing it to carry out binary arithmetic.

3.1.1   FET Logic Gates

The FET is the basic switching device which appears in the PIC data sheet in the equivalent circuits for various functional blocks. It is a transistor which works as a current switch; current flow through a semiconductor 'channel' is controlled by the voltage at the input 'gate'.

A single FET is shown in Fig. 3.1(a). Current flows through the channel when it is switched on by applying a positive voltage between the gate and 0 V. When the input voltage is zero, the channel has a high resistance to current flow, and the device is off. Some FETs operate with a negative voltage at the input to control the current flow.

A logical 'invert' operation is implemented by the FET circuit in Fig. 3.1(b). Assume that the FET is switched on with +5 V at input A. The channel will then have a low resistance allowing current to flow through the load resistor, R, causing a volt drop across it. This means that the voltage at F must fall, and for correct operation, F must be near zero volts when the FET is on. Thus the output is near 0 V (logic 0) when the input is +5 V (logic 1).

Conversely, the output is 'pulled up' to +5 V (logic 1) by R when the input is low (logic 0). There is then no current flow in the FET channel, and no voltage dropped across the resistor. The output must therefore be at the same voltage as the supply, +5 V.

The logic operation 'AND' requires the output of a gate to be HIGH only when all inputs are HIGH (see Table 3.1). 'NAND', the inverse operation, requires that the output is LOW only when all inputs are HIGH. This operation can be implemented as shown in Fig. 3.1(c). The output F is only low when both transistors are on. The AND function can be obtained by inverting the NAND output; this can be achieved by connecting the inverter circuit to the NAND output.

Similarly, the logic operation 'OR' requires the output of a gate to be HIGH when either input is HIGH (see Table 3.1). 'NOR', the inverse output, requires that the output is LOW when either input is HIGH. This operation can be implemented as shown in Fig. 3.1(d). The output F is low when either transistor is on. The OR function can then be obtained by inverting the NOR output, by connecting the inverter circuit.

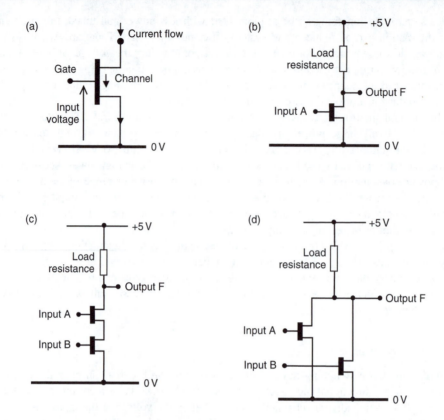

Figure 3.1 Field effect transistor logic gates. (a) Field effect transistor; (b) FET logic inverter; (c) Simplified NAND gate; (d) Simplified NOR gate.

Table 3.1 Logic table for one and two input gates

Inputs	Outputs					
	NOT	AND	OR	NAND	NOR	XOR
0	1	–	–	–	–	–
1	0	–	–	–	–	–
0 0	–	0	0	1	1	0
0 1	–	0	1	1	0	1
1 0	–	0	1	1	0	1
1 1	–	1	1	0	0	0

3.1.2 Logic Circuits

In real logic gates, the circuits are a little more complex. There are no actual resistors used because they waste too much power; instead, other FETs are used as 'active loads', which reduces the power which would be dissipated as heat in the resistors. The logic operations in Table 3.1 are all we need to make any logic or processor circuit.

Digital circuits are based on various combinations of these logic gates, fabricated on a silicon wafer. They can be supplied as discrete gates on small-scale ICs (SSI), or as complete logic circuits on large-scale ICs (LSI). Microprocessors are the most complex of all, containing thousands of gates and millions of transistors; these are called very large-scale ICs (VLSI).

3.1.3 Logic Gates

Whichever technology is used to fabricate the gates, the logical operation is the same. The symbols for logic gates used in most data sheets, including the PIC, conform to US standards, because that is where the chips are often designed. The basic set of logic devices are the AND gate, OR gate and NOT gate (or logic inverter), shown in Fig. 3.2.

There are three additional gates which can be made up from the basic set, the NAND gate, NOR gate and XOR (exclusive OR) gate. The NAND is just an AND gate followed by a NOT gate, and a NOR gate is an OR gate followed by a NOT gate. An XOR gate is similar to an OR gate (see Table 3.1). The inputs on the left accept logic (binary) inputs, producing a resulting output on the right. These logic values are typically represented by $+5\,\text{V}$ and $0\,\text{V}$, as we have seen. These gates, in various combinations, are used to make the control and data processing circuits in a microprocessor, microcontroller and supporting chips. Their functions have been summarised in the logic table, Table 3.1.

The logic table shows all the possible input combinations for one and two inputs. Obviously, the only possible inputs for the inverter are 1 and 0. The number of different inputs for the two input gates is four, that is, the total number of unique 2-bit codes. When specifying logic gate or circuit operation, all possible input combinations can be generated by counting up from zero in binary to the maximum allowed by the number of inputs. The resulting output which is obtained from each gate is then listed.

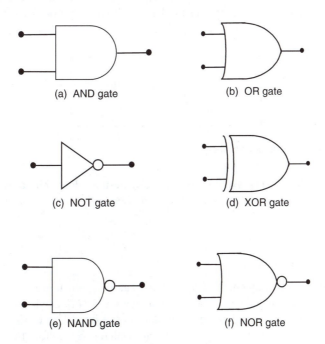

(a) AND gate (b) OR gate

(c) NOT gate (d) XOR gate

(e) NAND gate (f) NOR gate

Figure 3.2 Logic gate symbols (US standard).

The operation of logic circuits is shown in this way in IC data sheets, sometimes with 0 represented by L (low) and 1 by H (high). Note that only two inputs to each gate are shown here, but there can be more than two. The logical operation will be similar; for instance, a 3-input AND gate requires all inputs to be high to give a high output.

Variations may appear in data sheets. For instance, the circle representing logic inversion may be used at the input to a gate, as well as the output. It should always be possible to work out the logical operation from the basic logic symbol set. More detailed analysis and design of discrete logic circuits is provided in standard textbooks, and does not need to be covered here. Such discrete design principles are, in any case, less important now for the circuit designer due to the availability of microcontrollers such as the PIC which provide a software-based alternative to 'hard-wired' logic.

3.2   Combinational Logic

Logic circuits can be divided into two categories, combinational and sequential. Combinational logic describes circuits in which the output is determined only by the current inputs, and not by the inputs at some previous point in time. Circuits for binary addition will be used as examples of simple combinational logic. Binary addition is a basic function of the arithmetic and logic unit (ALU) in any microprocessor. A 4-bit binary addition is shown in Fig. 3.3, to illustrate the process required.

The process of binary addition (Fig. 3.3) is carried out in a similar way to decimal addition. The digits in the least significant column are added first, and the result 1 or 0 inserted in the 'Sum' row. If the sum is two (10_2), the result is zero with a carry into the next column. The carry is then added to the sum of the next column, and so on, until the last carry out is written down as the most significant bit of the result. The result can therefore have an extra digit, as in our example.

```
          1 1 1 1    (A)
     +    0 1 1 0    (B)
     =  1 0 1 0 1    (Sum)
  Carry: 1 1 1
```

Figure 3.3 Example of binary addition.

Having specified the process required, we can now design a logic circuit to implement this process. We will use a binary adder circuit for each column, feeding the carry bits forward as required.

3.2.1   Simple Binary Adder

The basic operation can be implemented using logic gates as shown in Fig. 3.4.

The two binary bits are applied at A and B, giving the result at F. Obviously, some additional mechanism is needed to store and present this data to the inputs, and this will be described later. This circuit is equivalent to a single XOR gate, which can therefore be used as our basic binary adder.

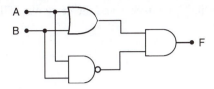

Figure 3.4 Binary adder logic circuit.

3.2.2 Full Adder

To add complete binary numbers, a carry bit must be generated from each bit adder, and added to the next significant bit in the result. This can be done by elaborating the basic adder circuit as shown in Fig. 3.5(a).

The required function of the circuit can be specified with a logic table, as shown in Fig. 3.5(b). To implement this logic function, the carry out (C_o) from each stage must be connected to the carry in (C_i) of the next, so that we end up with four full adders cascaded together. The overall

(a)

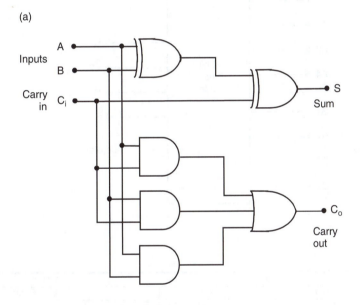

(b)

Input A	Input B	Carry in C_i	Carry out C_o	Sum S
0	0	0	0	0
0	0	1	0	1
0	1	0	0	1
0	1	1	1	0
1	0	0	0	1
1	0	1	1	0
1	1	0	1	0
1	1	1	1	1

Figure 3.5 Full adder circuit and logic table. (a) Full adder logic circuit; (b) Full adder logic table.

carry in must be applied to the C_i of stage 1 and the carry out will then be obtained from C_o of stage 4.

3.2.3 4-Bit Adder

A set of 4 full adders can be used to produce a 4-bit adder, or any other number of bits, by cascading one adder into the next. The PIC 16F84A ALU, for example, processes 8-bit data. As we are not particularly concerned with exactly how the logic is designed, as we can hide it inside a block, and then define the required logical inputs and the resulting outputs (See Fig. 3.6).

All possible input combinations must be correctly processed, and these can be specified by using a binary count in the input columns. The state of the output for each possible input

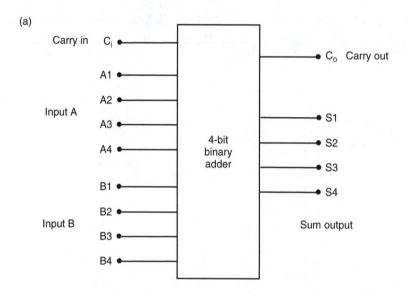

	INPUTS					OUTPUT				
	Input A				Input B							Output sum				
Row	A4	A3	A2	A1	B4	B3	B2	B1	C_i	C_o	S4	S3	S2	S1	Dec
0	0	0	0	0	0	0	0	0	0	0	0	0	0	0	0
1	0	0	0	0	0	0	0	0	1	0	0	0	0	1	1
2	0	0	0	0	0	0	0	1	0	0	0	0	0	1	1
3	0	0	0	0	0	0	0	1	1	0	0	0	1	0	2
4	0	0	0	0	0	0	1	0	0	0	0	0	1	0	2
5	0	0	0	0	0	0	1	0	1	0	0	0	1	1	3
6	0	0	0	0	0	0	1	1	0	0	0	0	1	1	3
etc.															etc.
509	1	1	1	1	1	1	1	0	1	1	1	1	0	1	30
510	1	1	1	1	1	1	1	1	0	1	1	1	1	0	30
511	1	1	1	1	1	1	1	1	1	1	1	1	1	1	31

Figure 3.6 4-bit binary full adder. (a) 4-bit adder block; (b) Logic table for 4-bit adder.

combination is then defined. With 2×4-bit inputs, plus the carry in, there are 512 possible input combinations in all, so the logic table only shows the first few and last rows, as examples.

In the past, logic circuits had to be designed using Boolean mathematics and built from discrete chips. Now, programmable logic devices (PLDs) make the job easier, as the required operation can be defined with a logic table or function statement. This is entered as a text file into a PC and converted into programming instructions which are sent to the chip, in much the same way that the PIC itself can be programmed.

3.3 Sequential Logic

Sequential logic refers to digital circuits whose outputs are determined by the current inputs AND the inputs which were present at an earlier point in time. That is, the sequence of inputs determines the output. Such circuits are used to make data storage cells in registers and memory, and counters and control logic in the processor.

3.3.1 Basic Latch

Sequential circuits are made from the same set of logic gates shown in Fig. 3.2. They are all based on a simple latching circuit made with two gates, where the output of one gate is connected with an input of the other, as shown in Fig. 3.7(a).

This circuit uses NAND gates, but NOR gates will work in a similar way. When both inputs, A and B, are low, both outputs must be high. This state is not useful here, so is called 'invalid'. When one input is taken high, the output of that gate is forced low, and the other output high.

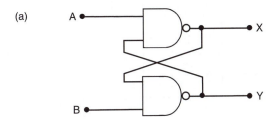

(a)

(b)

Time	Inputs		Outputs		Comment
	A	B	X	Y	
1	0	0	1	1	Invalid
2	0	1	1	0	X = 1
3	1	1	1	0	Hold X = 1
4	1	0	0	1	Reset X = 0
5	1	1	0	1	Hold X = 0
6	0	1	1	0	Set X = 1
7	1	1	1	0	Hold X = 1

Figure 3.7 Basic latch operation. (a) Basic latch circuit; (b) Sequential logic table for basic latch.

The latch is now set, or reset, depending on which output X or Y, is being used. In Fig. 3.7, X is taken as the output and is set high. This state is 'held' when the other input is taken high, and this gives us the data storage operation required. The output X can now be reset to zero by taking input B low. This state is held when B is returned high.

The sequence of events is shown in Fig. 3.7(b). At time slot 3 a data bit '1' is stored at X, while at time slot 7 data bit '0' is stored. Note that in the time slots when both inputs are high, output X can be high or low, depending on the sequence of inputs before that step was reached.

With additional control logic, the basic latch circuit can be developed to give two main types of circuit: the D-type ('Data') bistable or latch which acts as a 1-bit data store, and T-type ('Toggle') bistable which is used in counters. Such bistable (two stable states) devices are frequently referred to as 'flip-flops'. Different kinds of sequential circuits including counters and registers can be constructed from a general purpose device called a 'J-K flip-flop'.

Counters will be covered in the next chapter.

3.3.2 Data Latch

A basic sequential circuit block is a data latch, which is shown in Fig. 3.8(a). The input and output sequence can be represented on a logic table, Fig. 3.8(b). When the enable (EN) input is high, the output (Q) follows the state of the input (D). When the enable is taken low, the output state is held. The output does not change until the enable is taken high again. It is called a transparent latch, because the data goes straight through when the enable is high. There are other types of latches, called edge-triggered, which latch the data input at a specific point in

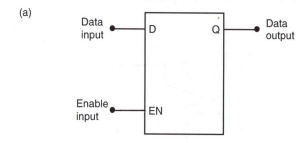

(a)

(b)

| | Inputs | | Output | |
Time	D	EN	Q	Comment
1	0	0	x	Output unknown
2	0	1	0	Output = Input 0
3	0	0	0	Data 0 latched
4	1	0	0	Data 0 held
5	1	1	1	Output = Input 1
6	1	0	1	Data 1 latched
7	0	0	1	Data 1 held
8	0	1	0	Output = Input 0

Figure 3.8 Data latch operation. (a) Data latch; (b) Sequential logic table for data latch.

time, when the enable (or 'clock') signal changes. This type of circuit block is used in registers and static RAM to store groups of, typically, 8 bits.

A timing diagram (Fig. 3.9) gives a pictorial view of the latch operating sequence, which may be easier to interpret than the logic table. It can also provide information about the precise timing of the signals, if required. This may be important, because there is always a delay between changes at the input and output of any gate. When designing high-speed circuits in particular, these timing characteristics must be carefully considered. However, for simplicity, time delays between the signal edges are not shown in Fig. 3.9.

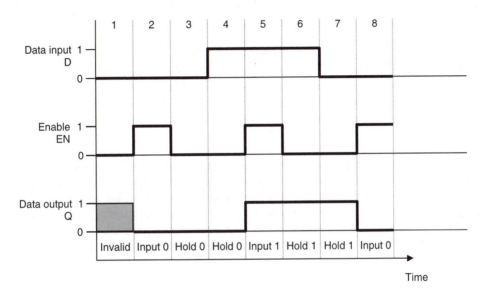

Figure 3.9 Data latch timing diagram.

These signals in the actual circuit can be displayed using an oscilloscope, if small timing delays are significant. If not, a logic analyser may be used, which operates like a multichannel digital oscilloscope allowing many signals to be displayed simultaneously; however, it may not record small time delays. The logic analyser works by sampling the signals at intervals, and can display the data in numerical form, as in the sequential logic table.

3.4 Data Devices

All data processing or digital control systems have circuits to carry out the following operations:

- data input
- data storage
- data processing
- data output
- control and timing

Data processing devices must be controlled in sequence to carry out useful work. In a microprocessor system, most of this control logic is built into the CPU and its support chips, but additional control circuits usually need to be designed for each specific system. In order to illustrate the principles of operation of microprocessor and microcontroller systems in simplified form, a set of basic logic devices will be used to make up a basic data processing circuit. These are shown in Fig. 3.10.

3.4.1   Data Input Switch

In Fig. 3.10(a), a switch (S) and resistor (R) are connected across a 5 V supply. If the switch is open, the data output is 'pulled up' to +5 V, via the resistor. If the switch is closed, the logic level at the data output must be zero, as it is connected directly to ground. The resistor is required to prevent a short circuit between the +5 V and 0 V supplies, while allowing the output to rise to +5 V when the switch is open.

This only works if a relatively small current is drawn by the load at the data output. This is usually not a problem, as digital inputs typically draw no more than a few microamps. If necessary, a capacitor may be connected across the switch for debouncing; if the switch contacts do not close cleanly, this ensures a smooth transition from high to low, and back.

3.4.2   Tri-State Gate

The tri-state gate (TSG) (Fig. 3.10(b)) is a digital device which allows electronic switching and routing of signals through a data processing system. It is controlled by the gate enable input (GE). When GE is active (in this example high), the gate is switched on, and data is allowed through, 1 or 0. When GE is inactive (low), the data is blocked, and the output goes into a high impedance (HiZ) state, which effectively disconnects it from the input of the following stage. The TSG may have an active low input, in which case the control input has a circular invert

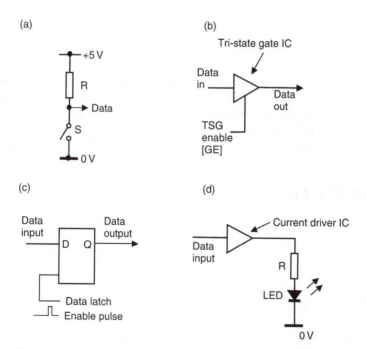

Figure 3.10   Data circuit elements. (a) Switch input; (b) Tri-state gate; (c) Data latch; (d) LED output.

symbol. TSGs can be obtained as individual gates in a small-scale integrated circuit chip, and are used as basic circuit building blocks within large-scale integrated circuits such as the PIC microcontroller.

3.4.3 Data Latch

A data latch (Fig. 3.10(c)) is a circuit block which stores one bit of data, as described above. If a data bit is presented at the input D (1 or 0), and the latch is 'clocked' by pulsing the latch enable input (0, 1, 0), the data appears at the output Q. It remains there when the input is removed or changed, until the latch is clocked again. Thus, the data bit is stored, and can be retrieved at a later time in the data processing sequence.

3.4.4 LED Data Display

An LED can provide a simple data display device. In Fig. 3.10(d) the logic level to be displayed (1 or 0) is fed to the current driver, which operates as a current amplifier and provides enough current (typically about 10 mA) to make the LED light up when the data is '1'. The resistor value controls the size of the current. Seven-segment and other matrix displays use LEDs to display decimal or hexadecimal digits by lighting up suitably arranged LED segments or dots.

3.5 Simple Data System

The way that data is transferred through a digital system using the devices described above is illustrated in Fig. 3.11. The circuit allows one data bit to be input at the switch (0 or 1), stored at the output of the latch and displayed on the LED.

The operational steps are as follows:

1. The data at D1 is generated manually at the switch ('0' = 0 V and '1'= +5 V).

2. When the TSG is enabled, the data becomes available at D2 (while the gate is disabled, the line D2 is floating, or indeterminate).

3. When the data latch is pulsed, level D2 is stored at its output, D3. D3 remains stored until new data is latched, or the system powered down.

4. While latched, the data at D3 is displayed by the LED (ON = '1'), via the current driver stage.

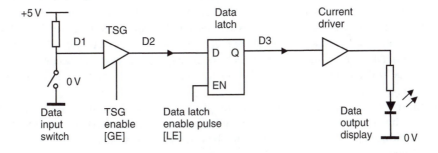

Figure 3.11   1-bit data system.

Table 3.2 1-bit system operating sequence

Operation	Switch	D1	GE	D2	LE	D3
Data input 1	Open	1	0	x	0	x
Input enable	Open	1	1	1	0	x
Latch data	Open	1	1	1	0-1-0	1
Input disable	Open	1	0	x	0	1
Data input 0	Closed	0	0	x	0	1
Input enable	Closed	0	1	0	0	1
Latch data	Closed	0	1	0	0-1-0	0
Input disable	Closed	0	0	x	0	0

Note that all active devices (gate, latch and driver) must be connected to the $+5\,V$ power supply, but these supply connections do not have to be shown in a block diagram or logic circuit. Table 3.2 details the control sequence, with the data states which exist after each operation. Note that 'x' represents 'don't know' or 'don't care' (it could be 1, 0 or floating).

3.6   4-Bit Data System

Data is usually moved and processed in parallel form within a microprocessor system. The circuit shown in Fig. 3.12 illustrates this process in a simplified way.

The function of the 4-bit system is to add two numbers which have been input at the switches. The two numbers A and B will be stored, processed and output on a seven-segment display which shows the output value in the range 0–F. The display has a built-in decoder which converts the 4-bit binary input into the corresponding digit pattern on the segments. To obtain the correct result, the two input numbers must add up to 15_{10} or less.

The common data bus is used to minimise the number of connections required, but it means that only one set of data can be on the bus at any one time, therefore, only one set of gates must be enabled at a time. The data destination is determined by which set of latches is operated when data is on the bus. The gates (data switches) and latches (data stores) must therefore be operated in the correct sequence by the control unit (See Table 3.3).

The 4-bit adder works as follows; for the moment, we will assume that these operations are carried out manually using suitable switches or push buttons to generate the control signals. The first number (6) is set up on the input switches, and the data input gate enable (DIGE) set active. This data word is now on the bus, and can be stored in latch A by pulsing the data A latch enable (DALE). Now the input switches are changed to generate the second number (5) which must be added to the first. This value now appears on the bus and can be stored in latch B by activating DBLE.

With the numbers stored at the outputs of the latches DA and DB, the result appears at the output of the binary adder, DO. If the data output gate is enabled (DOGE), the result will appear on the bus. However, the data input gate must be disabled first, so that there is no conflict on the bus. The result can then be stored and displayed by operating the data output latch enable (DOLE).

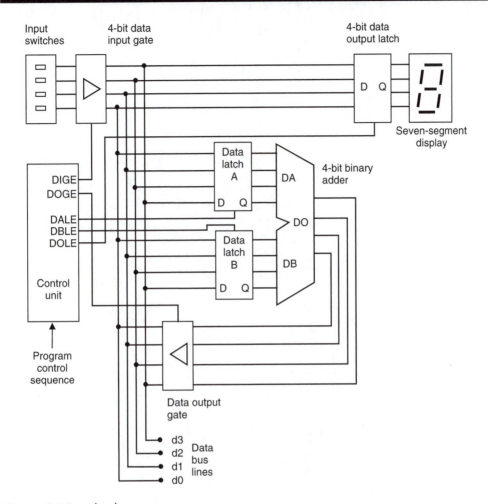

Figure 3.12 4-bit data system.

If this operating sequence can be automated, and we then have the makings of a microprocessor. The binary operating sequence produced by the control unit must be recorded and played
back in some way. This can be done by storing it in a ROM memory block, along with the
data to be input at the switches. Combining the 'instruction codes' (control switch operations)
with the 'operands' (input data) gives us a 'machine code program' as seen in Table 3.4.

The program described above has three instructions, of 9 bits in length – it is running a
simple processing system with a set of codes which are equivalent to the machine code program
in a microcontroller. The next step would be to replace the binary adder with a block that could
also subtract and carry out logical operations such as increment, shift, AND, OR and so on.
Different instruction codes would then set up the circuit to carry out all the required operations.
More latches could be added, forming registers within the processor. Better input and output
devices such as a keypad and multi-digit display would then give a usable system, as outlined
in Chapter 1. The means to program the ROM (a development system) completes our processor
system. This is how early calculator chips were developed, leading to microprocessors and
microcontrollers. The system outlined can be built (if you can get the obsolete components!).

Table 3.3 4-bit system operating sequence

	Input switches (4-bit binary)	DIGE (Data input gate enable)	DOGE (Data output gate enable)	DALE (Data A latch enable)	DBLE (Data B latch enable)	DOLE (Data output latch enable)	Display hex 0-F	Data bus (4-bit binary)	Operation
0	xxxx	0	0	0	0	1	X	xxxx	Ready for input
1	0110	0	0	0	0	1	X	xxxx	Set data input number A on switches
2	0110	1	0	0	0	1	6	0110	Enable data A onto bus by switching on input gates
3	0110	1	0	0–1–0	0	1	6	0110	Store data A in latch A by clocking it with a pulse
4	0110	0	0	0	0	1	X	xxxx	Disable input gates – no valid data on bus
5	0101	0	0	0	0	1	X	xxxx	Set data input number B on switches
6	0101	1	0	0	0	1	5	0101	Enable data B onto bus by switching on input gates
7	0101	1	0	0	0–1–0	1	5	0101	Store data B in latch B by clocking it with a pulse
8	0101	0	0	0	0	1	X	xxxx	Disable input gates – no valid data on bus
9	xxxx	0	1	0	0	1	B	1011	Enable result from ALU onto bus
10	xxxx	0	1	0	0	0	B	1011	Store result in output latch by clocking it with a pulse
11	xxxx	0	0	0	0	0	B	xxxx	Result displayed – ready for next input

Table 3.4 4-bit system 'machine code program'

'Instruction' code	'Operand'	Hex 'program'	Operation
1 0101	0110	15 6	Input and latch data A
1 0011	0101	13 5	Input and latch data B
0 1001	0000	09 0	Latch and display result

Summary

- MOS digital circuits are based on the field effect transistor acting as a current switch. These are combined to form logic gates on integrated circuits.

- The basic set of logic gates is AND, OR and NOT, from which all logic functions can be implemented. NAND, NOR and XOR form a useful additional set.

- Combinational logic gives outputs which depend only on the current input combination. Sequential logic outputs additionally depend on the prior sequence of inputs.

- Basic data system devices are the data latch used for bit storage and the TSG used to route data. Data input and output devices are also needed. Additional logic circuits provide sequential control.

Questions

1. Why is it necessary for battery-powered digital circuits to operate at a wide range of voltages?

2. Draw a simple logic inverter using a FET and a resistor, and then add to the circuit to provide the AND and OR logic functions.

3. Describe in one sentence the operation of (a) an OR gate and (b) an AND gate.

4. Representing a 1-bit full adder as a single block, with inputs A, B and C_i, and outputs S and C_o, draw a 4-bit adder consisting of four of these blocks, with the inputs and outputs shown in Fig. 3.6(a).

5. Construct a timing diagram for the sequential logic table shown in Fig. 3.7(b) to show how a basic latch works.

6. Draw a circuit with two input logic switches whose data can be stored in one of three different D-Type latches. Describe how the control logic must work to allow either input to be stored in any of the latches.

7. Modify the 4-bit system (Fig. 3.12) operating sequence so that only the final result is displayed.

1.   Construct the full adder circuit using the necessary logic chips and check that it works as described; then simulate it using a suitable schematic capture and simulation package and check that the function is accurately simulated.

2.   Investigate the operation of a suitable programmable logic device, and work out how it would be programmed to create a 4-bit adder.

3.   Investigate how the basic latch circuit could be used to 'debounce' the data input switch.

4.   Using suitable digital circuit simulator software, test the 4-bit system operation, if the components are available in the device libraries.

Chapter 4
Digital Systems

The basic set of digital devices described in Chapter 3 are enough to build working data systems, but they can be combined into common circuit blocks which are used in more complex digital designs. These have previously been available as discrete small-and medium-scale integrated circuits, and are essential elements within all microprocessor system chips and microcontrollers.

4.1 Encoder and Decoder

A digital encoder is a device which has a number of separate inputs and a binary output. An output binary number is generated corresponding to the numbered input which is active. A decoder is a device which carries out the inverse logical operation: a binary input code activates the corresponding output. Thus, if the binary code for 5 (101) is input, output 5 of the decoder goes active (usually low). An example of encoder and decoder operation is described below, where they are used to operate a keypad.

A set of switches are combined in a two-dimensional array to form a simple keyboard. These may have 12 keys (decimal) or 16 keys (hexadecimal). The decimal pad has digits 0–9, hash (#) and star (*), while the hex keypad has digits 0–F. A hex keypad is used in the example illustrated in Fig. 4.1. To read the keypad, an interface is needed which can detect when one of the buttons has been pressed. A software-based solution is explained later in Chapter 15, so a hardware solution will be described here.

Four select lines are output from a row decoder which are normally high (pull-up resistors can be attached to each line if necessary). When a binary input code is applied, the corresponding

(a)

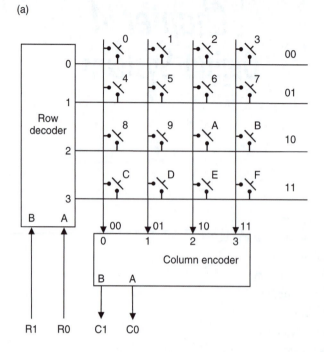

(b)

Inputs		Outputs			
B	A	0	1	2	3
0	0	0	1	1	1
0	1	1	0	1	1
1	0	1	1	0	1
1	1	1	1	1	0

(c)

Inputs				Outputs	
0	1	2	3	B	A
0	1	1	1	0	0
1	0	1	1	0	1
1	1	0	1	1	0
1	1	1	0	1	1

Figure 4.1   Keypad scanning using an encoder and decoder. (a) Hexadecimal key pad operation; (b) 2-bit decoder logic table; (c) 2-bit encoder logic table.

row select line goes low. A 2-bit binary counter can be used to drive the row decoder (see below for counters), which will generate each row select code in turn, continuously. If a switch on the active row is pressed, this low bit can be detected on the column line. The column lines, which are also normally high, are connected to a column encoder. This generates a binary code which corresponds to the input which has been taken low by connection to the row which is low.

Thus the combination of the row select binary code (R1, R0) and the column detect binary code (C1, C0) will give the number of the key which has been pressed. For instance, if key 9 is pressed, row 2 will go low when the input code is 10. This will take column 1 low, which will give the column code 01 out. The complete code is then 1001, which is 9 in binary. Encoders and decoders are combinational logic circuits which can be designed with any number of code bits, n, giving 2^n select lines.

4.2 Multiplexer, Demultiplexer and Buffer

These devices can be constructed from the same set of gates: two TSGs and a logic inverter, as seen in Fig. 4.2. All are important for the operation of bus systems, as outlined in Chapter 3.

A multiplexer is basically an electronic changeover switch, which can select data from alternative sources within the data system. A typical application is to allow two different signal sources to use a common signal path (bus line) at different times. In Fig. 4.2(a), input 1 or 2 is selected by the logic state of the select input. The logic inverter ensures that only one of the TSGs is enabled at a time. Conversely, a demultiplexer (Fig. 4.2(b)) splits the signal using the same basic devices. That is, it can pass data to alternative destinations from the bus.

The bidirectional buffer (Fig. 4.2(c)) is used to allow data to pass in one direction at a time along a data path, for example, on a bidirectional data bus. To achieve this, the TSGs are connected nose to tail, and operate alternately as in the multiplexer. When the control input is low, the data is enabled through from left to right, and when high, from right to left.

4.3 Registers and Memory

We have seen previously how a 1-bit data latch works. If the bidirectional data buffer (Fig. 4.2(c)) is added, data can be read from a data line into the latch, or written to the data line from it, depending on the data direction selected. We then have a register bit store. In Fig. 4.3(a), the data in/out line can be connected to the D input or Q output, depending on the state of the data direction select. If data is to be stored by the latch from the data line, latch enable is activated at the appropriate time.

If a set of these register elements are used together, a data word can be stored. A common data word size is 8 bits (1 byte), and most systems handle data in multiples of 8 bits. An 8-bit register, consisting of 8 data latches, is shown in Fig. 4.3(b). The register enable and read/write (data direction select) lines are connected to all the register bits, which operate simultaneously to read and write data to and from the 8-bit data bus.

4.4 Memory Address Decoding

A static RAM memory location operates in a similar way to the register. The memory device typically stores a block of 8-bit data bytes which are accessed by numbered locations (Fig. 4.4).

Each location consists of eight data latches which are loaded and read together. A read operation is illustrated; the data is being output from the selected location. A 3-bit code is needed to select one of the eight locations in the memory block, using an internal address decoder to generate the location select signal. The selected data byte is enabled out via an output buffer, which allows the memory device to be electrically disconnected when another device wants to use the bus.

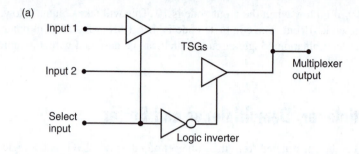

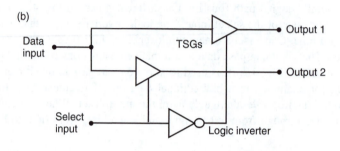

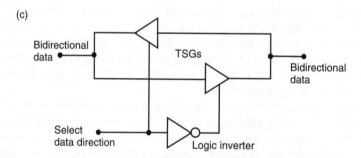

Figure 4.2 (a) 1-bit multiplexer; (b) 1-bit demultiplexer; (c) Bidirectional data buffer.

The number of locations in a memory device can be calculated from the number of address pins on the chip. In the example above, a 3-bit address provides eight unique location addresses (000_2–111_2). This number of locations can be calculated directly as $2^3 = (2 \times 2 \times 2) = 8$. Thus, the number of locations is calculated as 2 raised to the power of the number of address lines. Some useful values are listed in Table 4.1. Each memory location normally contains 1 byte. Table 2.5 was derived in a similar way and contains other significant values up to the largest 16-bit number, FFFF.

In a microprocessor system, program and data words (binary numbers) are stored in these memory locations, ready for processing in the CPU. The data transfer is implemented using a

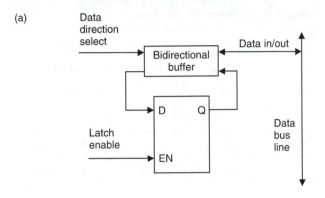

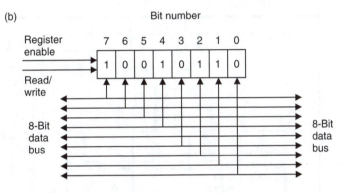

Figure 4.3 Register operation. (a) Data register bit operation; (b) 8-bit data register operation.

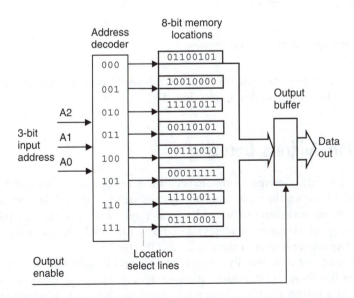

Figure 4.4 Memory device operation.

Table 4.1 Common Memory Sizes

Address lines	Locations (1 byte each)	Memory size
8	$2^8 = 256$	256 bytes
10	$2^{10} = 1024$	1 kb (kilobyte)
16	$2^{16} = 65536$	64 kb
20	$2^{20} = 1048576$	1 Mb (megabyte)
30	$2^{30} = 1073741824$	1 Gb (gigabyte)

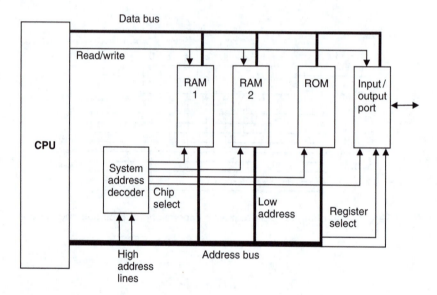

Figure 4.5 Microprocessor system addressing.

bus, which is a common set of lines which pass data between memory and registers, with TSGs directing the data to the correct destination (Fig. 4.5).

4.5 System Address Decoding

Although we are mainly concerned with microcontroller architecture, it is worth looking briefly at memory and I/O access in a conventional system, because it explains the process which occurs within the microcontroller chip and is important for an overview of microprocessor systems. It is a logical extension of address decoding within each memory chip. We will look at conventional system operation in more detail in Chapter 16.

As we have seen, there are usually several memory and input/output devices connected to a common data bus in the typical microprocessor system. Only one can use the data bus at any one time, so a system of chip selection is needed, so that the processor can 'talk' to the required peripheral chip.

Table 4.2 Typical memory map

Address range			
Lowest address	Highest address	Number of locations	Device
0000	7FFF	$8000_{16} = 32768_{10}$	RAM (32k)
8000	801F	$20_{16} = 32_{10}$	Parallel port registers
A000	A008	8	Serial port registers
C000	FFFF	$4000_{16} = 16384_{10}$	ROM (16k)

Figure 4.5 shows the basic connections in a microprocessor system which allows the CPU to read and write data to and from the memory and I/O devices. Let us assume that the CPU is reading a program instruction from ROM, although all data transfers are done in the same way.

The CPU program counter contains the address of the instruction; this is output as a binary code on the address bus. The system address decoder takes, in this system, the 2-bit code on the most significant address lines and sets one of four chip select lines active accordingly, which activates the chip to be accessed (ROM). The low order address lines are used, as described in section 4.4, to select the required location within the chip. Thus, the location select is a two-stage process, with external (system) and internal (chip) decoding of the address.

When the location has been selected, the data stored in it can be read (or written) via the data bus according to the setting of the read/write (R/W) line, generated by the CPU. To read from memory, the TSGs at the output of the selected device (ROM) are enabled, while all others connected to the bus are disabled, allowing the ROM data onto the bus lines. The data can then be read off the bus by the CPU, and copied into a suitable register (instruction register in this case). Note that ROM cannot be written and therefore does not need the R/W line connected. The I/O port only has a few addressable locations, its registers, so only a few of the address lines are needed for this device.

As a result of the design of the decoding system, the memory and I/O devices are allocated to specific ranges of addresses. The system can thus be tailored to a specific type of application by the hardware designer, with just the right amount and type of memory and I/O.

A typical memory map for a system with a 16-bit address bus (four hex digits) is shown in Table 4.2. The I/O is memory mapped, that is, the port registers are placed in the same address space as the memory. Notice that not all the available addresses have to be used.

4.6 Counters and Timers

A counter/timer register can count the number of digital pulses applied to its input. If a clock signal of known frequency is used, it becomes a timer, because the duration of the count is equal to the count value multiplied by the clock period. Like the data register, the counter/timer register is made from bistable units, but connected in 'toggle' mode, so that each stage drives the next. Each stage outputs one pulse for every two pulses which are input, so the output pulse frequency is half the input frequency for each stage (Fig. 4.6(a)). The counter/timer register can therefore be viewed as a binary counter or frequency divider, depending on the application.

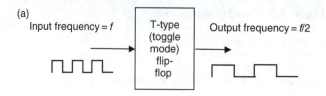

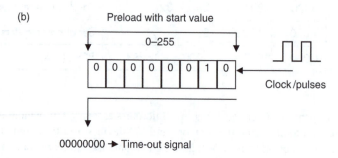

Figure 4.6 Counter/timer register operation. (a) Toggle mode stage; (b) 8-stage counter register.

Figure 4.6(b) shows an 8-bit counter/timer, with the input to the LSB at the right. The binary count stored increments each time the LSB is pulsed. Two pulses have been applied, so the counter shows binary 2. After 255 pulses have been applied, the counter will 'roll over' from 11111111 to 00000000 on the next pulse. A signal is output to indicate this, which can be used as a 'carry out' in counting operations or 'time out' in timing operations. In a microprocessor system, the 'time-out' signal typically sets a bit in a 'status' register to record this event. Optionally, an 'interrupt' signal may be generated, which forces the processor to carry out an 'interrupt service routine' to process the time-out event. Interrupts will be explained later.

If the clock pulse frequency is 1 MHz [1 megahertz], the period will be 1 μ s [1 microsecond], and the counter will generate a time-out signal every 256 μs. If the counter can be preloaded, we can make it time out after some other number of input pulses. For example, if preloaded with a count of 56, it will time out after 200 μs. In this way, known time intervals can be generated. In conventional microprocessor systems, the I/O ports often contain timers that the processor uses for timing operations. Most PICs have an 8-bit counter/timer, with a 'prescaler' that divides the input frequency by a factor of between 2 and 256 in order to extend its range; some have 16-bit counters, which allow longer intervals to be generated without a prescaler. PIC timer/counters are explained in more detail in Chapter 9.

4.7   Serial and Shift Registers

The general purpose data register, as described in Section 4.3, is loaded and read in parallel. A shift register is designed to be loaded or the data read out in serial form. It consists of a set of data latches which are connected so that a data bit fed into one end can be moved from one stage to the next, under the control of a clock signal. An 8-bit shift register can therefore store a data byte which is read in one bit at a time from a single data line. The data can then

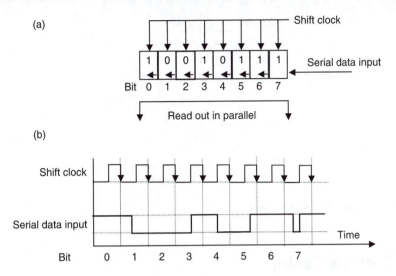

(a)

Shift clock

Serial data input

Bit   0   1   2   3   4   5   6   7

Read out in parallel

(b)

Shift clock

Serial data input

Time

Bit   0   1   2   3   4   5   6   7

Figure 4.7 Shift register operation. (a) Shift register; (b) Shift register signals.

be shifted out again, one bit at a time, or read in parallel. Alternatively, the register could be loaded in parallel and the data shifted out onto a serial output line.

In Fig. 4.7(a), the 8-bit shift register is fed data from the right. The shift clock has to operate at the same rate as the data, so that the register samples the data at the right time at the serial data input. This means that there must be agreed clock rates used to set up the shift register in advance. As each bit is read in, the preceding bits are shifted left to allow the next bit into the LSB. The timing diagram shows the data being sampled and shifted on the falling clock edge; note that only the state of the input at the sampling instant is registered, so the short negative-going pulse between bits 6 and 7 is ignored.

This type of register is used in microprocessor serial ports, where data is sent or received in serial form. In the PC, this could be the modem or network port, the keyboard input or VDU output.

4.8   Arithmetic and Logic Unit

The main function of any processor system is to process data, for example, to add two numbers together. The arithmetic and logic unit (ALU) shown in Fig. 4.8 therefore is an essential feature of any microprocessor or microcontroller.

A binary adder block has already been described, but this would be just one of the functions of an ALU. The ALU takes two data words as inputs and combines them together by adding, subtracting, comparing and carrying out logical operations such as AND, OR, NOT, XOR; these will be described in more detail in the next chapter. The operation to be carried out is determined by function select inputs. These in turn are derived from the instruction code in the program being executed in the processor. The block arrows used in the diagram indicate the parallel data paths, which carry the operands to the ALU and the result away. A set of data registers which store the operands are usually associated with the ALU, as seen in the 4-bit data system in Chapter 3.

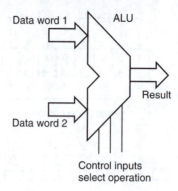

Figure 4.8 Arithmetic and logic unit.

4.9 Processor Control

The instruction decoder is a logic circuit in the CPU which takes the instruction codes from the program to control the sequence of operations. The decoder output lines, which are connected to the registers, ALU, gates and other control logic, are set up for a particular instruction to be carried out (e.g. add two data bytes).

The processor control block (Fig. 4.9) also includes timing control and other logic to manage the processor operations. The clock signal drives the sequence of events so that after a certain number of clock cycles, the results of the instruction are generated and stored in suitable register or back in memory.

The block diagram for the PIC 16F84A provided in the Appendix (Fig 1-1) shows a complete system. The data paths between each block show the data word size and the possible data transfer routes. The control lines which are set up by the instruction decode and control block are not shown, because it would make the diagram too complicated; but they are implicit in that the system cannot work without them. They connect to all parts of the processor, enabling data outputs at the source end, and operating data latches at the receiving end, of all data transfers.

We can now see more clearly the main difference between the microcontroller and microprocessor. In the conventional system, the bus system is external to the CPU, while it is internal to the microcontroller.

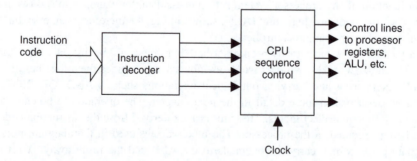

Figure 4.9 CPU control logic.

Summary

- The encoder generates a binary code corresponding to the active numbered input, and the decoder carries out the inverse operation activating the selected output according to the binary input.

- The multiplexer allows a selected input to be connected to a single output line, and the demultiplexer carries out the inverse operation connecting a single input line to a selected output line.

- A register or memory cell stores one bit of data using a data latch and bidirectional buffer.

- Numbered memory locations are accessed by decoding the address to generate a system device select and chip location select.

- Counters and timers use a counting register to count digital pulses, or measure time intervals using a clock input.

- Shift registers convert parallel to serial data, and back.

- The ALU provides data processing operations.

- The processor control signals are generated by the instruction decoder and timing circuits.

- The clock signal provides the timing reference signal for all processor operations.

Questions

1. Describe the process whereby an encoder and decoder could be used to scan a 4×32 key computer keyboard.

2. Two 8-bit registers, A and B, are connected to an 8-bit data bus via bidirectional buffers, so as to allow data to be stored and retrieved. Draw a block diagram and explain the sequence of signals required from the controller circuits to transfer the contents from A to B using data direction select and data latching signals.

3. A minimal microprocessor system, configured as shown in Fig. 4.5, has a 16-bit address bus. The two most significant lines, A14 and A15 are connected to the 2-bit decoder, which operates as specified in Fig. 4.1(b), with A14 = input A (LSB) and A15 = input B (MSB). The four select outputs are connected to memory and I/O chip select inputs as follows: 0 = RAM1, 1 = RAM2, 2 = ROM, 3 = I/O. The RAM1 chip is selected in the range of addresses from 0000 to 3FFF (hex). Work out the lowest address where each of the three remaining chips are selected.

4. Calculate the number of locations in a memory chip which has 12 address pins.

5. Calculate the time interval generated by an 8-bit timer preloaded with the value 11001110 and clocked at 125 kHz.

3.   4000, 8000, C000

4.   4096

5.   400 μs

Activities

1.   In a suitable TTL logic device data book or supplier's catalogue, look up the chip numbers and internal configuration of the medium scale ICs: 3 to 8-line decoder, octal D latch, octal bus transceiver, 8-bit shift register, 4-bit binary counter. Also identify the largest capacity RAM chip listed in your source.

2.   Refer to the PIC 16F8X data sheet, Fig 1-1 (block diagram). State the function of the following features: ROM program memory, RAM file registers 68 × 8, program counter, instruction register, instruction decode and control, multiplexer, ALU, W reg, I/O Ports, TMR0.

3.   Refer to the PIC 16F8X data sheet, Fig 4-1 (block diagram of pins RA3:RA0). Identify the following devices in the circuit diagram: FET, OR gate, TSG, transparent data latch, edge-triggered data latch. Describe how an input data bit would be transferred onto the internal data bus line from the I/O pin.

Chapter 5
Microcontroller Operation

5.1   **Microcontroller Architecture**
5.2   **Program Operations**

To understand the operation of a microcontroller requires some knowledge of both the internal hardware arrangement and the instruction set which it uses to carry out the program operations. We now have some knowledge of the digital circuit blocks which make up microsystems. In this chapter we will look at some common elements of microcontrollers and the basic features of machine code programs.

5.1   Microcontroller Architecture

The architecture (internal hardware arrangement) of a complex chip is best represented as a block diagram. This allows the overall operation to be described without having to analyse the circuit, which will be very complex, in detail. The PIC data sheet (see Appendix A) contains the definitive block diagram (Fig. 1-1) of the PIC 16F84A, but simplified versions will be used to help explain particular aspects of the chip operation. First, however, we will look at a general block diagram which shows some of the common features of microcontrollers (Fig. 5.1).

The block diagram shows a general microcontroller that can be considered in two parts, the program execution section and register processing section. This division reflects the PIC architecture, where the program and data are accessed separately. This arrangement increases overall program execution speed and is known as Harvard architecture.

The program execution section contains the program memory, instruction register and control logic which store, decode and execute the program. The register processing section has special registers used to set up the processor operations, data registers to store the current data, port registers for input and output, and the ALU to process the data. The timing and control block co-ordinates the operation of the two parts as determined by the program instructions and responds to external control inputs, such as the reset.

5.1.1   Program Memory

The control program is normally stored in non-volatile ROM. Microcontrollers which are designed for prototyping and short production runs have traditionally used Erasable Programmable ROM (EPROM) into which the program can be 'blown' using a suitable programming

Program execution section   **Register processing section**

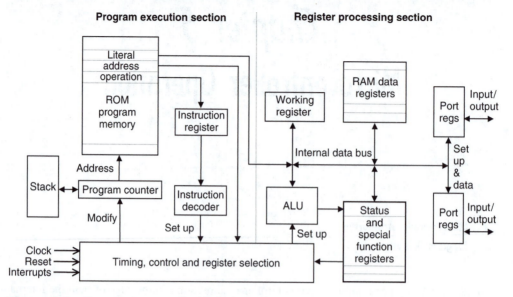

Figure 5.1   General microcontroller block diagram.

unit. Though EPROM can be erased and reprogrammed, the chip must be removed and placed under an ultraviolet lamp to clear an existing program, which is inconvenient.

More recently, microcontrollers which have flash ROM have become more common; these are generally more suitable for learning programming and prototyping. An existing program can be simply overwritten, and this can even be done while the chip is still in the application circuit. Usually, however, the chip is placed in a programming unit attached to the host computer for program downloading, prior to fitting it in the application board.

When the program is known to be correct, and will not need further modification, it can be downloaded into one-time programmable memory, which, as the name implies, cannot be erased or overwritten. For longer production runs, ready-programmed chips can be ordered from the manufacturer, which use mask progrmmed ROM, where the program is built in during chip fabrication.

5.1.2   *Program Counter*

The program counter is a register which keeps track of the program sequence by storing the address of the instruction currently being executed. The default start address of the program is usually zero; this is where the first instruction in the program will be stored unless the program author specifies otherwise. The program counter is therefore automatically loaded with zero when the chip is powered up or reset. In the PIC 16XXXX chips (any chip starting with 16), the program counter is file register 2.

As each instruction is executed, the program counter is incremented (increased by one) to point to the next instruction. Program jumps are achieved by changing the program counter to point to an instruction other than the next in sequence. For instance, if a branch back by three instructions is required, 3 is subtracted from the contents of the PC.

Sometimes, it is necessary to jump from address zero to the start of the actual program at a higher address, because special control words must be stored in specific low addresses. For instance, PIC 16XXXX devices use address 004 to store the 'interrupt vector', if interrupts are to be used. In this case, the main program should not be located at address zero, instead a jump to a higher address should be placed there. However, this problem can be ignored for programs which do not use interrupts, and our simple programs can be located at address zero. Interrupts will be explained in more detail later.

Associated with the program counter is the 'stack'. This is a temporary program counter store. When a subroutine is executed (see Section 5.2.3), a stack register temporarily stores the current address so that it can be recovered at a later point in the program. It is called a stack, because the addresses are restored to the PC in the reverse order to which they were stored, that is, 'last in, first out' (LIFO), like a stack of plates.

5.1.3 Instruction Register (IR) and Decoder

To execute an instruction, the processor copies the instruction code from program memory into the instruction register. It can then be decoded by the instruction decoder, which is a combinational logic block which sets up the processor control lines as required. These control lines are not shown explicitly in the block diagram, as they go to all parts of the chip.

In the PIC, the instruction code includes the operand, which may be a literal value or register address. For example, if a literal given in the instruction is to be loaded into the working register (W), it is placed on an internal data bus and the W register latch enable lines are activated by the timing and control logic. The internal data bus can be seen in the manufacturer's block diagram (Fig 1-1) in the data sheet.

5.1.4 Timing and Control

This sequential logic block provides overall control of the chip, and from it, control signals go to all parts of the chip to move the data around and carry out logical operations and calculations. A clock signal is needed to drive the program sequence; it is normally derived from a crystal oscillator, which provides an accurate, fixed frequency signal. There is always a maximum frequency of operation specified: PIC 16XXXX chips can operate at any frequency from a maximum of 20 MHz, down to zero.

The reset input can restart the program at any time by clearing the program counter to zero. If the program runs in a continuous loop, and there is no instruction to exit the loop, the reset may be needed. However, it is not essential to connect an active reset input because the program will start automatically at program ROM address zero, as long as the reset input (!MCLR) is connected in its active state, which is high. In most of the sample programs in this book, it is assumed that the chip would be switched off, then on again, to restart the program, and a reset switch is not required.

The only other way to stop or redirect a continuous loop is via an 'interrupt'. Interrupts are signals generated externally or internally, which force a change in the sequence of operations. If an interrupt source goes active in the PIC 16XXXX, the program will restart at address 004, where the sequence known as the 'interrupt service routine' (or a jump to it) must be stored. More details are provided in Chapter 9.

5.1.5 Working Register

In some microcontrollers and microprocessors, this is called the accumulator (A), but the name working register (W) used in the PIC system is a better description. It holds the data that the processor is working on at the current time, and most data has to pass through it. In the PIC, if a data byte is to be transferred from the port register to a RAM data register, it must be moved into W first. The working register or accumulator works closely with the ALU in the data processing operations.

5.1.6   Arithmetic and Logic Unit

This is a combinational logic block which takes one or two input binary words and combines them to produce an arithmetic or logical result. In the PIC, it can operate directly on the contents of a register, but if a pair of data bytes is being processed (for instance, added together), one must be in W. The ALU is set up according to the requirements of the instruction being executed by the timing and control block. Typical ALU/register operations are detailed later in this chapter.

5.1.7   Port Registers

Input and output in a microcontroller are achieved by simply reading or writing a port data register. If a binary code is presented to the input pins of the microcontroller by an external device (for instance, a set of switches), the data is latched into the register allocated to that port when it is read. This input data can then be moved (copied) into another register for processing. If a port register is initialised for output, the data moved to that register is immediately available at the pins of the chip. It can then be displayed, for example, on a set of LEDs. Each port has a 'data direction' register associated with its data register. This allows each pin to be set individually as an input or output before the data is read from or written to the port data register.

In the PIC 16F84A, there are two ports, A and B. Port A has five pins and Port B has eight. A '0' in the data direction register sets the port bit as an output, and a '1' sets it as an input. These port registers are mapped (addressed) as special function registers 5 and 6, respectively (Bank 0). In larger chips, additional ports may be available. For example, in the PIC 16F877, Ports C, D and E are available at register addresses 7, 8 and 9, respectively. The port data direction registers are mapped into a second register 'bank' (Bank 1) with addresses starting at 85_{16} for Port A, 86_{16} for Port B, and so on. These are accessed by special instructions which will be explained later.

5.1.8   Special Function Registers (SFRs)

These registers provide dedicated program control registers and processor status bits. In the PIC, the program counter, port registers and spare registers are mapped as part of this block. The working register is the only one that is not located in the main register block and is accessed by name, 'W', not by number.

A processor will also contain control registers whose bits are used individually to set up the processor operating mode or record significant results from those operations. Processors generally have a status register (SR) which will contain a zero (Z) flag. This bit is automatically set to 1 if the result of any operation is zero in the destination register (the register which receives the result). The carry (C) flag is another commonly used bit in the status register – it is set if the result of an arithmetic operation produces a carry out of the most significant bit of the destination register, that is, the register overflows. In the PIC 16XXXX, the status register is file register 3.

The status register bits are often used to control program sequence by conditional branching. Alternate sections of code are executed depending on the state of the status flag. In the PIC

Table 5.1 Selected PIC special function registers

File register address	Name	Function
Bank 0		
01	TMR0	Timer/counter allows external and internal clock pulses to be counted
02	PCL	Program counter stores the current execution address
03	STATUS	Individual bits record results and control operational options
05	PORTA	Bidirectional input and output bits
06	PORTB	Bidirectional input and output bits
0B	INTCON	Interrupt control bits
Bank 1		
85	TRISA	Port A data direction bits
86	TRISB	Port B data direction bits

system, this is achieved by an instruction which tests the bit and skips the next instruction conditionally. The test and skip instruction is generally followed by a jump instruction to take the execution point to another part of the program, or not, as the case may be. This will be explained more fully in the next section.

The most important SFRs are listed in Table 5.1.

5.2 Program Operations

We have seen in Chapter 2 that a machine code program consists of a list of binary codes stored in the microcontroller memory. They are decoded in sequence by the processor element, which generates control signals that set up the microcontroller to carry out the instruction. Typical operations are:

- load a register with a given number;
- copy data from one register to another;
- carry out an arithmetic or logic operation on a data word;
- carry out an arithmetic or logic operation on a pair of data words;
- jump to an alternative point in the program;
- test a bit or word and jump, or not, depending on the result of the test;
- jump to a subroutine, and return later to the same point;
- carry out a special control operation.

The machine code program must be made up only from those binary codes which the instruction decoder will recognise. These codes could be worked out manually from the instruction set given in the data sheet (Table 7-2). When computers were first developed, this was indeed how the program was entered, using a set of switches or keyboard. This is obviously time consuming and inefficient, and it was soon realised that it would be useful to have a software tool which would generate the machine code automatically from a program which was written in a more user-friendly form. Assembly language programming was therefore developed, when hardware had moved on enough to make it practicable.

Assembly language allows the program to be written using mnemonic ('designed to aid the memory') code words. Each processor has its own set of instruction codes and corresponding mnemonics. For example, a commonly used instruction mnemonic in PIC programs is 'MOVWF', which means move (actually copy) the contents of the working register (W) to a file register which is specified as the operand. The destination register is specified by number (file register address), such as 0C (the first general purpose register in the PIC 16F84). The complete instruction, with its machine code equivalent, is:

008C   MOVWF   0C

There are two main types of instruction:

1. *Data processing operations*

   - MOVE   copy data between registers.
   - REGISTER   manipulate data in a single register.
   - ARITHMETIC   combine register pairs arithmetically.
   - LOGIC   combine register pairs logically.

2. *Program sequence control operations*

   - UNCONDITIONAL JUMP   jump to a specified destination.
   - CONDITIONAL JUMP   jump, or not, depending on a test.
   - CALL   jump to a subroutine and return.
   - CONTROL   miscellaneous operations.

Together, these types of operations allow inputs to be read and processed, and the results stored or output, or used to determine the subsequent program sequence.

5.2.1   Single Register Operations

The processor operates on data stored in registers, which typically contain 8 bits. The data can originate in three ways:

1. A literal (numerical value) provided in the program;

2. An input via a port data register;

3. The result of a previous operation.

This data can be processed using the set of instructions defined for that processor. Table 5.2 shows a typical set of operations which can be applied to a single register. The same binary number is shown before processing, and then after the operation has been applied to the register.

As an example of how these operations are specified in mnemonic form in the program, the assembler code to increment a PIC register is:

0A06   INCF   06

Register number 06 happens to be Port B data register, so the effect of this instruction can be seen immediately at I/O pins of the chip. The corresponding machine code instruction is

Table 5.2 Single register operations

Operation	Before		After	Comment
CLEAR	0101 1101	→	0000 0000	Reset all bits to zero
INCREMENT	0101 1101	→	0101 1110	Increase binary value by one
DECREMENT	0101 1101	→	0101 1100	Decrease binary value by one
COMPLEMENT	0101 1101	→	1010 0010	Invert all bits
ROTATE LEFT/RIGHT	0101 1101	→	1011 1010	Shift all bits left by one place, replace MSB in LSB
SHIFT LEFT/RIGHT	0101 1101	→	0010 1110	Shift all bits right by one place, losing the LSB
CLEAR BIT	0101 1101	→	0101 0101	Reset bit (3) to 0
SET BIT	0101 1101	→	1101 1101	Set bit (7) to 1

0A86 in hexadecimal, or 00 1010 1000 0110 in binary (14 bits). As you can see, it is easier to recognise the mnemonic form!

5.2.2 Register Pair Operations

Table 5.3 shows basic operations that can be applied to pairs of registers. Normally, the result is retained in one of the registers, which is referred to as the destination register. A binary code to be combined with the contents of the destination register is obtained from the source register. The source register contents remain unchanged after the operation.

The meaning of each type instruction is explained below, with an example from the PIC instruction set. In the PIC, there is an option to store the result in W, the working register, if that is the source. Note also that the PIC does not provide moves directly between file registers, all data moves are via W.

Move

It is the most commonly used instruction in any program and simply moves data from one register to another. It is actually a copy operation, as the data in the source register remains unchanged until overwritten or the processor is reset

```
080C    MOVF  0C, W
```

This instruction moves the contents of register 0C (12_{10}) into the working register.

Arithmetic

Add and subtract are the basic arithmetic operations, carried out on binary numbers. Some processors also provide multiply and divide in their instruction set, but these can be created if necessary by using shift, add and subtract operations.

```
078C      ADDWF  0C
```

This instruction adds the contents of W to register 0C.

Table 5.3 Operations on register pairs

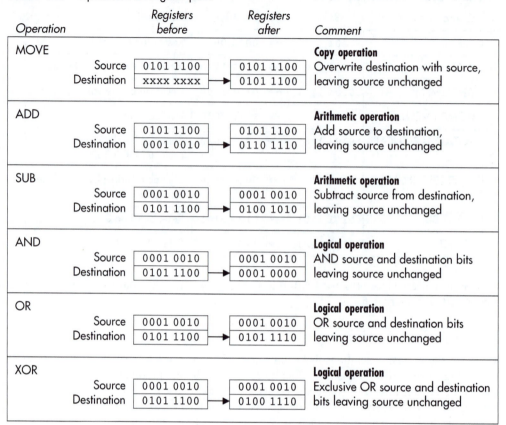

Operation		Registers before	Registers after	Comment
MOVE				**Copy operation**
	Source	0101 1100	0101 1100	Overwrite destination with source,
	Destination	xxxx xxxx	0101 1100	leaving source unchanged
ADD				**Arithmetic operation**
	Source	0101 1100	0101 1100	Add source to destination,
	Destination	0001 0010	0110 1110	leaving source unchanged
SUB				**Arithmetic operation**
	Source	0001 0010	0001 0010	Subtract source from destination,
	Destination	0101 1100	0100 1010	leaving source unchanged
AND				**Logical operation**
	Source	0001 0010	0001 0010	AND source and destination bits
	Destination	0101 1100	0001 0000	leaving source unchanged
OR				**Logical operation**
	Source	0001 0010	0001 0010	OR source and destination bits
	Destination	0101 1100	0101 1110	leaving source unchanged
XOR				**Logical operation**
	Source	0001 0010	0001 0010	Exclusive OR source and destination
	Destination	0101 1100	0100 1110	bits leaving source unchanged

Logic

Logical operations act on the corresponding pairs of bits in a literal, or source register, and destination. The result is normally retained in the destination, leaving the source unchanged. The result in each bit position is obtained as if the bits had been fed through the equivalent logical gate (see Chapter 3).

```
3901          ANDLW   01
```

This instruction carries out an AND operation on the corresponding pairs of bits in the binary number in W and the binary number 00000001, leaving the result in W. In this example, the result is zero if the LSB in W is zero. This type of operation can be used for bit testing if the processor does not provide a specific instruction.

5.2.3 Program Control

As we have already seen, the microcontroller program is a list of binary codes in the program memory which are executed in sequence. The sequence is controlled by the program counter. Most of the time, the PC is simply incremented by one to proceed to the next instruction. However, if a program jump (branch) is needed, the PC must be modified, that is, the address of the next instruction must be loaded into the PC, replacing the existing value. This new address can be given as a jump instruction operand (absolute addressing), or calculated from the current address, for example, by adding a number to the current PC value (relative branch).

The PC is cleared to zero when the chip is reset or powered up for the first time, so program execution starts at address 0000. The clock signal then drives the execution sequence forward. During the execution cycle, the program counter is incremented to 0001, so that the processor is ready to execute the next instruction. This process is repeated unless there is a jump instruction.

On the question of terminology, 'jump' and 'branch' are two terms for describing sequence control operations, but the ways in which they work are slightly different. A 'branch' is made relative to the current address, by adding to the current value in the PC. A 'jump' uses absolute addressing, that is, the contents of the PC are replaced with the destination address. Because the PIC is a RISC processor, it does not provide branch instructions in its basic instruction set, but such operations can be created from the available instructions, if required. Therefore, the program counter can be modified directly using a register processing operation to create a relative jump or branch.

The jump instructions must have a destination address as the operand. This can be given a numerical address, but this would mean that the instructions would have to be counted up by the programmer to work out this address. So, as we will see later in the program examples, a destination address is usually specified in the program source code by using a recognisable label, such as 'again', 'start' or 'wait', in the same way that mnemonics are used to represent the binary machine code instructions. The assembler program then replaces the label with the actual address when the assembler code is converted to machine code.

Program sequence control operations are illustrated in Figs 5.2, 5.3 and 5.4.

Jump

The unconditional jump (Fig. 5.2) forces a jump to another point in the program every time it is executed. This is carried out by replacing the contents of the program counter with the address of the destination instruction, in this case, 005. Execution then continues from the new address. Note that the code for GOTO is 28h combined with the destination address 05h, giving the instruction code 2805h (h for hexadecimal).

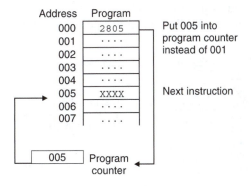

Figure 5.2 Unconditional jump.

The unconditional jump is often used at the very end of a program to go back to the beginning of the sequence, and keep repeating it.

```
start     first instruction
          . . . . . . . . . . . . . . . . .
          . . . . . . . . . . . . . . . . .
          GOTO start
```

The label 'start' is placed in the first column of the program code, to differentiate it from the instruction mnemonics, which must be placed in the second column, as we will see. The label and its reference must match exactly. The label is replaced by the corresponding address by the assembler when creating the machine code for the GOTO instruction.

Conditional Jump

The conditional jump instruction is required for making decisions in the program. Instructions to change the program sequence depending on, for instance, the result of a calculation or a test on an input are an essential feature of any microprocessor instruction set.

In Fig. 5.3, the code 1885 tests an input bit of the PIC and skips the next instruction if it is zero ('0'). Instruction YYYY (representing any valid instruction code) is then executed. If the bit is not zero (i.e. '1'), the instruction 2807 is executed, which causes a jump to address 007, and instruction ZZZZ is executed next. This is called Bit Test and Skip, and is the way that conditional branches are achieved in the PIC.

In PIC assembly language, this program fragment looks like this:

```
          . . . .
          . . . .
          BTFSC 05,1     ; Test bit 1 of file register 5
          GOTO dest1     ; Execute this jump if bit = 1
          . . . .        ; otherwise carry on from here
          . . . .
          . . . .
dest1     . . . .        ; branch destination
```

The PIC is an RISC processor, designed with a minimal number of instructions, so the conditional branch has to be made up from two simpler instructions. The first instruction tests a bit in a register and then skips (misses out) the next instruction, or not, depending on the result. This next instruction is usually a jump instruction (GOTO or CALL). Thus, program

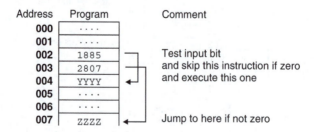

Figure 5.3 Conditional jump.

execution continues either at the instruction following the jump, if the jump is skipped, or at the jump destination.

In complex instruction set processors, conditional jump instructions which test specific bits in the CPU status register are usually available. When the CPU operates on some data in a register, status register bits record certain results, such as whether the result was zero or not. In this case, if the result of a previous instruction was zero, a 'zero flag' is set to 1 in the status register.

Pseudocode (structured program description) for a typical use of the conditional jump, a delay routine, would look like this:

```
            Allocate 'Count' Register
            ....
            ....
            Load 'Count' register with literal XX
  again     Decrement 'Count' register
            Test 'Count' register for zero
            If not zero, jump to label 'again'
            Next Instruction
            ....
```

This software timing loop simply causes a time delay in the program, which is useful, for instance, for outputting signals at specific intervals. A register is used as a down counter by loading it with a number, XX, and decremented it repeatedly until it is zero. A test instruction then detects that the zero flag has gone active, and the loop is terminated. Each instruction takes a known time to execute, therefore the delay can be calculated.

Subroutine

Subroutines are used to carry out discrete program functions. They allow programs to be written in manageable, self-contained blocks, which can then be executed as required. The instruction CALL is used to jump to a subroutine, which must be terminated with the instruction RETURN.

CALL has the address of the first instruction in the subroutine as its operand. When the CALL instruction is decoded, the destination address is copied to the PC, as for the GOTO instruction. In addition, the address of the next instruction in the main program is saved in the 'stack'. In the PIC, this is a special block of RAM memory used only for this purpose, but in conventional processors, a part of main RAM memory may be set aside for this purpose. The return address is 'pushed' onto the stack when the subroutine is called, and 'popped' back into the program counter at the end of the routine, when the RETURN instruction is executed.

In Fig. 5.4, the subroutine is a block of code whose start address has been defined by label as 0F0. The CALL instruction at address 002 contains the destination address as its operand. When this instruction is encountered, the processor carries out the jump by copying the destination address (0F0) into the program counter. At the same time, the address of the next instruction in the main program (003) is pushed onto the stack, so that the program can come back to the original point after the subroutine has been executed.

One advantage of using subroutines is that the block of code can be used more than once in the program, but only needs to be typed in once. In the PIC program SCALE (see Chapter 12),

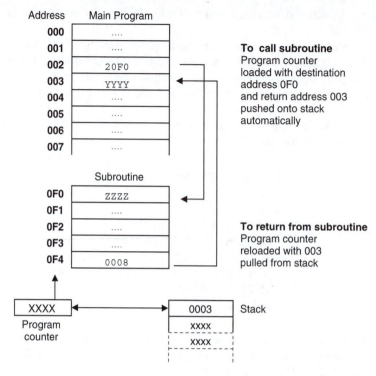

Figure 5.4 Subroutine call.

a delay loop is written as a subroutine. It is a counting loop which just uses up time to give a delay between output changes, which is 'called' twice within a loop which sets an output high, delays, sets the output low, and delays again before repeating the whole process. The same program also contains an example of direct modification of the program counter (labelled PCL) to create a data table.

Pseudocode for a delay loop written as a subroutine would be as follows:

```
; Program DELTWICE *********************************

                Allocate 'Count' Register
                ....
                ....
                Load 'Count' register with value XX
                CALL 'delay'
                Next Instruction
                ....
                ....
                Load 'Count' register with value YY
                CALL 'delay'
                Next Instruction
                ....
```

```
              . . . .
              END of Program

; Subroutine DELAY *********************************

delay         Decrement 'Count' register
              Test 'Count' register for zero
              If not zero, jump to label 'delay'
              RETURN from subroutine

; End of code ****************************************
```

Note that the 'delay' routine is called twice, but using a different delay value in the 'Count' register. Thus, the same code can be used to give different delay times. Notice also that we have started using comments in our pseudocode to identify the functional blocks as the program gets more complex.

Summary

- The typical microcontroller contains a program execution section, and a register processing section.

- The program counter steps through the program ROM addresses, and the instructions are decoded and executed.

- Data is transferred via port registers, stored in RAM/registers and processed in the ALU.

- Special function registers hold control, setup and status information.

- Instructions move or process data, or control the execution sequence.

- The content of the data registers is manipulated as single data words, or using register pairs.

- Program jumps can be unconditional or conditional, using bit testing or status bits to determine the sequence.

- Subroutines are distinct program blocks which operate using call, execute and return.

Questions

1. Outline the sequence of program execution in a microcontroller, describing the role of the program ROM, program counter, instruction register, instruction decoder, and timing and control block.

2. A register is loaded with the binary code 01101010. State the contents of the register after the following operations on this data: (a) clear, (b) increment, (c) decrement, (d) complement, (e) rotate right, (f) shift left, (g) clear bit 5, (h) set bit 0.

3. A source register is loaded with the binary code 01001011, and a destination register loaded with 01100010. State the contents of the destination register after the following operations: (a) MOVE, (b) ADD, (c) AND, (d) OR, (e) XOR.

4. In a microcontroller program, a subroutine starts at address 016F and ends with a 'return' instruction at address 0172. A 'call subroutine' instruction is located at address 02F3. Assuming that the microcontroller has one complete instruction in each address, list the changes in the contents of the program counter and stack between the time of execution of the instruction before the call and the instruction following the call. Indicate an unknown value as XXXX.

5. Write a pseudocode program for the process by which two numbers, say 4 and 3, could be multiplied by successive addition. Use the register instructions Clear, Move, Add, Decrement, Test for Zero and Jump if Zero to Label.

2. (a) 00000000, (b) 01101011, (c) 01101001, (d) 10010101, (e) 00110101, (f) 11010100, (g) 01001010, (h) 01101011.

3. (a) 01001011, (b) 10101101, (c) 01000010, (d) 01101011, (e) 00101001.

4.
```
PC      Stack
....    ....
02F2    XXXX    Instructions
02F3    XXXX    before Call
016F    02F4    Subroutine Start
0170    02F4    Subroutine
0171    02F4    instructions..
0172    02F4    Return
02F4    XXXX    Instructions
02F5    XXXX    after Call
....    ....
```

5.
```
Allocate registers A,B,C
Clear register A
Move 4 into register B
Move 3 into register C
Loop1     Add B to A
          Decrement C
          Test C for zero
          Jump back to 'Loop1' if C not zero
Finished with product in A
```

1. Study the PIC 16F8X block diagram (Appendix A, Fig. 1-1), and identify the features described in Section 5.1.

2. Study the PIC instruction set (Appendix A, Table 7-2) and allocate the instructions to the following categories: Move, Arithmetic, Logic, Jump and Control. Make a list of instructions, organised in these categories, with a description and an example for each showing the required syntax (one line per instruction). This can then be used as a handy instruction reference when programming.

Part B
The PIC Microcontroller

Chapter 6
A Simple PIC Application

6.1 **Hardware Design**
6.2 **Program Execution**
6.3 **Program BIN1**
6.4 **Assembly Language**

A very simple machine code program for the PIC will now be developed, avoiding complicating factors as far as possible. A simplified internal architecture will be used to explain the execution of the program, and the program will then be developed further, with new programming techniques being added at each step. Since the core architecture and programming methods are similar for all PIC microcontrollers, this serves as an introduction to the whole family.

The specification for the application is as follows. The circuit will output a binary count to eight LEDs, under the control of two push button inputs. One input will start the output sequence when pressed. The sequence will stop when the button is released, retaining the current value on the display. The other input will clear the output (all LEDs off), allowing the count to resume from zero.

6.1 Hardware Design

We need a microcontroller which will provide two inputs and eight outputs, which will drive the LEDs without additional interfacing, and has reprogrammable flash memory to allow the program to be developed in stages. An accurate clock is not required, so a crystal oscillator is not necessary. The PIC 16F84A meets these requirements; it is a basic device, so we will not be distracted by unused features.

6.1.1 PIC 16F84A Pinout

The PIC 16F84A microcontroller is supplied in an 18-pin DIL (dual in line) chip. Simplified pin labelling, taken from the data sheet (see Appendix A), is shown in Fig. 6.1. Some of the pins have dual functions which will be discussed later. The suffix 'A' in the chip number indicates the enhanced version of the chip, which, as far as we are concerned, is functionally identical to the original 16F84, except that it can run at 20 MHz.

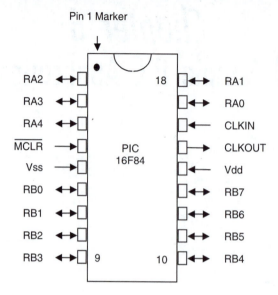

Figure 6.1 Pin-out of PIC 16F84.

The chip has two ports, A and B. The port pins allow data to be input and output as digital signals, at the same voltage levels as the supply which is connected to Vdd and Vss. CLKIN and CLKOUT are used to connect clock circuit components, and the chip then generates a fixed frequency clock signal which drives all its operations. !MCLR ('NOT Master CLeaR') is a reset input, which can (optionally) be used to restart the program. Note that the active low operation of this input is indicated by a bar over the pin label. An exclamation mark at the beginning of the pin label means the same thing. In many applications, this input does not need to be used, but it *must* be connected to the positive supply rail to allow the chip to run.

A summary of the pin functions is provided in Table 6.1. Port B has eight pins, so we will assign these pins to the LEDs and initialise them as outputs. Port A has five pins, of which two can be used for the input switches. A resistor and capacitor will be connected to the CLKIN pin to control the clock frequency. !MCLR must be connected to +5 V.

6.1.2   BIN Hardware Block Diagram

The hardware arrangement required for the application can be represented in a simplified form as a block diagram (Fig. 6.2). The main parts of the hardware and relevant inputs and outputs should be identified, together with the direction of signal flow. The nature of the signals may be described with labels or illustrated with simple diagrams. The power connections need not be shown; it is assumed that suitable supplies are available for the active components. The idea is to outline the basic hardware arrangement without having to design the circuit in detail at this stage.

Port A (5 bits) and Port B (8 bits) give access to the data registers of the ports, the pins being labelled RA0 through to RA4, and RB0 through to RB7 respectively. The two push button switches will be connected to RA0 and RA1, and a set of LEDs connected to RB0–RB7. The switches will later be used to control the output sequence. RA1 will be programmed to act

Table 6.1 PIC 16F84 pins arranged by function

Pin	Label	Function	Comment
14	Vdd	Positive supply	+5 V nominal, 3–6 V allowed
5	Vss	Ground supply	0 V
4	!MCLR	Master clear	Active low reset input
16	CLKIN	Clock input	Connect RC clock components to 16
15	CLKOUT	Clock output	Connnect crystal oscillator to 15 and 16
17	RA0	Port A, Bit 0	Bidirectional Input/Output
18	RA1	Port A, Bit 1	Bidirectional Input/Output
1	RA2	Port A, Bit 2	Bidirectional Input/Output
2	RA3	Port A, Bit 3	Bidirectional Input/Output
3	RA4	Port A, Bit 4	Bidirectional Input/Output + TMR0 Input
6	RB0	Port B, Bit 0	Bidirectional Input/Output + Interrupt Input
7	RB1	Port B, Bit 1	Bidirectional Input/Output
8	RB2	Port B, Bit 2	Bidirectional Input/Output
9	RB3	Port B, Bit 3	Bidirectional Input/Output
10	RB4	Port B, Bit 4	Bidirectional Input/Output + Interrupt Input
11	RB5	Port B, Bit 5	Bidirectional Input/Output + Interrupt Input
12	RB6	Port B, Bit 6	Bidirectional Input/Output + Interrupt Input
13	RB7	Port B, Bit 7	Bidirectional Input/Output + Interrupt Input

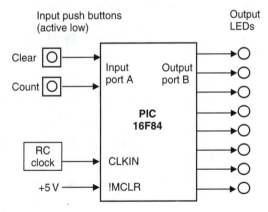

Figure 6.2 Block diagram of BIN hardware.

as a 'run' input, enabling the binary count, while RA0 will provide a 'reset' input to restart the output sequence. However, these inputs will not be used in the first program, BIN1. The connections required are shown in Table 6.2.

The block diagram can now be converted into a circuit diagram. The input and output circuits have already been introduced in Section 3.4. The clock components are the only additional parts needed, and the configuration and values for these are obtained from the data sheet. The circuit diagram is shown in Fig. 6.3.

Table 6.2   PIC 16F84A pin allocation for BIN application

Pin	Connection
Vss	0 V
Vdd	+5 V
!MCLR	+5 V
CLKIN	CR clock circuit
CLKOUT	Not connected (n/c)
RA0	Reset switch
RA1	Count switch
RA2	n/c
RA3	n/c
RA4	n/c
RB0	LED bit 0
RB1	LED bit 1
RB2	LED bit 2
RB3	LED bit 3
RB4	LED bit 4
RB5	LED bit 5
RB6	LED bit 6
RB7	LED bit 7

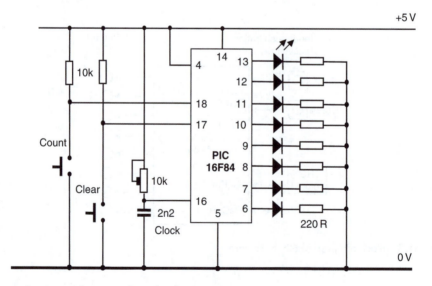

Figure 6.3   Circuit diagram of BIN hardware.

6.1.3   BIN Circuit Operation

Active low switch circuits, consisting of normally open push buttons and pull-up resistors, are connected to the control inputs. The resistors ensure that the inputs are high when the buttons are not pressed. The outputs are connected to LEDs in series with current-limiting resistors.

The PIC outputs are capable of supplying enough current (up to 20 mA) to drive LEDs directly, making the circuit relatively simple. The external clock circuit consists of a capacitor (C) and resistor (R) in series; the value of C and R multiplied together will determine the chip clock rate. The resistance in this circuit has been made variable, and the values shown should allow the clock frequency to be adjusted to 100 kHz. The reset input (!MCLR) must be connected to the positive supply (+5 V) to allow the chip to run. Other unused pins can be left open circuit, but unused I/O pins should be programmed as inputs.

6.2 Program Execution

The program for the chip is created using the PIC development system software on a PC and downloaded via a serial data link. This process will be described in more detail later, but for now we will assume that the program is in memory.

A block diagram showing a simplified program execution model for the PIC 16F84 is shown in Fig. 6.4. The binary program, shown in hexadecimal, is stored in the program memory. The instructions are decoded one at a time by the instruction decoder, and the required operations set up in the registers by the control logic. The file registers are numbered from 00 to 4F, with the first 12 registers (00–0B) being reserved for specific purposes. These are called special function registers (SFRs). The rest may be used for temporary data storage, and are called general purpose registers (GPRs). Only GPR1 is shown in Fig. 6.4.

6.2.1 Program Memory

The program memory is a block of flash ROM, which means it is non-volatile, but can be easily re-programmed. The program created in the host computer is downloaded via port register pins RB6 and RB7 when the chip is placed in its programming unit and set to program mode by supplying +14 V at the !MCLR pin. It is possible to write the program directly into the

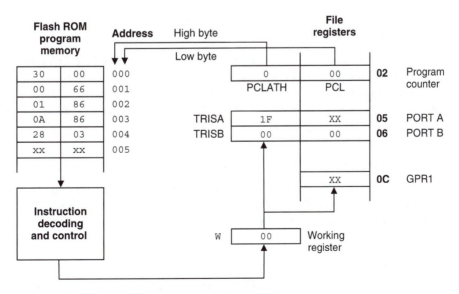

Figure 6.4 PIC 16F84 simple program execution model.

programming software in hexadecimal form, but it is normally created using assembly language. This will be described later.

The 14-bit codes are loaded into memory starting at address 000. When the chip is powered up, the program counter resets automatically to 000, and the first instruction is fetched from this address, copied to the instruction register in the control block, decoded and executed.

6.2.2   Program Counter: File Register 02

The program counter keeps track of the program execution by holding the address of the current instruction. It is automatically incremented to point to the next instruction during the execution cycle. If there is a jump in the program, the program counter is modified by the jump instruction, so that it then points to the required jump destination.

6.2.3   Working Register: W

This is the main data register (8 bits) used for holding the data that is currently being worked on. It is separate from the file register set and is therefore referred to as W in the PIC program. Literals (values given in the program) must be loaded into W before being moved to another register, or used in a calculation. Most data movements have to be via W, in two stages, since direct moves between file registers are not available in the instruction set.

6.2.4   Port B Data Register: File Register 06

The 8 bits stored in the Port B data register will appear on the LEDs connected to pins RB0–RB7, if the port bits are initialised as outputs. The data direction is set as output by placing a data direction code in the register TRISB. A '0' in TRISB sets the corresponding pin in the port register as an output (0 = Output). A '1' sets it to input (1 = Input). In this case, 00000000 (binary) will be placed in TRISB to set all bits as outputs, but any combination of inputs and outputs can be used.

6.2.5   Port A Data Register: File Register 05

The least significant five bits of file register 05 are connected to pins RA0–RA4, the other three being unused. This port will be used later to read the push buttons. If not initialised as outputs, the PIC I/O pins automatically become inputs, i.e. TRISA = xxx11111. We will use this default setting for Port A. However, the state of these inputs will have no effect unless the program uses them; the first program BIN1 will not.

6.2.6   General Purpose Register 1: File Register 0C

The first general purpose register will be used later in a timing loop. It is the first of a block of 68 such registers, numbered 0C–4F. They may be allocated by the programmer as required for temporary data storage, counting and so on.

6.2.7   Bank 1 Registers

The main registers such as the program counter and port data registers are in a RAM block called register bank 0, while TRISA, TRISB and PCLATH are in a separate block, bank 1.

Bank 0 can be directly addressed, meaning that data can be moved into them using a simple 'move' instruction.

Unfortunately, this is not the case with bank 1 registers. Special instructions are needed to load them, and there are two ways to do this. The first way is a simple method which we will use initially. It requires the 8-bit code to be loaded to be placed in W first, and then moved into the bank 1 register using the TRIS command. Later, we will use the recommended method, using bank selection.

PCLATH stands for Program Counter Latch High. This stores the most significant two bits of the 10-bit program counter, which also cannot be accessed directly.

6.3   Program BIN1

The simple program called BIN1, introduced in Chapter 2, is listed as Program 6.1. The program consists of a list of 14-bit binary machine code instructions, represented as 4-digit hex numbers. If bits 14 and 15 are assumed to be zero, the codes are represented by hex numbers in the range 0000–3FFF. The program is stored at addresses 000–004 (5 instructions) in program memory.

Program 6.1   BIN1 machine code

Memory address	Machine code instruction	Meaning
000	**3000**	Load working register (W) with number 00
001	**0066**	Store W in Port B direction code register
002	**0186**	Clear Port B data register
003	**0A86**	Increment Port B data register
004	**2803**	Jump back to address 0003 above

6.3.1   Program Analysis

The explanation of the program instructions must be related to the internal hardware of the PIC 16F84, as shown in Fig. 6.4.

Address 0000:   Instruction = 3000

The code 3000 means move (copy) a literal (number given in the program) into the working register (W). All literals must be placed initially in W before transfer to another register. The literal, which is zero in this case, can be seen in the code as the last two digits, 00.

Address 0001:   Instruction = 0066

This means copy the contents of W to the Port B data direction register (TRISB). W now contains 00, which was loaded in the first instruction. This code will set all 8 bits of TRISB to zero, making all bits output. The file register address of Port B (6) is given as the last digit of the code.

These first two instructions are required to initialise Port B for output, using the TRIS command to load the bank 1 register called TRISB in the register set.

Address 0002:   Instruction = 0186

This instruction will clear file register 6 (last digit), which means that it will set all bits in the Port B data register (PORTB) to zero. Operations can be carried out directly on the port data register, and the result will appear immediately on the LEDs. On start-up, the register bits default to '1', switching the LEDs on. When the 'clear' instruction is executed, they will go out.

Address 0003:   Instruction = 0A86

Port B data is modified; the binary value is increased by 1 and this value will be seen on the LEDs.

Address 0004:   Instruction = 2803

This is a jump instruction, which causes the program to go back and repeat the previous instruction. This is achieved by the instruction overwriting the current program counter contents with the value 03, the destination address, which is given as the last two digits of the instruction code.

6.3.2 Program Execution

BIN1 is a complete working program, which initialises and clears Port B and then keeps incrementing it. The last two instructions, increment Port B and jump back, will repeat indefinitely with the value being increased by 1 each time. In other words, Port B data register will act as an 8-bit binary counter. When it reaches FF, it will roll over to 00 on the next increment operation. If you study the binary count given in Table 2.5, you can see that the least significant bit is inverted each time the binary count is incremented. RB0 will thus be toggled (inverted) every time the increment operation is repeated. The next bit, RB1, will toggle at half this rate, and so on, with each bit toggling at half the frequency of the previous bit. The MSB therefore toggles at 1/128 of the frequency of the LSB. The output pattern generated is shown in Fig. 6.5.

An instruction in the PIC takes 4 clock cycles to complete, unless it causes a jump, in which case, it will take 8 clock cycles (or two instruction cycles). The repeated loop in BIN1 will

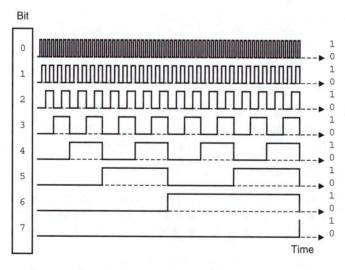

Figure 6.5 Waveforms produced by program BIN1 at Port B.

therefore take $4 + 8 = 12$ clock cycles, and thus it will take 24 cycles for the RB0 to go low and high, giving the output period of the LSB. With the clock component values as indicated on the circuit diagram in Fig. 6.3, the clock can be set to run at 100 kHz, so RB0 would then flash at 100 kHz/24 = 4.167 kHz, and RB7 will then flash at 4167/128 = 32.5 Hz. This is too fast to see unaided, but it is possible to reduce the clock speed by increasing the value of the capacitor C in the clock circuit. Alternatively, the outputs can be displayed on an oscilloscope. We will see later how to slow the outputs down without changing the clock.

The frequencies generated are actually in the audio range, and they can be heard by passing them to a small loudspeaker or peizo buzzer. This is a handy way of checking quickly that the program is working, and also immediately suggests a range of PIC applications – generating signals and tones at known frequencies by adjusting the clock rate or using a crystal oscillator. Again, we will come back to this idea later and see how to generate audio outputs or a tone sequence to make a tune like a mobile phone ring tone.

6.4 Assembly Language

It should be apparent that writing the machine code manually for any but the most trivial applications is going to be a bit tedious. Not only do the actual hex instruction codes have to be worked out, but so do jump destination addresses and so on. In addition, the codes are not easy to recognise or remember.

6.4.1 Mnemonics

For this reason, microcontroller programs are normally written in assembly language, not machine code. Each instruction has a corresponding mnemonic defined in the instruction set in the data sheet. The main task of the assembler program supplied with the chip is to convert a source code program written in mnemonic form into the required machine code. The mnemonic form of the program BIN1 is shown in Program 6.2.

The instructions can now be written as recognisable (when you get used to them!) code words. The program can be typed into a text editor, spaced out as shown, using the tab key to place the code in the correct columns. Note that the first column (column 0) must be kept blank – we will see why later. The instruction mnemonics are placed in column 1, and the operands (data to be operated on) in column 2. The operand 00 is the data direction code for the port initialisation, 06 is the file register number of the port data register, and 03 is the jump destination address, line 3 of the program. The PIC instructions are all 14 bits long, so each line of source code becomes a 14-bit code, which we have already seen.

Program 6.2 Mnemonic form of program BIN1

Top left of edit window

Line number	Column 0	Column 1	Column 2	Column 3
0		MOVLW	00	
1		TRIS	06	
2		CLRF	06	
3		INCF	06	
4		GOTO	03	
5		END		

The meaning of the mnemonics is as follows:

```
0    MOVLW   00    Move Literal 00 into W
1    TRIS    06    Move W into TRISB to set Port B as outputs
2    CLRF    06    Clear file register 06 (Port B)
3    INCF    06    Increment file register 06 (Port B)
4    GOTO    03    Jump to address 03 (back to previous instruction)

     END     End of source code - this is not an instruction!
```

The END statement is an 'assembler directive'; it tells the assembler that this is the end of the program, and is not converted into an actual instruction. When entering the program, there must be space before and after each instruction mnemonic, and it is advisable to lay out the program in columns as shown to improve its readability.

6.4.2 Assembly

The source code program could be created using a general purpose text editor, but is normally created within a dedicated software package such as MPLAB, the PIC integrated development environment (IDE), which contains the assembler as well as a text editor. The source code text is entered and the assembler invoked from the menus. The assembler program analyses the source code, character by character, and works out the binary code required for each instruction. The terminology can be confusing here; the assembly language application program (source code) is created in the text editor, while the software tool which does the conversion is the assembler program.

The source code is saved on disk as a text file called PROGNAME.ASM, where 'progname' represents any suitable filename. This is then assembled by the assembler program MPASM.EXE, which creates the machine code file PROGNAME.HEX. This appears as hexadecimal code when listed. At the same time, PROGNAME.LST, the list file is created which contains both the source and hex code, which may be useful later on when debugging (fault finding) the program. Further information on using MPLAB will be given later.

6.4.3 Labels

The mnemonic form of the program with numerical operands can now be further improved. The operands can be input in a more easily recognisable form, in the same way that the mnemonics represent the instruction codes. The assembler is designed to recognise labels. A label is a word which represents an address, register or literal. Examples used below are 'again', 'portb', and 'allout'.

The jump destinations are defined by label, by simply placing the label at the beginning of the destination line, and using a matching label as the jump instruction operand. When the program is assembled, the assembler notes the numerical address of the instruction where the label was found, and replaces the label, when found as an operand, with this address. Register and literal labels, on the other hand, must be 'declared' at the beginning of the program, and the assembler will then substitute the numerical operand for the label when it is found in the source code.

The program BIN1 can thus be re-written using labels as shown in BIN2 source code (Program 6.3). The literal value 00 and the port register address 06 have been replaced with labels which are assigned at the beginning of the program. These are 'equate' statements, which

Program 6.3 BIN2 source code using labels

Edit window

```
allout      EQU         00
portb       EQU         06

            MOVLW       allout
            TRIS        portb

            CLRF        portb
again       INCF        portb
            GOTO        again

            END
```

allow the numbers which are to be replaced in the source code to be declared. In this case, the label 'allout' will represent the Port B data direction code, while the data register address itself, 06, will be represented by the label 'portb'. 'EQU' is another example of an assembler directive, which is an instruction to the assembler program and will not be translated into code in the executable program.

Note that lower case is used for the labels, while upper case is used for the instruction mnemonics and assembler directives. Although this is not obligatory, this convention will be used because the instruction mnemonics are given in upper case in the instruction set. The labels can then be distinguished by using lower case. The jump destination label is simply defined by placing it in column 0 of the line containing the destination instruction. The 'GOTO label' instruction then uses a matching label. Initially, labels will be limited to six characters; they must start with a letter, but can contain numbers, e.g. 'loop1'.

The programs BIN1 and BIN2 are functionally identical, and the machine code will be the same.

6.4.4 Layout and Comments

A final version of BIN2 (Program 6.4) includes comments in the program to explain the action of each line, and the overall program. As much information as possible should be provided; when learning programming, comments help the learner to retain information, and when developing real applications, it will help with future modifications and upgrading (software maintenance).

Comments must be preceded with a semicolon (;), which tells the assembler to ignore the rest of that line. Comments and information can thus occupy a whole line, or can be added after each instruction in column 3. A minimal header has been added to BIN2, with the source code file name, author and date, and a comment added to each line. Blank lines can be used without a comment 'delimiter' (the semicolon); these are used to break up the source code into functional sections, and thus make the operation of the program easier to understand. In BIN2.ASM, the first block contains the operand label equates, the second the port initialisation and the third the output sequence. The layout of the program is very important in showing how it works.

We now have a program that can be entered into a text editor, assembled and downloaded to the PIC chip. The exact way of doing this will vary with the version of the PIC software and programming hardware that you use.

Program 6.4 BIN2 source code with comments

```
;       BIN2.ASM          M.Bates                        11-10-03
;
;       Outputs a binary count at Port B
; ....................................................................

allout  EQU     00        ; Define Data Direction Code
portb   EQU     06        ; Declare Port B Address

        MOVLW   allout    ; Load W with DDC
        TRIS    portb     ; Set Port B as outputs

        CLRF    portb     ; Switch off LEDs
again   INCF    portb     ; Increment output
        GOTO    again     ; Repeat endlessly

        END               ; Terminate source code
```

Summary

- A block diagram can be used to outline the hardware, and the circuit designed from it.

- The PIC 16F84 program is stored in flash ROM, at addresses from 000. The instructions are decoded and executed by the processor control logic.

- The CPU registers are modified according to the program, and the sequence can be modified by the instructions.

- The PIC 16F84 has 14-bit instructions, containing both the operation code and operand.

- The program is written using assembler mnemonics and labels to represent the machine code instructions and operands.

- Layout and comments are used to document the program operation.

Questions

1. State the 4-digit hex code for the instruction INCF 06.

2. State the 2-digit hex code for the instruction MOVLW.

3. What is the meaning of the least significant two digits in the PIC machine code instruction 2803?

4. Why must the instruction mnemonic be in the second column of the source code?

5. Give two examples of a PIC assembler directive. Why are they not represented in the machine code?

6. What are the numerical values of the labels 'allout' and 'again' in BIN2?

Answers

1. 0A86

2. 30

3. Jump destination

4. Labels go in first column

5. EQU, END

6. 00, 03

Activities

1. Check the machine code for BIN1 against the information given in the PIC instruction set in the data sheet, so that you could, if necessary, work out a program entirely in machine code. Modify the machine code program by deleting the 'Clear Port B' operation and changing the 'Increment Port B' to 'Decrement Port B'. What would be the effect at the output when the program was run?

2. Construct the circuit shown in Fig. 6.3 using a suitable hardware prototyping method. Refer to Chapter 12 if necessary. A socket must be used for the PIC chip. Enter the machine code for BIN1 directly into the programming software memory buffer and download to the chip. Run the program in the hardware or a simulated circuit. Feed the outputs to a small loudspeaker with a 220R current-limiting resistor in series. Use an oscilloscope to measure the clock and output frequencies. Confirm the relationship between the clock frequency and the output frequencies. Increase the capacitor value to 220 nF, which should make the MSB flash at a visible rate. Predict the output frequency (Hint: the rate is proportional to the product of RC clock components).

3. Enter the program BIN2, using labels, into the text editor, assemble and test as above. Check that the machine code and function is identical to BIN1.

4. Display or print out the list file BIN2.LST and check that the machine code generated is the same as BIN1. Note that there is no machine code generated for comment lines or assembler directives. See Table 7.4 for a list file example.

Chapter 7
PIC Program Development

We have seen how to start developing PIC application hardware and software, and can now take a closer look at some of the software tools available, and how each is used in the program development process. The program BIN2 will be further developed using the same hardware as described in Chapter 6.

This chapter will describe features of the standard PIC development system which is currently available, but hardware and software support to application developers are being continuously developed by Microchip and independent suppliers. The Internet provides ready access to the most recent information on the range of PIC chips and support software available at any given time. The manufacturer's website can be found at www.microchip.com.

Since the available software tools are continuously updated, a definitive tutorial in using a particular version would soon be out of date. At the time of writing, MPLAB Version 6 is the most recently released version of the PIC IDE, but the reader will need to refer to the manufacturer's documentation for details concerning the use of any particular version. The intention at this stage is to outline how to assemble and test demonstration programs BIN3 and BIN4 in general terms. More details on debugging programs are provided in Chapter 11.

The flowchart in Fig. 7.1 gives an overview of the program development process. The starting point is the specification for the program, which describes how the application will function when complete. This must then be analysed by the software designer so that the required program can be derived from it, taking into account the features of the instruction set of the microcontroller. The program algorithm describes the process whereby the correct outputs are obtained from the given inputs. Various software design techniques are available to outline the program, including flowcharts and pseudocode, which we will use here. These must represent the program processes and their sequence in a consistent way, which can then be converted to source code.

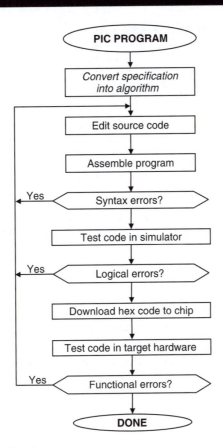

Figure 7.1 PIC program development process.

The program source code is developed from the program algorithm, by filling in the details and converting each program block to assembler code. The program must be saved on disk as it is developed; it is a good idea to always have copies on different disks (floppy, hard disk or network drive) in case of disk failure – this always happens when you least expect it! The source code text file is called PROGNAME.ASM, where PROGNAME is the application name, such as BIN1. Successive versions of a program can be numbered: BIN1, BIN2, etc.

The program can then be assembled by calling up the assembler utility; this is called MPASM. It converts the source code into machine code and creates additional files to help with debugging (fault finding) the program. If a mistake has been made in the individual instruction (e.g. misspelling a mnemonic), it will be reported in an error message window and an entry added to the error file on disk. This must then be corrected in the source code and the program re-assembled until it is free of syntax errors.

The program can then be tested for correct operation before it is downloaded to the chip using the simulator MPLAB SIM. The program is loaded and executed on-screen, and checked step by step for the correct logical operation, by monitoring the changes in the registers, and checking the timing if necessary. Simulated inputs are also needed. If a logical error is found, the source code must be re-edited, re-assembled and the simulation repeated.

When the logical errors have been removed, the program can be downloaded to the chip and it should work first time, if the hardware has been correctly designed. Final testing can

Table 7.1 Components of MPLAB development system

Software tool	Tool function	Files produced or used	File description
Text editor	Used to create and modify source code text file	PROGNAME.ASM	Source code text file
Assembler	Generates machine code from source code, reports syntax errors, generates list and symbol files	PROGNAME.HEX PROGNAME.ERR PROGNAME.LST PROGNAME.COD	Executable machine code Error messages List file with source and machine code Symbol and debug information
Simulator	Allows program to be tested in software before downloading	PROGNAME.HEX PROGNAME.COD	
Programmer	Downloads machine code to chip	PROGNAME.HEX	

then compare the finished circuit function with the specification, and, hopefully, no further debugging should be necessary at this stage.

The main software tools and files created and used by MPLAB during the development process are listed in Table 7.1.

7.1 Program Design

There are national and company standards for specifying engineering designs which should be applied in commercial work. The design rules for different types of products will vary; for instance, a military application will be designed to a higher standard of reliability and more rigorously tested and documented than a commercial one. Our designs here are artificial in that they are intended to illustrate features of the PIC microcontroller rather than meet a user's requirement. Nevertheless, we can follow the design process through the main steps.

7.1.1 Application Design

The first step in the design process is to specify the functions and performance required by the application. In the real world, this needs to be done in some detail so that the overall design, development and production costings and timescales can be predicted, as well as establishing the market or customer requirements. For our purposes, the minimal specification given in Chapter 6 will suffice.

The next step is to design the hardware on which the application program will run. A block diagram which shows the user interface (input and output) requirements is a good starting point. The interfacing of the microcontroller is generally based on a limited number of standard devices, such as push buttons, keypad, LED indicators, LCD (liquid crystal display), relays and so on. The circuit design techniques required will not be covered here, but we must ensure that the demands on the microcontroller are within its specification. For example, the maximum current available at the standard PIC output is about 20 mA, sufficient to drive an LED, but not a relay.

The microcontroller itself must be selected by specifying the requirements such as:

- number of inputs and outputs
- program memory size
- data memory size (number of spare file registers)
- program execution speed
- other special interfaces (e.g. analogue inputs, serial ports)

The hardware configuration for the BINx applications has already been described in Chapter 6. We have established that the instruction set and programming features of the microcontroller selected are suitable. If further features were required, the existing hardware design could be modified. If the microcontroller selected was then found to be lacking in some way, for example, not enough I/O pins, another microcontroller, or other types of hardware such as a conventional microprocessor system, must be considered. However, it is easier to stay within one family of processors, and most manufacturers supply a range of chips, from which the most suitable can be selected.

7.1.2   Program Specification

The operational requirements of the application must be clearly specified in advance. In the commercial environment, a customer may do this, or if the application is a more speculative venture, the requirements of the potential market must be analysed. A specification must then be written in a way that lends itself to conversion into a software product using the language and tools available. Each programming language offers a different combination of features which must be matched to the user requirement as closely as possible. Similarly, the hardware system type must be selected to suit the application, before attempting the detailed circuit design. Choosing the most suitable microprocessor or microcontroller is clearly crucial. To make this choice, one needs a knowledge of the whole range of options. Chapter 14 provides a starting point for investigating and comparing different solutions.

For so-called embedded applications (the controller is built into the application circuit), the main choice of language is between assembler and a high level language (HLL) such as 'C'. C allows such features as screen graphics, file handling and complex calculations to be more easily included in an application. For example, a maths function such as 'Sin x' is provided in C; this would require a much more complex calculation in assembly language. On the other hand, assembly language code is generally faster and requires less memory. Of course, ultimately all languages are converted into machine code to run on the selected processor. The HLL requires more program memory because each statement is converted into several machine code instructions.

High level languages are normally used to develop applications for conventional processor systems, especially if using the Windows/Intel PC as a standard hardware platform. However, for less complex applications, a suitable microcontroller would be used, programmed in assembler language. The PIC family contains an expanding range of devices, which offer a combination of different features, for example, built-in serial communication ports or analogue to digital converters (ADCs). The more powerful 18XXXX series of PICs is typically programmed in 'C' (see Chapter 16).

7.1.3   Program Algorithm

The specification for the demonstration application is as follows. The circuit will output a binary count to eight LEDs, under the control of two push button inputs. One input will start the output sequence when pressed. The sequence will stop when the button is released, retaining the current value on the display. The other input will clear the output (all LEDs off), allowing the count to resume from zero.

The BIN3 specification is much less demanding than would normally be the case for real software product. The frequency of operation of the output could, for instance, be specified. As the specification is not very specific, it should be easy to meet!

A flowchart is useful for clarifying the algorithm, particularly when learning, as it provides a pictorial representation. A flowchart for BIN3 is shown in Fig. 7.2. The program title is placed in the start symbol at the top of the flowchart, and the process required defined as a sequence of blocks. Each flowchart box will contain a description of the action at each stage, using different shaped boxes for processes (rectangle), input and output (sloping) and decisions (pointed). The decision box has two outputs, to represent a conditional branch in the program. This decision box should contain a question with the answer yes or no, and the active selection labelled Yes or No as appropriate; only one needs to be labelled. The jump destinations are also labelled; these same labels will be used in the program as address labels. Software design techniques, including flowcharts, will be covered in more detail later.

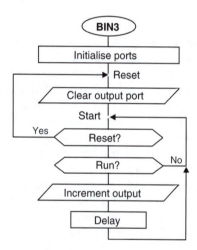

Figure 7.2   Flowchart for BIN3.

7.2   Program Editing

The program is written using the instruction set of the processor selected. This is provided with the hardware data sheet, for the 16F84A, Section 7 of the data sheet. A summary is provided in Table 7-2 of the data sheet. The source code, that is, the assembly code program, must be entered into a suitable text editor, usually the MPLAB edit window. We will not go into the details of using a text editor, as it is assumed that the reader is familiar with using a word processor.

The MPLAB text editor has limited editing features, because it is only used for creating plain text files. The typeface 'Courier' is used because each character occupies the same space, unlike proportionally spaced typefaces such as 'Arial' and 'Times Roman'. Displayed in this way, the text lines up vertically as well as horizontally, so the program can be laid out consistently in columns using tab stops, making it easier to understand. The tab spacing should be set to 8 characters for the programs in this book.

When a new application is started, a separate folder should be created to contain the source code file, and all the other files that will be created. Name the folder with the application name, e.g. BIN3. When the source code file has been opened, enter the source code filename (e.g. BIN3.ASM) at the top of the file, and immediately save it in the folder. This ensures that the required filepath is checked for correct operation before any further source text is entered. When saving on floppy, there is a high risk of disk failure and possible operating system crash, resulting in the loss of the edit file. Avoid disaster by keeping at least two copies of the source code on different drives.

7.2.1 Instruction Set

Table 7.2 is a more user-friendly form of the PIC 16F84 instruction set organised by function. An example is given with each instruction so that the exact syntax can be seen. More detailed information is provided in the data sheet. Other PIC chips have additional instructions, but they all use the same basic set.

7.2.2 BIN3 Source Code

In program BIN3 the same instructions are used as in BIN2 (Chapter 6), with additional statements to read the switches and control the output. Program 7.1 is the result.

Firstly, note the general layout and punctuation required. The program header block contains as much information as is necessary at this stage. These comments are preceded by a semicolon on each line to indicate to the assembler that this text is not part of the program. Assembler directives such as EQU and END are also not part of the program proper, but used to define labels and the end of the program source code. The labels 'porta', 'portb' and 'timer' refer to file registers 05, 06 and 0C, respectively; 'inres' and 'inrun' are input bit labels representing the push buttons. The program uses 'Bit Test and Skip' instructions followed by 'GOTO label' for conditional jumping.

At this stage, the reader can type the source code into the editor without full analysis in order to practise use of the editor. The instructions are placed in the first three columns, and the comments can be left out to save time. Labels go in the first column, instruction mnemonics in the second, and the instruction operands in the third. The source code text file should be saved as BIN3.ASM in a suitably named directory or folder on disk.

7.2.3 Syntax

'Syntax' refers to the way that words are put together to create meaningful statements, or a series of statements. In programming, the syntax rules are determined by the assembler which will be used to create the machine code, in our case, MPASM.EXE. The assembler must be provided with source code which it can convert into the required machine code without any ambiguity, that is, only one meaning is possible. This is why the assembler syntax rules are very strict.

7.2.4 Layout

The program layout should be in four columns, as described in Table 7.3. Each character then occupies the same space, and the columns are correctly aligned. The label, command and

Table 7.2 PIC 16F84 instruction set by functional groups

PIC 16F84 INSTRUCTION SET BY FUNCTIONAL GROUPS

F = Any file register (specified by number or label), example is 0C
W = Working register, W
L = Literal value (follows instruction), example is 0F9
* = Use of these instructions not now recommended by manufacturer

	Operation	Example	
Move			
	Move data from F to W	MOVF	0C,W
	Move data from W to F	MOVWF	0C
	Move literal into W	MOVLW	0F9
Register			
	Clear W (reset all bits and value to 0)	CLRW	
	Clear F (reset all bits and value to 0)	CLRF	0C
	Decrement F (reduce by 1)	DECF	0C
	Increment F (increase by 1)	INCF	0C
	Swap the upper and lower four bits in F	SWAPF	0C
	Complement F value (invert all bits)	COMF	0C
	Rotate bits Left through Carry Flag	RLF	0C
	Rotate bits Right through Carry Flag	RRF	0C
	Clear (reset to zero) the bit specified (e.g. bit 3)	BCF	0C,3
	Set (to 1) the bit specified (e.g. bit 3)	BSF	0C,3
Arithmetic			
	Add W to F	ADDWF	0C
	Add F to W	ADDWF	0C,W
	Add L to W	ADDLW	0F9
	Subtract W from F	SUBWF	0C
	Subtract W from F, placing result in W	SUBWF	0C,W
	Subtract W from L, placing result in W	SUBLW	0F9
Logic			
	AND the bits of W and F, result in F	ANDWF	0C
	AND the bits of W and F, result in W	ANDWF	0C,W
	AND the bits of L and W, result in W	ANDLW	0F9
	OR the bits of W and F, result in F	IORWF	0C
	OR the bits of W and F, result in W	IORWF	0C,W
	OR the bits of L and W, result in W	IORLW	0F9
	Exclusive OR the bits of W and F, result in F	XORWF	0C
	Exclusive OR the bits of W and F, result in W	XORWF	0C,W
	Exclusive OR the bits of L and W	XORLW	0F9
Test and Skip			
	Test a bit in F and Skip next instruction if it is Clear (=0)	BTFSC	0C,3
	Test a bit in F and Skip next instruction if it is Set (=1)	BTFSS	0C,3
	Decrement F and Skip next Instruction if it is now Zero	DECFSZ	0C
	Increment F and Skip next Instruction if it is now Zero	INCFSZ	0C
Jump			
	Go To a Labelled Line in the Program	GOTO	start
	Jump to the Label at the start of a Subroutine	CALL	delay
	Return at the end of a Subroutine to the next instruction	RETURN	
	Return at the end of a Subroutine with L in W	RETLW	0F9
	Return from Interrupt Service Routine to next instruction	RETFIE	

Control

No Operation – delay for 1 cycle	NOP	
Go into Standby Mode to save power	SLEEP	
Clear Watchdog Timer to prevent automatic reset	CLRWDT	
Load Port Data Direction Register from W*	TRIS	06
Load Option Control Register from W*	OPTION	

The result of arithmetic and logic operations can generally be stored in W instead of the file register by adding ',W' to the instruction. General purpose register 1, address 0C, represents all file registers (00–4F). Literal value 0F9 represents all values 00–FF. Bit 3 is used to represent file register bits 0–7. For MOVE instructions data is copied to the destination but retained in the source register.

Program 7.1 BIN3 source code

```
;       BIN3.ASM                M. Bates          12-10-03
;.................................................................
;
;       Slow output binary count is stopped, started
;       and reset with push buttons.
;
;       Processor = 16F84        Clock = CR, 100kHz
;       Inputs: RA0, RA1         Outputs: RB0 – RB7
;
; ****************************************************************

; Register Label Equates ..........................................

porta   EQU     05              ; Port A Data Register
portb   EQU     06              ; Port B Data Register
timer   EQU     0C              ; Spare register for delay

; Input Bit Label Equates..........................................

inres   EQU     0               ; 'Reset' input button = RA0
inrun   EQU     1               ; 'Run' input button = RA1


; ****************************************************************

; Initialise Port B (Port A defaults to inputs).....................

        MOVLW   00              ; Port B Data Direction Code
        TRIS    portb           ; Load the DDR code into F86

; Start main loop .................................................

reset   CLRF    portb           ; Clear Port B

start   BTFSS   porta,inres     ; Test RA0 input button
        GOTO    reset           ; and reset Port B if pressed
                                              continued ...
```

```
         BTFSC        porta,inrun        ; Test RA1 input button
         GOTO         start              ; and run count if pressed

         INCF         portb              ; Increment count at Port B

         MOVLW        0FF                ; Delay count literal
         MOVWF        timer              ; Copy W to timer register
down     DECFSZ       timer              ; Decrement timer register
         GOTO         down               ; and repeat until zero

         GOTO         start              ; Repeat main loop always
         END                             ; Terminate source code
```

Table 7.3   Layout of assembler source code

Column 1	Column 2	Column 3	Column 4
Label	**COMMAND**	**Operand/s**	**; Comment**
Label EQUated to a value, or to indicate a program destination address for jumps.	Mnemonic form of the instruction for the processor to carry out a specific operation. Only mnemonics specified in the instruction set may be used.	The data or register contents to be used in the instruction. Registers are usually represented by a label. Some instructions do not need an operand.	Explanatory text to the right of a semicolon on any line of code helps the programmer and user to understand the program. It has no effect on the operation of the program. Full line comments may also be used between program blocks.

operand columns should be set to a width of 8 characters, with the maximum label length of 6 characters, leaving a minimum of two clear spaces between columns (longer labels can be used, but a different form of the program layout must then be used). The tab key is normally used to place the text in columns, and the tab spacing can be adjusted if necessary.

7.2.5   Comments

Comments are not part of the actual program, but are included to help the programmer and user understand how the program works. Comments are preceded by a semicolon (;), which can be placed at the beginning of a line to indicate a comment which relates to a whole program block (functional set of statements), or at the start of Column 4 for line comment. The comment and line are terminated with a line return ('Enter' key).

A standard header block is recommended (see Program 7.1). For simple programs, the first line should at least contain the source code file name, the author and date, and/or version number. A program description should also be provided in the header, and for more complex programs, the processor type, hardware setup and other relevant information.

7.2.6 Creating a Project

MPLAB is designed to work using a named project to keep track of the application files created. A project file can be created which records information about the project such as the location of the application files and the window configuration used for testing. When the project is re-opened, the windows re-appear as they were last set up. It is not essential to create a project for our simple applications, which can be assembled by selecting 'Quickbuild' (Ver. 6) or 'Build Node' (Ver. 5) but earlier versions of MPLAB required it, and it will be needed for more complex applications.

7.3   Program Structure

Structured programming means constructing the program, as far as possible, from discrete blocks. This makes the program easier to write and understand, more reliable and easier to modify at a later date.

7.3.1   BIN4 Source Code

Program BIN3 (Chapter 6) is unstructured, in that the program instructions are essentially executed in the order given in the source code. An equivalent 'structured' program, BIN4, is listed as Program 7.2.

The main difference between BIN3 and BIN4 is that the program now has the delay code as a 'subroutine'. The subroutine is inserted before the main program block, and assembled first. It is then 'called' from the main program by label. The subroutine can be created as a self-contained program block, and re-used in the program as necessary. It can be called as many times as required, which means that the block of code only needs to be written once. It can also be saved as a separate file and re-used in another program.

A program flowchart has been given for BIN3 (Fig. 7.2). The same flowchart describes BIN4, but the delay routine can now be expanded as a separate subroutine flowchart (Fig. 7.3). In addition, the delay time is loaded prior to the subroutine execution, so the same delay routine could be used to provide different delay times. The use of flowcharts in program design will be more fully examined in Chapter 10.

7.4   Program Analysis

The program BIN4 will now be analysed in some detail as it was designed to contain examples of most of the basic PIC syntax. A sample instruction is given in each case.

7.4.1   Label Equates

```
timer    EQU    0C
```

The use of labels in place of numbers makes programs easier to write and understand, but we have to 'declare' those labels at the beginning of the program. In assembly code, the assembler directive EQU is used to assign a label to number, which can be a literal, file register number or individual register bit. In BIN4, 'porta' and 'portb' are the port data registers (05 and 06) and 'timer' is the first spare register (0C), which will be used as a software counter. The labels 'inres' and 'inrun' will represent Bit 0 and Bit 1 of Port A; they are simply given the numerical value 0 and 1.

Program 7.2 BIN4 source code

```
;       Source File:    BIN4.ASM
;       Author:         M. Bates
;       Date:           15-10-03
;       ..............................................................
;       Program Description:
;
;       Slow output binary count is stopped, started
;       and reset with push buttons. This version uses a
;       subroutine for the delay....
;
;       Processor:      PIC 16F84
;
;       Hardware:       BIN Demo System
;       Clock:          CR ~100kHz
;       Inputs:         Push Buttons RA0, RA1 (active low)
;       Outputs:        LEDs (active high)
;
;       WDTimer:        Disabled
;       PUTimer:        Enabled
;       Interrupts:     Disabled
;       Code Protect:   Disabled
;
; ********************************************************************

; Register Label Equates.........................................

porta   EQU     05              ; Port A Data Register
portb   EQU     06              ; Port B Data Register
timer   EQU     0C              ; Spare register for delay

; Input Bit Label Equates........................................

inres   EQU     0               ; 'Reset' input button = RA0
inrun   EQU     1               ; 'Run' input button = RA1

; ********************************************************************

; Initialise Port B (Port A defaults to inputs)..................

        MOVLW   b'00000000'     ; Port B Data Direction Code
        TRIS    portb           ; Load the DDR code into F86
        GOTO    reset

; 'delay' subroutine.............................................

delay   MOVWF   timer           ; Copy W to timer register
down    DECFSZ  timer           ; Decrement timer register
        GOTO    down            ; and repeat until zero
        RETURN                  ; Jump back to main program
```

```
; Start main loop.........................................

reset    CLRF     portb                 ; Clear Port B Data

start    BTFSS    porta,inres           ; Test RA0 input button
         GOTO     reset                 ; and reset Port B if pressed
         BTFSC    porta,inrun           ; Test RA1 input button
         GOTO     start                 ; and run count if pressed

         INCF     portb                 ; Increment count at Port B
         MOVLW    0FF                   ; Delay count literal
         CALL     delay                 ; Jump to subroutine 'delay'

         GOTO     start                 ; Repeat main loop always
         END                            ; Terminate source code
```

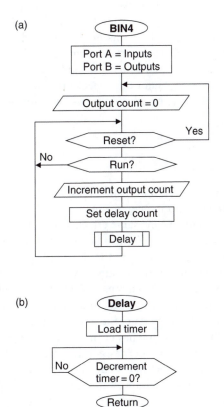

Figure 7.3 Flowcharts for program BIN4. (a) Main routine; (b) Subroutine.

7.4.2   Port Initialisation

```
TRIS      portb
```

Port B is used as the output for the 8-bit binary count. The data direction must be set up using the TRIS command, which loads the port data direction register with the data direction code. In this example, the code is given in binary, b'00000000'. This is useful, especially if the port bits are to be set as a mixture of inputs and outputs; the binary code identifies the data direction for each bit individually. This code is loaded into W using MOVLW, and the TRIS command follows.

The TRIS instruction is still available as a simple way of initialising the ports, but the manufacturers recommend an alternative method which involves bank selection, and will be covered later. Hopefully, TRIS will continue to be supported in by the MPASM assembler, as it is easier for beginners.

7.4.3   Program Jumps

```
GOTO      start
```

The 'GOTO label' command is used to make the program jump to a line other than the one following. In BIN4, 'GOTO reset' skips over the following DELAY routine, to start the main loop. We will come back to the reason for this in a moment. There is another unconditional jump at the end of the program, 'GOTO start', which makes the main loop repeat endlessly. Other 'GOTO label' instructions are used with 'Test and Skip' instructions to create conditional branches. In this program, the input buttons are checked using this type of instruction and the program branches, or not, depending on whether it has been pressed.

7.4.4   Bit Test and Skip if Set/Clear

```
BTFSS     porta,inres
```

The input button connected to Port A, bit 0 is tested using the above instruction, which means 'Bit Test File (register bit) and Skip the next instruction if it is Set (=1)'. Without labels, the instruction' 'BTFSS 05,0' would have the same effect. The buttons are connected 'active low', meaning that the input goes from '1' to '0' when the button is pressed. If the button connected to RA0 is not pressed, the input will be high, that is, set. The following instruction, 'GOTO reset' is therefore skipped, and the next executed. When the button is pressed, the 'GOTO reset' is executed, and the CLRF instruction repeated, clearing the previous count.

BTFSC means 'Bit Test and Skip if Clear'; it works in the same way as BTFSS, except that the logic is reversed. Thus, 'BTFSC porta,inrun' tests bit 1 of Port A register and skips the following 'GOTO start' if the 'run' button has been pressed. The program will then proceed to increment the output count. If button is not pressed, the program waits by jumping back to the 'start' line. The combined effect is that the count runs when the 'run' button is pressed, and the count is reset to zero if the 'reset' button is pressed.

7.4.5   Decrement/Increment Register and Skip If Zero

```
DECFSZ    timer
```

The other instructions for conditional branching allow a register to be incremented or decremented and then checked for a zero result. This is a common requirement for counting and timing applications, and in the delay routine in BIN3, a register 'timer' is loaded with the maximum value FF and decremented. If the result is not yet zero, the jump 'GOTO down' is executed. When the register reaches zero, the GOTO is skipped and the subroutine ends. In BIN4, the timer value is set up before the delay subroutine is called.

7.4.6 *Subroutine Call and Return*

The main elements of the subroutine call structure are shown below:

```
main   ......              ; start main program
       ......
       CALL delay          ; jump to subroutine
       ......              ; return to here

delay ......               ; subroutine start
       ......
       ......
       RETURN              ; subroutine ends
```

In this program, the subroutine provides a delay by loading a register and counting down to zero. The delay is started using the 'CALL delay' instruction, when the program jumps to the label 'delay' and runs from there. CALL means 'jump and come back to the same place after the subroutine', so the return address has to be stored for later recall in a special memory block called the 'stack'.

The address of the instruction following (in this case 'GOTO start') is saved automatically on the stack as part of the execution of the CALL instruction. The subroutine is terminated with the instruction 'RETURN', which does not require an operand because the return destination address is automatically pulled from the stack and replaced in the program counter. This takes the program back to the original place in the main program. The stack can store up to eight return addresses, so multiple levels of subroutine can be used. The return addresses are pushed onto and pulled from the stack in order, so if a CALL or RETURN is missed out of the program, a stack error will occur.

7.4.7 *End of Source Code*

```
END
```

The source code must be terminated with assembler directive END so that the assembly process can be stopped in an orderly way, and control returned to the host operating system.

7.5 Program Assembly

The assembler program (MPASM) takes the source code text and decodes it character by character, line by line, starting at the top left. In MPLAB, the correct processor type must first be selected via the configuration menu, as there is some variation in valid syntax between processors. Then, in the project menu, select the option to assemble a single file. It is not necessary, at this stage, to create a project.

When the assembler runs, the corresponding 14-bit binary machine code for each line in the source code is generated, until the END directive is detected. The binary code created is automatically saved as a file called BIN4.HEX in the same folder as the source code.

7.5.1   Syntax Errors

If there are any syntax errors in the source code, such as spelling, layout, punctuation or failure to define labels properly, error messages will be generated by the assembler. These will be displayed in a separate window, indicating the type of error and line number. You must note the messages and line numbers, or print out the error file, BIN4.ERR. Then go back and re-edit the source code and make the necessary changes. The error is sometimes on a previous line to the one indicated, and sometimes one error can generate more than one message. Warnings and information messages can usually be ignored. There is more details about error messages in Chapter 11.

You may receive the following messages:

```
Warning[224] C:\MPLAB\BOOKPRGS\BIN4.ASM 65 : Use of this
instruction is not recommended.
```

```
Message[305] C:\MPLAB\BOOKPRGS\BIN4.ASM 81 : Using default
destination of 1 (file).
```

The first warning will be caused by using the instruction TRIS, which the manufacturer warns may not be supported in future (OPTION is also not recommended). However, it is used here because it simplifies the initialisation of the ports. The message about the 'default destination' is caused by the simplified syntax used in these programs, where the file register is not explicitly specified as destination in instructions where the result can be placed either in the file register or in the working register (see Section 9.4.1). The assembler assumes that the file register is the destination by default.

When all errors have been eliminated, and the program successfully assembled, the machine code can be inspected by viewing program memory. Note that the source code labels are not reproduced, as the program code has been 'disassembled' from the machine code. That is, the hex file has been converted back to mnemonic form so that it can be checked against the original.

7.5.2   List File

A program 'list file' BIN4.LST is also produced by the assembler, which contains the source code, the machine code, error messages and other information all in one listing (Table 7.4). This is useful for analysing the program and assembler operations, and debugging the source code.

The list file header shows the assembler version used and source file details. The column headings are then given:

LOC: Memory location addresses at which the machine code
                will be stored
VALUE: The numerical value with which equate labels will be
                replaced
OBJECT CODE: Machine code produced for each instruction
LINE: Line number of list file
SOURCE TEXT: Source code including comments

At the end of the list file, additional information is provided:

SYMBOL TABLE: Lists all the equate and address labels allocated
MEMORY USAGE MAP: Shows the locations occupied by the object code

Note that there is no machine code produced by the lines which are occupied by a full line comment. The actual program starts to be produced at line 00040. The machine code for the first

Table 7.4 BIN4 list file

```
MPASM 01.21 Released     BIN4.ASM      24-10-03     15:04:14

LOC OBJECT CODE    LINE SOURCE TEXT
  VALUE
                00001 ;
                00002 ;      BIN4.ASM     M. Bates     24-10-03
                00003 ;      ...........................................
                00004 ;
                00005 ;      Output binary sequence is stopped, started
                00006 ;      and reset with input buttons...
                00007 ;
                00008 ;      Processor:    PIC 16F84
                00009 ;
                00010 ;      Hardware:     PIC Demo System
                00011 ;      Clock:        CR ~100kHz
                00012 ;      Inputs:       Push Buttons RA0, RA1 (active low)
                00013 ;      Outputs:      LEDs (active high)
                00014 ;
                00015 ;      WDTimer:      Disabled
                00016 ;      PUTimer:      Enabled
                00017 ;      Interrupts:   Disabled
                00018 ;      Code Protect: Disabled
                00019 ;
                00020 ;      Subroutines: DELAY
                00021 ;      Parameters:  None
                00022 ;
                00023 ; ****************************************************
                00024
                00025 ;      Register Label Equates ......................
                00026
  0005          00027 porta EQU    05            ; Port A Data Register
  0006          00028 portb EQU    06            ; Port B Data Register
  000C          00029 timer EQU    0C            ; Spare register for delay
                00030
                00031 ;      Input Bit Label Equates .....................
                00032
  0000          00033 inres EQU    0             ; 'Reset' input button=RA0
  0001          00034 inrun EQU    1             ; 'Run' input button = RA1
                00035
                00036 ; ****************************************************
                00037
                00038 ;      Initialise Port B (Port A defaults to inputs).....
                00039
  0000 3000     00040         MOVLW b'00000000' ; Port B Data Direction Code
  0001 0066     00041         TRIS   portb      ; Load the DDR code into F86
                00042
  0002 2808     00043         GOTO   reset      ; Jump to start of main
                00044
                00045 ;      Define DELAY subroutine......................
                00046
  0003 30FF     00047 delay  MOVLW  0xFF        ; Delay count literal
  0004 008C     00048        MOVWF  timer       ; is loaded into spare reg.
                00049
                                                        continued...
```

Table 7.4   continued

```
0005 0B8C   00050 down   DECFSZ   timer         ; Decrement timer register
0006 2805   00051        GOTO     down          ; and repeat until zero then
0007 0008   00052        RETURN                 ; return to main program
            00053
            00054
            00055 ;    Start main loop ...............................
            00056
            00057
0008 0186   00058 reset  CLRF     portb         ; Clear Port B Data
            00059
0009 1C05   00060 start  BTFSS    porta,inres   ; Test RA0 input button
000A 2808   00061        GOTO     reset         ; and reset Port B
            00062
000B 1885   00063        BTFSC    porta,inrun   ; Test RA1 input button
000C 2809   00064        GOTO     start         ; and run count if pressed
            00065
            00066
000D 0A86   00067        INCF     portb         ; Increment count at Port B
000E 2003   00068        CALL     delay         ; Execute delay subroutine
000F 2809   00069        GOTO     start         ; Repeat main loop
            00070
            00071
            00072        END                    ; Terminate source code
```

```
SYMBOL TABLE
  LABEL                          VALUE

__16C84                          00000001
delay                            00000003
down                             00000005
inres                            00000000
inrun                            00000001
porta                            00000005
portb                            00000006
reset                            00000008
start                            00000009
timer                            0000000C


MEMORY USAGE MAP ('X' = Used, '-' = Unused)

0000 : XXXXXXXXXXXXXXXX ---------- ---------- ----------
0040 : ---------- ---------- ---------- ----------

All other memory blocks unused.


Errors    :    0
Warnings  :    0
Messages  :    2
```

instruction is shown in column 2 (3000), and the address where it will be stored in the chip when downloaded is shown in column 1 (0000). The whole program will occupy locations 0000–000F (16 instructions).

If we study the machine code, we can see how the labelling works; for example, the last instruction 'GOTO start' is encoded as 2809, and the 09 refers to address 09 in column 1, the location with the label 'start'. The assembler program has replaced the label with the corresponding numerical address for the jump destination. Similarly, the label 'porta' is replaced with its file register number 05 in the instruction code to test the input, 1C05.

The label values are listed again in the symbol table. These values will be used by the simulator to allow the user to display the simulated registers by label. The amount of program memory used, 16 locations (0000–000F), is shown in graphical format in the memory usage map, and finally a total of errors, warnings and messages given. If there are fatal errors, which prevent successful assembly of the program, the list file will not be produced.

7.6   Program Simulation

The BIN4.HEX file could now be downloaded to the PIC chip and executed; it would run correctly, because the program given here has already been tested. However, when a program is first developed, it is quite likely that 'logical' errors will be found. Logical errors prevent the program from work correctly; that is, the program executes but it does not necessarily carry out the right operations in the right order.

If this is so, the source code must be analysed again to try to find the errors. In complex programs, this process might have to be repeated many times, making it time-consuming and inefficient if the program has to be downloaded for testing in the hardware each time. This is where a software simulator comes in – it allows the program to be 'run' on the host PC, as if it were being executed in the chip, but without having to download to the actual hardware. It can then be checked for logical errors and the source code changed and re-tested much more quickly and easily.

Using the simulator MPLAB SIM (Fig. 7.4), the program can be run and stopped at will; this allows the effect of the program on the registers to be checked at critical points. For example, in BIN4 we would check to see that Port B has been incremented after the execution of the main loop, because this is the primary function of the program.

Suppose that in developing BIN4 we had failed to analyse the switching logic correctly, and instead of the instruction BTFSS (Bit Test and Skip if Set) at the line labelled 'start', BTFSC (Bit Test and Skip if Clear) had been entered. The program would assemble successfully, but when tested, would not run correctly. While the 'inres' switch was not pressed, the program would jump back to the 'reset' line instead of going to the second switch ('inrun') test to allow the output sequence to be started. This fault could be detected in the simulator, by running the program, stepping through the start sequence and simulating the switch inputs.

BIN4 can be tested as follows. Figure 7.4 shows the relevant windows in a screen shot from MPLAB 6.

7.6.1 Single Stepping

To test a program in the edit window (source code debugging), make sure it is saved and assembled correctly. Enable the simulator mode as required by the MPLAB version in use, so that the simulator toolbar buttons are showing. Operate the 'Run' button; nothing appears to happen, but when the 'Stop' button is operated, the current execution point is indicated in the source code window.

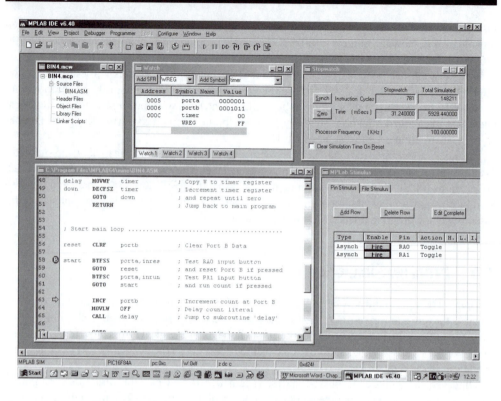

Figure 7.4   MPLAB simulation windows for debugging program BIN4.

The program can now be executed one instruction at a time using the 'Step Into' button, and the sequence examined. The program should loop through the reset sequence. The program can be restarted from the top at any time by clicking on the 'Reset' button.

7.6.2   Input Simulation

We now need to simulate the action of the push buttons in the hardware which are used to start and stop the output sequence. The simplest way to do this is to use 'asynchronous input' found under the dubugger, stimulus menu. This displays a set of screen buttons which can be assigned to any input. Assign a button to RA0 and RA1, and set to toggle mode; the toggle option will make the input change over each time the button is pressed.

The inputs can now be toggled to allow the program to proceed from the reset loop. Set both inputs high initially; taking RA1 low will allow the main loop to proceed, and operating RA0 will execute the reset loop. Unfortunately, the state of the input is not indicated in the simulator stimulus window, so file register 05, Port A, must be displayed (preferably in binary) in order to confirm the changes on the inputs.

7.6.3   Register Display

View the file registers, or special function registers, to check the effect of the program on the output register 06, Port B. The changes at the inputs can also be checked, and any intermediate changes in internal registers tracked. Registers can also be displayed selectively using a watch window, where only those registers affected by the program are seen.

7.6.4 Step Over

Once the program has entered the delay loop, single stepping is not so useful, because the program is simply executing the same simple sequence over and over. A short cut around the delay subroutine is needed. One way to do this is to use the 'Step Over' button; when the subroutine is reached in the main loop, it is executed at full speed and the single stepping mode re-entered upon return from the subroutine.

7.6.5 Breakpoints

Another technique for executing some parts of the program at full speed is the use of breakpoints. For example, if part of a large program is known to be correct, it can be skipped and single stepping started at a later point in the program. In BIN4, a break point can be set at the start of the main loop by right-clicking with the mouse over the program line labelled 'start', and selecting 'Set Breakpoint' from the menu. The program can then be run from the start, and it will stop at the breakpoint. Run again, and a complete loop will be completed at full speed and Port B should increment.

7.6.6 Stopwatch

The program timing can be checked using the stopwatch feature. This displays the total number of instructions executed and the time elapsed, calculated from the processor clock frequency. For BIN4, a CR clock is assumed, operating at 100 kHz. The processor frequency must be set to this value in the simulator. Then run the program to the break point at 'start', zero the clock and run again. The stopwatch will display the total time for one cycle.

The frequency of the output can now be predicted. Two program loop cycles will cause the low output bit RB0 to be toggled up and down once, giving one full output cycle. Therefore, we can double the loop time to give the output period, and calculate the reciprocal to give the frequency at RB0. The period at RB7 will be 128 times longer, and the frequency 128 times higher.

From the stopwatch readings:

```
Number of instructions executed per loop = 777
Processor frequency                      = 100 kHz
Loop time                                = 31.08 ms
```

Therefore:

```
Output period at RB0     = 2 × 31.08      = 62.16 ms
Output period at RB7     = 0.06216 × 128  = 7.96 s
Output frequency at RB0  = 1/0.06216      = 16.1 Hz
```

This shows that changes in the higher order output bits will be clearly visible using this clock frequency with the maximum delay loop count (FF). The frequency at RB1 will be about 8 Hz, RB2 4 Hz, RB3 2 Hz and RB4 1 Hz and so on.

More information about using MPLAB for debugging is given in Chapter 11.

7.7   Program Downloading

After testing in the simulator for correct operation, the machine code program can now be blown into the Flash ROM on the chip. The program is downloaded via a serial or parallel link into RB6 (clock) and RB7 (data) on the chip, while a programming voltage of about 14 V is applied to MCLR. There are two methods for program downloading, outlined below.

7.7.1   Programming Unit

In this case, a programming unit must be plugged into the serial port of the PC (COM1 or COM2), power connected and the chip inserted into the socket on the programmer. This is usually a ZIF socket which is opened to allow the IC to be dropped in, and closed to clamp the pins (Fig. 7.5). The chip orientation must be carefully checked, as inversion could probably damage the chip, as the supplies will be reversed. Anti-static precautions should be observed, since the PIC is a CMOS device (conductive bench cover and earthed wrist strap). However, the device has not been found to be too sensitive in practice.

The programming unit is enabled from the main menu bar. If the programming unit has been correctly connected, a programming dialogue should open, with the hex code to be downloaded also visible. If not, the COM port may need to be changed.

Before downloading, the chip configuration options need to be selected (Fig. 7.6).

Figure 7.5   PIC programming unit.

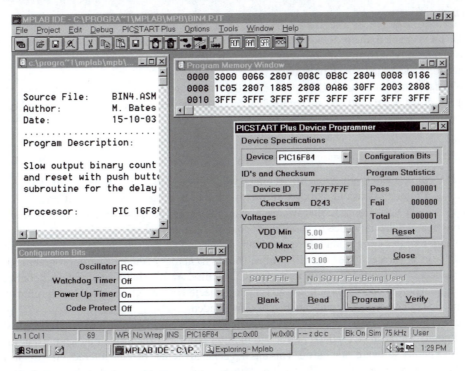

Figure 7.6 MPLAB (version 5) program downloading windows.

Oscillator (Clock): RC

The main options here are 'RC' and 'XT'. RC must be selected for the oscillator configuration used in the BIN hardware. XT will be used later for crystal-oscillator-clocked applications. The program will not run in the hardware if the wrong type of oscillator has been selected, so check this carefully.

Watchdog Timer (WDT): Off

The watchdog timer is an internal timer which automatically restarts the program if it is not cleared back to zero within 18 ms, using the instruction CLRWDT. This can be used to stop the controller getting 'stuck' in a loop, due to an undetected program bug, or an input condition which has not been predicted in testing. For applications not using this feature, WDT must be switched off, or the program will reset repeatedly, preventing normal operation.

Power Up Timer (PuT): On

Mains-derived power supplies may take some time to reach the correct value (+5 V) when first switched on. The power up timer is an internal timer which delays the start of program execution until the power supply is at the correct voltage and stable. This helps to ensure that the program starts correctly every time. Always enable the power up timer unless there is a good reason not to do so.

Code Protect (CP): Off

If this bit is enabled, the program cannot be read back into MPLAB and copied or manipulated. This is normally only necessary for commercial applications to prevent software piracy, so code protection can normally be switched off.

When the configuration bits have been set, click on the program button, and the progress of the program download should be displayed. When complete, the chip can be removed to the application circuit.

7.7.2 In-Circuit Programming and Debugging

The PIC chip can alternatively be programmed while fitted in the application circuit. This is very useful as it minimises risk of damage, allows the chip to be programmed with different software versions after completion of the hardware and even allows remote re-programming via a communications link. All that is needed is an on-board connection to RB6, RB7 and MCLR.

In addition, this connection allows in-circuit debugging (ICD), which is a very useful feature available in more recently designed PIC chips. The 16F877 is one of these; it can be programmed and the application debugged in circuit using the ICD facility. An ICD hardware module is connected to the serial port of the PC; its output lead is then connected to pins RB6 and RB7 of the chip in the application circuit. The program can then be downloaded.

If the chip is now set to run in ICD mode, the debugging features available MPLAB (single step, breakpoints, etc.) can be used to control and monitor program execution in the chip itself, rather than in a purely software simulation. This has the great advantage that the interaction with the real hardware can be monitored, and hence the hardware and software verified at the same time. See Chapter 14 for more details.

7.8 Program Testing

The application circuit should be checked for correct on-board connections before inserting the microcontroller and other active devices (this is easier if the chips are in sockets). Static hardware testing is important if it is a newly constructed circuit, and essential if it is newly designed. When the hardware has been thoroughly checked, insert the microcontroller (ensuring correct orientation) and power up!

In a commercial product, a test schedule must be devised and correct operation to that schedule confirmed and recorded. The test procedure should check all possible input sequences, not just the correct ones, if the design is to be foolproof. It is, in fact, quite difficult to be sure that complex programs will always be 100% reliable, as it is often not feasible to predict every possible operating sequence. A outline test procedure for BIN4 is suggested in Table 7.5.

The program should start immediately on power up. If it does not function correctly, when tested against the original specification, a fault finding process needs to be followed, as outlined below.

1. Hardware checks
   (a) +5 V on MCLR, Vdd, 0 V on Vss,
   (b) clock signal on CLKIN,
   (b) input changes on RB0, RB1.

2. Software Checks
   (a) simulation correct,
   (b) correct clock selected,

Table 7.5 Basic test schedule for BIN4

	Test	Correct operation	Checked
1	Check PIC connections	Correct orientation and pins	
2	Power up	LEDs off	
3	Clock frequency	100 kHz	
4	Press RUN	Count on LEDs	
5	Release RUN	LED count halted	
6	Press and release RESET	LEDs off	
7	Press RUN	Count on LEDs from zero	

```
(c)  WDT off, PuT on, CP off,
(d)  program verified.
```

More suggestions on hardware and software testing are given in Chapter 11.

Summary

- The development process consists of application specification, hardware selection and design, and program development and testing.

- The program is converted from a software design to assembler source code, FILENAME.ASM, using the instruction format defined for the assembler.

- The assembler converts the source code text into object code, FILENAME.HEX. Any syntax errors detected must be corrected.

- A list file, FILENAME.LST, is created which lists the source code, object code, label and memory allocation.

- The simulator allows the machine code to be tested without downloading to the actual target system. Logical errors can be detected and corrected at this stage.

- The program can then be downloaded and tested in the target hardware, using a test schedule developed from the specification.

Questions

1. Place the following program development steps in the correct order: Test in Hardware, Simulate, Assemble, Edit Source Code, Download.

2. Suggest two advantages of using 'C' as the programming language.

3. State two advantages of using subroutines.

4.   State the instruction for incrementing the register 0F.

5.   In which register must a port data direction code be placed prior to using the TRIS instruction?

6.   How could you halve the delay time in BIN4?

7.   Explain how a switched input on RA4 of the PIC 16F84 is simulated in MPLAB.

8.   State the configuration bit settings which should normally be selected when downloading a simple application to a PIC chip. Why is it generally desirable to enable the power on timer?

4.   INCF 0F

5.   W

6.   Delay Count = 80h

1.   Download from www.microchip.com or otherwise obtain the supporting documentation for MPLAB. Study the tutorial in the User's Guide and the help files supplied with MPLAB as necessary to familiarise yourself with editing, assembling and simulating an application program. Start up MPLAB, create a source code file for BIN3, and enter the assembler code program, leaving out the comments. Assemble, correct any errors and simulate. Check that the Port B (F6) file register operates as required.

2.   Construct a prototype circuit and test the program to the test schedule given in Table 7.5. Refer forward to Chapter 12 if necessary.

3.   Modify the program as BIN4 and confirm that its operation is essentially the same.

4.   Modify the program to scan the output, that is, move one lit LED up and down the display, repeating indefinitely. Use the rotate instructions, and subtraction from 1 and 80h to check if the bit is at one end.

Chapter 8
PIC 16F84 Architecture

8.1 **Block Diagram**
8.2 **Program Execution**
8.3 **Register Set**

An overview of programming the PIC microcontroller has been provided in Chapter 7; we can now look at the PIC internal hardware arrangement in more detail. We will use the 16F84 as a reference, since it has all the essential elements, without some of the more advanced features such as analogue inputs and serial ports found on larger chips such as the 16F877, which we will look at later. The other members of the PIC family are based on the same architecture, with elements added, removed or modified according to the combination of features provided in each chip.

The key reference is Fig. 1.1 in the PIC 16F84A Data Sheet, the 'PIC 16F84A Block Diagram'. The data sheet contains all the details of the internal architecture discussed in this chapter. Refer back to Chapter 4 for a description of the function of elements such as registers, ALU, multiplexer, decoder, program counter and memory.

8.1 Block Diagram

A somewhat simplified internal architecture (Fig. 8.1) has been derived from the block diagram given in the data sheet. Some features seen in the manufacturer's diagram have been left out because they are not important at this stage. The functional blocks of the chip are shown, with the main address paths identified as block arrows. The 8-bit data paths are shown in an alternative style as single arrows in this diagram. The timing and control block has control connections to all other blocks, which set up the processor operations, but they are not all shown explicitly in order to keep the diagram as clear as possible.

The file register set contains various control and status registers, as well as the port registers and the program counter. The most commonly used are the ports (A and B), status register (STATUS), real-time clock counter (TMR0) and interrupt control (INTCON). There are also a number of spare general purpose registers (GPRs) which can be used as data registers, counters and so on. The file registers are numbered 00–4F, but are usually given suitable labels in the program source code. File registers also give access to a block of EEPROM, a non-volatile data memory.

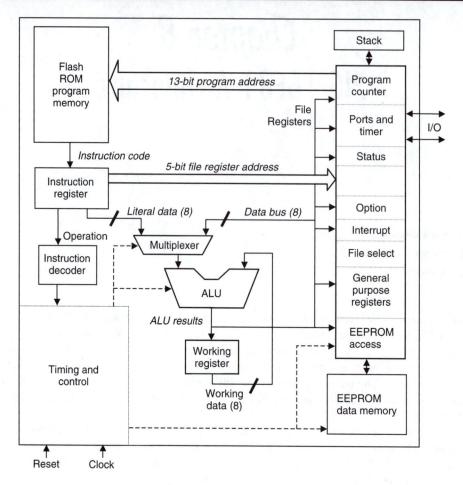

Figure 8.1 PIC 16F84 internal architecture.

8.1.1 Clock and Reset

A clock circuit is connected to the timing and control block to drive all the operations of the chip. For applications where precise timing is not required, a simple external resistor and capacitor network controls the frequency of the internal oscillator. Typically, relatively low frequencies are used (<1 MHz) with an RC clock. For more precise timing, a crystal oscillator is used (see Fig. 12.7); a convenient frequency is 4 MHz, because each instruction takes four clock cycles to execute, that is, 1 μs. The exact program execution timing can then be more easily calculated, and the hardware timer used for accurate signal generation and measurement. With the high-speed oscillator option selected, the processor can be clocked at up to 20 MHz, giving a minimum 200 ns instruction execution period, and a maximum instruction execution rate of 5 MIPs (millions of instructions per second).

The timing and control circuits contain start up timers which means that the reset input !MCLR can simply be connected to Vdd, the positive supply, to enable the processor. In earlier processors, an external reset circuit was often needed to ensure a smooth start up. An external reset button or control signal can still be connected to !MCLR if required. The controller program can then be restarted by pulsing the reset input low.

8.1.2 Harvard Architecture

It can be seen in the block diagrams that the memory and file register address lines are separate from the data paths within the processor. This is referred to as Harvard architecture; it improves the speed of processor operation because data and addresses do not have to share the same bus lines. It also allows 'pipelining' of the instruction execution; as one instruction is executed, the next is being fetched from program memory, thus doubling the overall execution rate. The reduced size of the instruction set also speeds up decoding and the short data path length in a single chip design reduces data transmission time. All these features contribute to high-speed operation, compared with some other microcontrollers which use a 'conventional' architecture.

8.2 Program Execution

The program consists of a block of 14-bit codes which contain both operation code and operand in a fixed length instruction. This machine code program is normally created using a host PC, and downloaded, as outlined in Chapter 7. We are not too concerned here exactly how the downloading is carried out; all we need to know for now is that the program is received in serial form from the host PC via the I/O port pin, RB7, and written to the program memory while a programming voltage applied to !MCLR. RB6 acts as a serial data clock.

8.2.1 Program Memory

Since the program memory is flash ROM, an existing program can be replaced by simply overwriting with a new program. Up to 1024 (1k) instructions can be stored in program memory. The program counter (PCL) holds the current address, and is reset to 0000 when the chip is powered up or reset.

The user program must therefore start at address 0000, but the first instruction can be GOTO the start of the program at some other labelled address. This is essential when using interrupts, as we shall see later, because the interrupt service routine (or GOTO ISR) must be placed at address 0004. The address is fed to the program memory via a 13-bit address bus. As the 16F84 memory only contains 1k locations, the actual address required is only 10 bits, but a standard 13-bit address is used in order to maintain compatibility with other PIC processors with more program memory, and other future products.

8.2.2 Instruction Execution

This section contains the instruction register, instruction decoder, and timing and control logic. The 14-bit instructions stored in program memory are copied to the instruction register for decoding. The instruction decoder logic converts the instruction input into settings for all the internal control lines. The 14-bit instruction contains both the operation code and operand. The instruction decoder will only use the operation code bits, while the operand provides a literal, file register address or program address when the instruction is executed.

If, for example, the instruction is MOVLW (Move a Literal into W), the control lines will be set up to feed the literal operand to W via the multiplexer and ALU. If the instruction is MOVWF, the control lines will be set up to copy the contents of W to the specified file register. The operand will be the address of the file register (00–4F) required. If we look at the 'move' instruction codes quoted in the Data Sheet, Table 7-2, we can see the difference in the code structure:

```
MOVLW k   = 11  00xx  kkkk  kkkk
MOVWF f   = 00  0000  1fff  ffff
MOVF f,d  = 00  1000  dfff  ffff
```

In the MOVLW instruction, the operation code is the high 4 bits (1100), 'x' are 'don't care' bits, and 'k' represents the literal bits, the low byte of the instruction. In the MOVWF instruction, the operation code is 0000001 (7 bits) and 'f' indicates the file register address bits. Only 7 bits are used for the register address, allowing a maximum of $2^7 (= 128)$ registers to be addressed. In fact, the 16F84 has only 80 registers in all, but 7 bits are still needed to address this number.

In the MOVF instruction the operation code is 001000, and the file register address is needed as before to identify the data source register. However, there is one bit, d, which controls the data destination. This bit must be 0 to direct the data into W, the usual operation. For example, to move an 8-bit data word from file register 0C to W requires the syntax MOVF 0C,W.

8.2.3 Data Processing

The ALU can add, subtract or carry out logical operations on single data bytes or pairs of numbers. These operations are carried out in conjunction with the data multiplexer and working register. The multiplexer allows new data to be fed from the instruction (a literal) or a register. This may be combined with data from W, or register data manipulated in a single register operation. W is used in register pair operations as the temporary store. Final results are stored back in the file registers.

8.2.4 Jump Instructions

If a GOTO instruction is decoded, the program counter will be loaded with the program memory address of the jump destination given as the instruction operand. The program label used in the source code will have been replaced by the destination address by the assembler. For conditional branching (making decisions), any file register bit can be used by a 'Bit Test and Skip' instruction, which is then followed by a GOTO or CALL instruction.

If a CALL instruction is decoded, the destination address is loaded into the PC in the same way as for the GOTO, but in addition, the next address to the CALL is stored on the stack (the return address). The subroutine is then executed until a RETURN instruction is encountered. The return address is then pulled from the stack and placed in the PC, allowing program execution to pass back to the original point. The stack works on a last in, first out (LIFO) basis, with the last address stored being the first to be recovered.

In conventional processors, the stack can be modified directly as it is typically located in main memory, but in the PIC this is not the case – stack operation is entirely automatic.

8.3 Register Set

All the file registers are 8-bits wide. They are divided into two main blocks – SFRs, which are reserved for specific purposes, and GPRs which can be used for temporary storage of any data byte. The file register set is shown in Fig. 8.2 in numerical order.

Address	Page 0	Page 1	Address
0	IND0		
1	**TMR0**	**OPTION**	81
2	**PCL**		
3	**STATUS**		
4	FSR		
5	**PORTA**	**TRISA**	85
6	**PORTB**	**TRISB**	86
7			
8	EEDATA	EECON1	88
9	EEADR	EECON2	89
A	PCLATH		
B	**INTCON**		
C	**GPR1**		
D	**GPR2**		
E	**GPR3**		
F	**GPR4**		
10	**GPR5**		
.	|		
.			
.	GPRs		
.			
.	|		
4F	**GPR68**		

Figure 8.2 PIC 16F84 file register set.

The registers in Page 0 (file addresses 00–4F) can be directly addressed, and it is recommended that the register labels given in Fig. 8.2, which match the data sheet, are used as the register labels. These labels are also used by default in MPLAB. Standard header files can be included in your programs which define all the register names.

Special instructions are available to access the Page 1 registers. We have already used the instruction TRIS to access the data direction registers TRISA and TRISB. In a similar way, we will use the instruction OPTION to access the option register; this will be used later to set up the hardware timer, TMR0. Alternatively, a register bank select bit in the status register can be used to access Page 1 file registers; this method is recommended for more advanced programming (see Section 9.4.2). Note that 'Page 0' means the same as 'Bank 0'.

8.3.1 Special Function Registers

The operation of the SFRs is summarised below, with the emphasis on those which are used most frequently. The functions of all the registers are detailed in the chip data sheet. The shaded registers in Fig. 8.2 either do not exist, or are repeated at addresses 80–CF (Page 1).

PCL Program Counter Low Byte
    File Register Number = 02

The program counter contains the address of (points to) the instruction currently being executed, and counts from 000 to 3FF unless there is a jump (GOTO or CALL). The PCL register contains only the low 8 bits (00–FF) of the whole program counter, with the high 2 bits (00–03) stored in the PCLATH register (address 0A). We only need to worry about the high bits if the program is longer than 255 instructions in total, which is not the case for any of the demonstration programs, and then only if the program counter is being modified directly.

The PC is automatically incremented during the instruction execution cycle, or the contents replaced entirely for a jump.

PORTA Port A Data Register
      File Register Number = 05

Port A has 5 I/O bits, RA0–RA4. Before use, the data direction for each pin must be set up by loading the TRISA register with a data direction code (see below). If a bit is set to output, data moved to this register appears at the output pins of the chip. If set as input, data presented to the pins can be acted on immediately, or stored for later use by moving the data to a spare register. Examples of this have already been seen in earlier chapters. In the 16F84, RA4 can alternatively be used as an input to the counter timer register (TMR0) for counting applications. The use of the hardware timer will be covered in Chapter 9. The PORTA register bit allocation is shown in Table 8.1. In other PIC chips, most port pins will have at least two uses, depending on how the control registers are set up.

All registers are read and written in 8-bit words, so we sometimes need to know what will happen with unused bits. When the Port A data register is read within a program (MOVF), the 3 unused bits will be seen as '0'. When writing to the port, the high 3 bits are simply ignored. When used as outputs, the port lines are able to provide up to 20 mA of current (except RA4), in or out, which is enough to drive our LEDs in the demonstration circuit. An equivalent circuit for each port pin is given in the data sheet, Section 5.

TRISA Port A Data Direction Register
      File Register Number = 85

The data direction of the port pins can be set bit by bit by loading this register with a suitable binary code, or the hex equivalent. A '1' sets the corresponding port bit to input, while a '0' sets it to output. Thus, to select all bits as inputs, the data direction code is 1111 1111 (FFh),

Table 8.1 PIC 16F84 port bit functions

Register bit	Chip pin label	Function
Port A		
0	RA0	Input or Output
1	RA1	Input or Output
2	RA2	Input or Output
3	RA3	Input or Output
4	RA4/T0CKI	Input or Output or Input to TMR0
5	–	None
6	–	None
7	–	None
Port B		
0	RB0/INT	Output or Input or Interrupt Input
1	RB1	Output or Input
2	RB2	Output or Input
3	RB3	Output or Input
4	RB4	Output or Input + Interrupt on change
5	RB5	Output or Input + Interrupt on change
6	RB6	Output or Input + Interrupt on change
7	RB7	Output or Input + Interrupt on change

and for all outputs is 0000 0000 (00h). Any combination of inputs and outputs can be set by loading the TRIS register with the required binary code.

When the chip is powered up, these bits default to '1', so it is not necessary to initialise for input, only for output. This makes sense, because if the pin is incorrectly wired up, it is more easily damaged if set to output. For instance, if the pin is accidentally grounded, and then driven to a high state, the short circuit current is likely to damage the output circuit. If set as an input, no damage would be done.

The data direction register TRISA is loaded by placing the required code in W and then using the instruction TRIS 05 or TRIS 06 for Port A and Port B, respectively. Alternatively all file registers with addresses 80–CF can be addressed directly, using the page selection bits in the option register, and this may be seen in later programs.

PORTB Port B Data Register
File Register Number = 06

Port B has the full set of eight I/O bits, RB0–RB7. If a bit is set to output, data moved to this register appears at the output pins of the chip. If set as input, data presented to the pins can be read at this address. The data direction is set in TRISB, as described above, and all bits default to input on power up. The PORTB register bit allocation is shown in Table 8.1.

Bit 0 of Port B has an alternate function; it can be initialised, using the Interrupt Control Register (INTCON), to allow the processor to respond to a change at this input with an interrupt sequence. In this case, the processor is forced to jump to a predefined interrupt service routine (ISR) upon completion of the current instruction (see Section 9.3). The processor can also be initialised to provide the same response to a change on any of the bits RB4–RB7.

TRISB Port B Data Direction Register
File Register Number = 86

As for Port A, the data direction can be set bit by bit by loading this register with a suitable binary code, or the hex equivalent, where '1' (default) sets an input, and '0' sets an output (must be initialised). The program instruction 'TRIS 06' moves the data direction code from W to TRISB register.

STATUS Status (or Flag) register
File Register Number = 03

Individual bits in the status register record information about the result of the previous instruction. Probably the most commonly used is the zero flag, bit 2; when the result of any operation is zero, this zero flag bit is set to '1'. It is used by the Decrement/Increment and Skip if Zero instructions, and can be used by the Bit Test and Skip instructions, to implement conditional branching of the program flow. The status register bit functions are shown in Table 8.2. We can leave consideration of the rest of these for the moment; the data sheet and more advanced programming references will provide more information.

TMR0 Timer Zero Register
File Register Number = 01

A timer/counter register counts the number of pulses applied to a clock input; the binary count can be read from the register when the count is finished. TMR0, being an 8-bit register, can count up to 255 pulses. For external inputs, the pulses are applied at pin RA4. When used as timer, the internal clock is used to supply the pulses. If the processor clock frequency is known, the time taken to reach a given count can be calculated. When the counter rolls over from FF

Table 8.2 STATUS register bit functions

Bit	Label	Name	Function
0	C	Carry flag	Set if register operation causes a carry out of bit 8 of the result (8-bit operations)
1	DC	Digit carry flag	Set if register operation causes a carry out of bit 3 of the result (4-bit operations)
2	Z	Zero flag	Set if the result of a register operation is zero
3	PD	Power down	Cleared when the processor is in Sleep mode
4	TO	Time out	Cleared when Watchdog Timer (WDT) times out
5	RP0	Register bank	RP0 selects File Registers 00–7F or 80–FF
6	RP1	Select bits	RP1 not used
7	IRP		IRP not used

to 00, an interrupt flag (see INTCON below) is set, if enabled. This allows the processor to check if the count is complete, or to be alerted when a set time interval has elapsed, even if it is doing something else at the time. The timer register can be read and written directly, so a count can be started at a preset value. The timer zero label refers to the fact that other PICs have more than one timer/counter register, but the 16F84 has only one. More details on using the TMR0 are given in Chapter 9.

OPTION Option Register
          File Register Number = 81

Table 8.3 details the option register bit functions. The counter/timer operation is controlled by OPTION register bits 0–5. When used as a timer, the processor clock signal is used to increment

Table 8.3 OPTION register bit functions

Bit	Label	Name	Function
0	PS0	Prescaler rate Select bit 0	These 3 bits form a 3-bit code to select one of 8 prescale values for the counter/timer TMR0 or WDT
1	PS1	Prescaler rate Select bit 1	
2	PS2	Prescaler rate Select bit 2	
3	PSA	Prescaler Assignment	Assigns prescaler to WDT or TMR0
4	T0SE	Timer zero Source edge select	Select rising or falling edge trigger for T0CKI input at RA4
5	T0CS	Timer zero Clock source select	Select timer/counter input as RA4 or internal clock
6	INTEDG	Interrupt edge select	Select rising or falling edge trigger for RB0 interrupt input
7	RBPU	Port B pull-up enable	Enable pull-ups on Port B pins so input data defaults to '1'

the counter register. If a crystal clock is in use, the timing will be very accurate. Prescaling can be selected to increase the maximum time interval; this means dividing down the timer input frequency by a factor of 2, 4, 8, 16, 32, 64, 128 or 256. As the case with the TRISA and TRISB registers, the OPTION register has to be accessed using a special instruction, namely 'OPTION'. The alternative method, which is recommended by the manufacturers, uses bank selection. There is more on using the timer in Chapter 9.

INTCON   Interrupt Control Register
          File Register Number = 0B

The INTCON bit functions are given in Table 8.4. An interrupt is a signal which causes the current program execution to be suspended, and an ISR to be carried out. An interrupt can be generated by an external device, via port B, or from the timer. However, in all cases, the ISR must start at address 004 in the program memory. If interrupts are in use, an unconditional jump from address zero, the program start address, to a higher start address, is needed. The INTCON register contains three interrupt flags and five interrupt enable bits, and these must be set up as required during the program initialisation by writing a suitable code to the INTCON register. Program 9.2 demonstrates the use of interrupts.

Other SFRs

Registers EEDATA, EEADR, EECON1 and EECON2 are used to access the non-volatile ROM data area (see Table 8.5). PCLATH acts as a holding register for the high bits (12:8) of the program counter. The file select register (FSR) acts as a pointer to the file registers. It can be used with IND0, which gives indirect access to the file register selected by FSR. This is useful

Table 8.4   Interrupt control register (INTCON) bit functions

Bit	Label	Name	Function
0	RBIF	Port B change Interrupt flag	Set when any one of RB4–RB7 changes state
1	INTF	RB0 pin Interrupt flag	Set when RB0 detects interrupt input
2	T0IF	Timer overflow Interrupt flag	Set when timer TMR0 rolls over from FF to 00
3	RBIE	Port B change Interrupt enable	Set to enable Port B change interrupt
4	INTE	RB0 pin Interrupt enable	Set to enable RB0 interrupt
5	T0IE	Timer overflow Interrupt enable	Set to enable timer overflow interrupt
6	EEIE	Data EEPROM write Interrupt enable	Set to enable interrupt on completion of write operation to non-volatile data memory
7	GIE	Global interrupt Enable	Enable all interrupts which have been selected

Table 8.5　Other PIC 16F84 registers

NUM	Name	Function
00 04	**INDF** **FSR**	File register memory indirect addressing for block access
0A	**PCLATH**	Program counter high byte
08 09	**EEDATA** **EEADR**	Data EEPROM indirect addressing for block access
88 89	**EECON1** **EECON2**	Data EEPROM read and write control

for a block read or write to the GPRs, for example, saving a set of data read in at a port. More information on this is given in Section 9.4.3.

8.3.2　General Purpose Registers (GPR1–GPR68)

The GPRs are numbered 0C–4F. They are also referred to as SRAM registers, because they can be used as a small block of static RAM for storing blocks of data, such as a data table of values read in at a port at intervals. We have already seen an example of using the GPR1 (address 0C) as a counter register in a delay loop. The register was labelled 'timer', preloaded with a value, and decremented until it reached zero. This is a common type of operation, and not only used for timing loops. For example, a counting loop can be used for doing an output a certain number of times. We could have used any of the GPRs for this function because they are all operationally identical; however, we do need to declare a different name for each when using more than one.

Summary

- The 16F84 internal architecture can be represented as a block diagram showing the main functional blocks, which are: program ROM, execution logic, data processing, file registers and data EEPROM.

- The following features of PIC chips enhance performance: Harvard/RISC architecture, instruction pipelining, high clock rate, single chip system.

- The program memory stores up to 1024 14-bit instructions. The program execution starts at address 0000.

- The 14-bit instruction contains operation code and operands, which can vary in length.

- The ALU processes data from the instruction, registers or W.

- Jump instructions modify the program counter to change the execution sequence.

- The file register set contains SFRs and GPRs.

- The most important of the SFRs (with their address/number) are the timer (01), program counter (02), status register (03), Port A (05), Port B (06) and interrupt control (0B).

- The GPRs are a block of registers that can be used separately, or in blocks, to store temporary data, act as counters, and so on.

Questions

1. State the function of the following blocks within a PIC microcontroller: program memory, program counter, instruction decoder, ALU, W.

2. Why is it not necessary to initialise a PIC port for input?

3. State the main functions of the ALU, and the three sources of its data input.

4. Why is the stack needed for subroutine execution?

5. State the function of the following PIC file registers: PORTA, TRISA, TMR0, PCLATH, GPRxx.

6. State the function of the register bits: STATUS, 2; INTCON, 1; OPTION, 5.

7. Which port pin gives access to TMR0?

8. What is the default destination of a 'move' operation?

Activities

1. Refer to Table 8.6. Complete the logic table to show the binary code present on the internal data connections and in the registers during or after each instruction cycle while the program BIN1 is executed. Copy this table and complete the additional columns to the right for each of the remaining four instructions. The first is given as a guide.

Table 8.6   Question 1

Instruction number	1	2	3	4	5
Address	0000				
Instruction	MOVLW 00				
Machine code	3000				
Program address bus (13 bits)	0 0000 0000 0000				
File register address (5 bits)	X XXXX				
Instruction code register	0011 0000				

Table 8.6 continued

Instruction number	1	2	3	4	5
Literal bus	0000 0000				
Data bus (8 bits)	XXXX XXXX				
Working register	0000 0000				
Port B data register	XXXX XXXX				
Port B data direction register	XXXX XXXX				

2. Study the PIC 16F84A Data Sheet (see Appendix A), Section 4.0. Study the block diagram of the internal circuit connected to pin RA0. The FETs at the output form a complementary pair of switches, a P-type and an N-type. The P-FET is on when its gate is low. The N-FET is on when its gate is high. For output, the TRIS latch is loaded with the data direction bit 0, and the data is loaded into the data latch from the data bus.

(a) Draw a logic table to represent the operation of the output logic when the TRIS latch is clear ($Q = 0$), that is, the pin is set as an output.

(b) Extend the logic table and prove that P and N are both off when the pin is initialised for input.

(c) By referring to Chapter 3 if necessary, describe how a data bit is read onto the data bus when the pin is set for input.

(d) What are the functions of the output FETs in the operation of the I/O pin?

Chapter 9
Further Programming Techniques

Now that the basic programming methods have been introduced, we can look at some more advanced techniques. Sample programs demonstrating use of the timer, interrupts and data table are included in this chapter.

9.1   Program Timing

The microcontroller program execution is driven by the clock signal generated by an internal oscillator whose frequency is controlled by either an external RC or crystal (XT) network. This signal is divided into four internal clocks (Q1–Q4) which run at a quarter of the oscillator frequency (F_{osc}/4). These provide four separate pulses during each cycle to trigger the processor operations. These include fetching the instruction code from the program memory, and copying it to the instruction register. The instruction code is then used by the decoder to set up the control lines to carry out the required process. The four clocks are used to operate the data gates and latches within the MCU to complete the data movement and processing.

This instruction timing is illustrated in Fig. 9.1. Note that, if the CR clock option is used, an output instruction clock signal at F_{osc}/4 is available at the CLKOUT pin to operate external circuits synchronously. It can also be used in hardware testing to check that the clock is running, and to measure its frequency.

The result of this clocking scheme is that each instruction takes four clock cycles to execute, unless a jump (GOTO or CALL) occurs. These will take eight clock cycles, because the program counter contents have to be replaced, and this takes an extra instruction cycle.

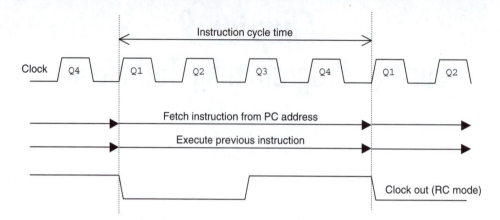

Figure 9.1   PIC instruction timing.

9.1.1   Pipelining

The instruction fetch and execute cycles can be carried out simultaneously, because the data is being transferred on separate data paths (see Fig. 8.1). While one instruction is being executed, the next is being fetched from the program memory into the instruction register. This overlapping of execution stages is called 'pipelining', with the PIC having a two-stage pipeline. The CISC microprocessors such as Pentium use more elaborate pipelining to break the instruction processing into multiple stages, and thereby boost performance.

9.1.2   Execution Time

We can now predict how long a particular sequence will take to execute. A clock rate of 4 MHz is a convenient default value because it is the maximum operating frequency in standard XT mode (see the PIC 16F84A data sheet, Table 6-1), and also gives an instruction execution rate of 1 MIP (millions of instructions per second) and an instruction cycle time of 1 μs.

A delay loop is shown in Table 9.1. The move instructions take one cycle each, and the DECFSZ instruction is then repeated 254 times. The GOTO takes two cycles, because each time the GOTO is executed, the RETURN is pre-fetched, and then not executed, so a cycle is wasted. On the 255th loop, the register becomes zero and the GOTO is skipped, and the RETURN executed. This also takes two cycles, because of another wasted pre-fetch cycle, but is only executed once per delay sequence. The total loop time can then be calculated, by totalling the time taken for each instruction and the loop. As we can see, this comes to 768 μs, at 4 MHz. This figure can be confirmed if the program containing the loop is run in the simulator, using the stopwatch, with the clock frequency set to 4 MHz.

The block execution time for a section of code can thus be predicted before testing in simulator or hardware. Alternatively, the timing can be checked and modified using the simulator. Incidentally, NOP (No OPeration) is useful here. For time critical sequences, NOP may be used to insert a delay of one instruction cycle, that is, four clock cycles; it has no other effect. Using this, a delay of 1 ms can be created using the delay loop with the count set to 249 and a NOP in the loop to make the loop execution time 4 μs. The total loop time is then (249 × 4 μs) plus a few cycles for the loop initialisation and return.

Table 9.1 Sequence execution time

Label	Instruction	Operand	Time (cycles)
delay	MOVLW	0xFF	1
	MOVWF	timer	+1
down	DECFSZ	timer	+ (1 × 255)
	GOTO	down	+ (2 × 254) +1
	RETURN		+2
			Total 768

Clock frequency	=	4 MHz
Instruction frequency	=	1 MHz
Instruction period	=	1 µs
Total delay time	=	768 µs

9.2 Hardware Counter/Timer

Accurate event timing and counting is often needed in microcontroller programs. For example, if we have a sensor on a motor shaft which gives one pulse per revolution of the shaft, the number of pulses per second will give the shaft speed. Alternatively, the interval between pulses can be measured, using a timer, to obtain the speed by calculation. A process for doing this would be:

1. wait for pulse,

2. read and reset the timer,

3. restart the timer,

4. process previous timer reading,

5. go to 1.

If an independent hardware timer is used to make the measurement, the controller program can carry on with other operations, such as processing the timing information, controlling the outputs and checking the sensor input, while the timer keeps an accurate record of the time elapsed. The motor application in Chapter 13 uses this technique.

9.2.1 Using TMR0

The special file register 01 in the 16F84 is called timer zero (TMR0); it is an 8-bit counter/timer register which, once started, runs independently. This means it can count inputs or clock pulses concurrently with (at the same time as) the main program execution. The counter/timer can also be set up to generate an interrupt when it has reached its maximum value, so that the main program does not have to keep checking it to see if a particular count has been reached. A block diagram of TMR0 and its associated hardware and control registers is shown in Fig. 9.2.

As an 8-bit register, TMR0 can count from 00 to FF (255). The operation of the timer is set up by moving a suitable control code into the OPTION register. The counter is then clocked by an external pulse train, or, more usually, from the chip oscillator. When it reaches its maximum value, FF, and is incremented again, it 'rolls over' to 00. This register 'overflow' is recorded by

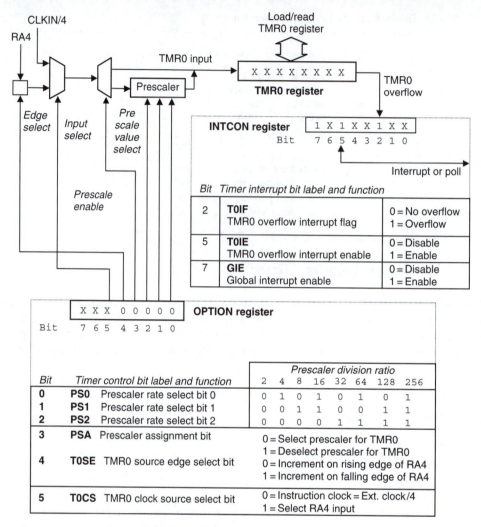

Figure 9.2 Hardware counter/timer setup and operation.

the INTCON (interrupt control) register, bit 2 (T0IF), going to '1' (assuming that it has been previously enabled and cleared). This condition can be checked by bit testing in the program, or can trigger an interrupt (Section 9.3).

9.2.2 Counter Mode

The simplest mode of operation of TMR0 is counting pulses applied to RA4, which has the alternate name T0CKI, Timer Zero Clock Input. These pulses could be input manually from a push button, or, more likely, would be produced by some other signal source, such as the sensor on the motor shaft mentioned above. If the sensor produces one pulse per revolution of the shaft, and one of the PIC outputs controls the motor, the microcontroller could be programmed to rotate the shaft by a set number of revolutions. If the motor were geared down, a positioning system could be designed to move the output through a set angle, in a robot, for example.

In order to increase the range of this kind of measurement, the prescaler allows the number of pulses received by the TMR0 register to be divided by a factor of 2, 4, 8, 16, 32, 64, 128 or 256. The ratio is selected by loading the least significant three bits in the OPTION register as follows: 000 selects divide 2, 001 divide by 4 and so on up to 111 for divide by 256. TMR0 can also be pre-loaded with a value, and the overflow detected when it has been 'topped up' by a set number of pulses.

9.2.3 Timer Mode

The internal clock is selected by setting the OPTION register, bit 5, to 0. To use TMR0 as an accurate hardware timer, a crystal oscillator must be used as the chip clock source. A convenient crystal frequency is 4 MHz, because it is divided by four before it is fed to the input of TMR0, giving a pulse frequency of 1 MHz. The counter would then be clocked every 1 μs exactly, and would take 256 μs to count from zero to zero again. Again, by preloading with a suitable value, a smaller time interval could be selected, with time out indicated by the timer interrupt flag. For example, by preloading with the value 156 (9C), the overflow would occur after 100 μs. Alternatively, the time period measured can be extended by selecting the prescaler. The maximum timer period would then be 512 μs, 1024 μs and so on to 65.536 ms. Crystals are also available in frequencies that are more conveniently divisible by 2. For example, a 32.768 kHz crystal frequency will produce a time-out every 1.0000 s, if the prescale value of 32 is selected.

In Fig. 9.1, TMR0 is set up with xxx00000$_2$ in the option register, selecting the internal clock source, with a prescale value of 2. The INTCON register has been set up with the timer interrupt enabled and the timer overflow interrupt flag has been set (overflow has occurred).

9.2.4 TIM1 Timer Program

Program TIM1, which demonstrates the use of the timer, is listed as Program 9.1. It is designed to increment a binary output once per second. The program uses the same demonstration BIN hardware as the previous programs, with eight LEDs displaying the contents of Port B. An adjustable CR clock is used, set to give a frequency of 65536 Hz (approximately). This frequency is divided by four, and is then divided by 64 in the prescaler, giving an overall frequency division of $4 \times 64 = 256$. The timer register is therefore clocked at $65536/256 = 256$ Hz. The timer register counts from zero to 256 and so overflows every second. The output is then incremented; it will take 256 s to complete the 8-bit binary output count.

9.2.5 Timing Problems

Each instruction in the program takes four clock cycles to complete, with jumps taking eight cycles. If the program sequence is studied carefully, extra time is taken in completing the program loop before the timer is restarted. In this application, it will cause only a small error, but in other applications it may be significant. Also notice that the program has to keep checking to see if the time-out flag has been set by the timer overflowing. It is more efficient to allow the processor to carry on with some other process while the timer runs, and allow the time-out condition to interrupt the main program when it has finished.

9.2.6 More Timers

Because they are so useful, some larger PIC chips have more than one timer/counter. The 16F877, for example, has three. In addition to Timer 0 (TMR0), it has Timer 1, a 16-bit counter

Program 9.1   TIM1 source code

```
; ****************************************************************
;         TIM1.ASM         M. Bates         6/1/99         Ver 1.2
; ****************************************************************
;
;   Minimal program to demonstrate the hardware timer operation.
;
;   The counter/timer register (TMR0) is initialised to zero and diven
;   from the instruction clock with a prescale value of 64.
;
;   T0IF is polled while the program waits for time out.
;   When the timer overflows, the Timer Interrupt Flag (T0IF) is set. The
;   output LED binary display is then incremented. With the clock adjusted
;   to 65536 Hz, the LSB LED flashes at 1 Hz.
;
;          Processor:        PIC 16F84
;
;          Hardware:         PIC BIN Demo Hardware
;          Clock:            CR = 65536 Hz (approx)
;          Outputs:          RB0 - RB7: LEDs (active high)
;          WDTimer:          Disabled
;          PUTimer:          Enabled
;          Interrupts:       Disabled
;          Timer:            Internal clock source
;                            Prescale = 1:64
;          Code Protect:     Disabled
;
;          Subroutines:      None
;          Parameters:       None
;
; ****************************************************************


; Register Label Equates.....................................

TMR0      EQU       01              ; Counter/Timer Register
PORTB     EQU       06              ; Port B Data Register (LEDs)
INTCON    EQU       0B              ; Interrupt Control Register

T0IF      EQU       2               ; Timer Interrupt Flag


; ****************************************************************


; Initialise Port B (Port A defaults to inputs)................

          MOVLW     b'00000000'     ; Set Port B Data Direction
          TRIS      PORTB

          MOVLW     b'00000101'     ; Set up Option register
          OPTION                    ; for internal timer/64

          CLRF      PORTB           ; Clear Port B (LEDs Off)
```

```
; Main output loop .............................................

next      CLRF    TMR0                ; clear timer register
          BCF     INTCON,T0IF         ; clear timeout flag

check     BTFSS   INTCON,T0IF         ; wait for next timeout
          GOTO    check               ; by polling timeout flag

          INCF    PORTB               ; Increment LED Count
          GOTO    next                ; repeat forever...

          END                         ; Terminate source code
```

(using two registers in cascade) which provides an accurate count up to 64 535. It can also operate with its own independent oscillator, which can operate with a 32.768 kHz crystal, which can be used to give an accurate time interval up to 2 s. Timer 2 is another 8-bit counter which is designed to be used in generating a pulse-width-modulated output signal. This can be used to drive motors and other loads which require a variable power input. Obviously, additional control registers are needed to setup and operate these extra timers.

9.3 Interrupts

Interrupts are generated by an internal or external asynchronous (not linked to the program timing) event, and the interrupt signal can be received at any time during the execution of the main process. For example, when you hit the keyboard or move the mouse on a PC, an interrupt signal is sent to the processor from the keyboard interface to request that the key be read in, or the mouse movement transferred to the screen. The code which is executed as a result of the interrupt is called the 'interrupt service routine' (ISR). When the ISR has finished its task, the process which was interrupted must be resumed as though nothing has happened. This means that any information being processed at the time of the interrupt may have to be stored temporarily, so that it can be recalled later. The program counter is saved automatically on the stack, as when a subroutine is called, so that the program can return to the original execution point after the ISR has been completed. This system allows the CPU to get on with other tasks without having to keep checking all the possible input sources.

9.3.1   Interrupt Setup

A block diagram detailing the 16F84 interrupt system is given in Fig. 9.3. The PIC has four possible interrupt sources:

1. RB0 can be selected as an edge-triggered interrupt input by setting INTCON,4 (INTE), with the active edge selected by OPTION,6 (INTEDG).

2. RB7–RB4 can be selected to trigger an interrupt if any of them changes state, by setting INTCON,3 (RBIE).

3. TMR0 overflow interrupt can be selected by setting INTCON,5 (T0IE).

4. Completion of an EEPROM (non-volatile data) write operation can be used to trigger the interrupt.

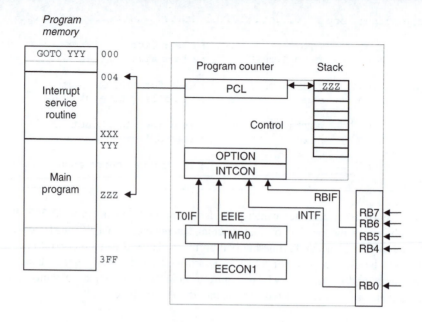

Interrupt control bit functions

	Bit	Label	Function	Settings
INTCON	0	**RBIF**	Port B (4:7) Interrupt flag	0 = No change 1 = Bit change detected
	1	**INTF**	RB0 Interrupt flag	0 = No interrupt 1 = Interrupt detected
	2	**T0IF**	TMR0 overflow Interrupt flag	0 = No overflow 1 = Overflow detected
	3	**RBIE**	Port B (4:7) Interrupt enable	0 = Disabled 1 = Enabled
	4	**INTE**	RB0 Interrupt enable	0 = Disabled 1 = Enabled
	5	**T0IE**	TMR0 overflow Interrupt enable	0 = Disabled 1 = Enabled
	6	**EEIE**	EEPROM write complete interrupt enable flag	0 = Disabled 1 = Enabled
	7	**GIE**	Global interrupt enable	0 = Disabled 1 = Enabled

| **OPTION** | 6 | **INTEDG** | RB0 interrupt Active edge select | 0 = Falling edge 1 = Rising edge |

Figure 9.3 Interrupt setup and operation.

If interrupts are required, the interrupt source must be enabled in the INTCON (interrupt control) register. Then, the global interrupt enable bit, which enables all interrupts, must be set (INTCON,7) and finally the specific interrupt bit must be set. Note that, although there are four interrupt sources, they will all call an ISR at location 0004. If more than one interrupt source is to be used, a mechanism for identifying which is active must be included in the application program. There is no hardware interrupt priority system, as is available in more complex processors.

9.3.2 Interrupt Execution

Interrupt execution is also illustrated in Fig. 9.3. Each interrupt source has a corresponding flag, which is set if the interrupt event has occurred. For example, if the timer overflows, T0IF (INTCON,2) is set. When this happens, and the interrupt is enabled, the current instruction is completed and the next program address is saved on the stack. The program counter is then loaded with 004, and the routine found at this address is executed. Alternatively, location 004 can contain a 'GOTO addlab' (address label) if the ISR is to be placed elsewhere in program memory. If interrupts are to be used, a GOTO must also be used at the reset vector address, 000, to redirect the program counter to the start of the main program at a higher memory address, because the ISR (or GOTO addlab) will occupy address 004. The ISR must be created and allocated to address 004 (ORG 004) as part of the program source code.

The ISR must be terminated with the instruction RETFIE (return from interrupt). This causes the original program address to be pulled from the stack, and program execution resumes at the instruction following the one which was interrupted. It may be necessary to save other registers as part of the ISR, so that they can be restored after the interrupt. This is called 'context saving'. This is illustrated in INT1 program below by saving and restoring the contents of Port B data register as part of the ISR.

9.3.3 INT1 Interrupt Program

A demonstration program, Program 9.2, illustrates the use of interrupts. The BIN hardware must be modified to run this program, with the push buttons connected to RB0 and RA4. This is necessary because only Port B pins can be used for external interrupts.

The program outputs the same binary count to Port B, as seen in the BINx programs, to represent its normal activity. This process is then interrupted by RB0 being pulsed manually. The interrupt service routine causes all the outputs to be switched on, and then waits for the button on RA4 to be pressed. The routine then terminates, restores the value in Port B data register and returns to the main program at the original point. The program structure and sequence can be represented by the flowcharts in Fig. 9.4.

Program 9.2 INT1 interrupt program

```
;********************************************************
;     INT1.ASM     M. Bates     12/6/99     Ver 2.1
;********************************************************
;
;     Minimal program to demonstrate interrupts.
;
;     An output binary count to LEDs on PortB, bits 1-7
;     is interrupted by an active low input at RB0/INT.
;     The Interrupt Service Routine sets all outputs high,
;     and waits for RA4 to go low before returning to
;     the main program.
;     Connect push button inputs to RB0 and RA4
;
;     Processor:          PIC 16F84
;     Hardware:           PIC Modular Demo System
;                         (reset switch connected to RB0)
```

continued...

```
;             Clock:              CR ~100kHz
;             Inputs:             Push Buttons
;                                 RB0 = 1 = Interrupt
;                                 RA4 = 0 = Return from Interrupt
;             Outputs:            RB1 – RB7: LEDs (active high)
;
;             WDTimer:            Disabled
;             PUTimer:            Enabled
;             Interrupts:         RB0 interrupt enabled
;             Code Protect:       Disabled
;
;             Subroutines:        DELAY
;             Parameters:         None
;
;****************************************************************


; Register Label Equates....................................

PORTA     EQU      05             ; Port A Data Register
PORTB     EQU      06             ; Port B Data Register
INTCON    EQU      0B             ; Interrupt Control Register
timer     EQU      0C             ; GPR1 = delay counter
tempb     EQU      0D             ; GPR2 = Output temp. store

; Input Bit Label Equates ..................................

intin     EQU      0              ; Interrupt input = RB0
resin     EQU      4              ; Restart input = RA4
INTF      EQU      1              ; RB0 Interrupt Flag

;
;****************************************************************


; Set program origin for Power On Reset.......................

          org      000            ; Program start address
          GOTO     setup          ; Jump to main program start

; Interrupt Service Routine at address 004.....................

          org      004            ; ISR start address

          MOVF     PORTB,W        ; Save current output value
          MOVWF    tempb          ; in temporary register

          MOVLW    b'11111111'    ; Switch LEDs 1-7 on
          MOVWF    PORTB

wait      BTFSC    PORTA,resin    ; Wait for restart input
          GOTO     wait           ; to go low

          MOVF     tempb,w        ; Restore previous output
          MOVWF    PORTB          ; at the LEDs
          BCF      INTCON,INTF    ; Clear RB0 interrupt flag
          RETFIE                  ; Return from interrupt
```

```
;  DELAY subroutine.........................................

delay     MOVLW     0xFF          ; Delay count literal is
          MOVWF     timer         ; loaded into spare register
down      DECFSZ    timer         ; Decrement timer register
          GOTO      down          ; and repeat until zero then
          RETURN                  ; return to main program


;  Main Program ************************************************

;  Initialise Port B (Port A defaults to inputs)..................

setup     MOVLW     b'00000001'   ; Set data direction bits
          TRIS      PORTB         ; and load TRISB

          MOVLW     b'10010000'   ; Enable RB0 interrupt in
          MOVWF     INTCON        ; Interrupt Control Register

;  Main output loop ...........................................

count     INCF      PORTB         ; Increment LED display
          CALL      delay         ; Execute delay subroutine
          GOTO      count         ; Repeat main loop always

          END                     ; Terminate source code

;  *************************************************************
```

The program is in three parts: the main sequence which runs the output count, the delay subroutine which controls the speed of the output count and the interrupt service routine. The delay process in the main program is implemented as a subroutine, and expanded in a separate flowchart. The ISR must be shown as a separate chart because it can run at any time within the program sequence. In this particular program, most of the time is spent executing the software delay, so this is the process which is most likely to be interrupted.

The interrupt routine is placed at address 004. The instruction 'GOTO setup' jumps over it at run time to the initialisation process at the start of the main program. The interrupt and delay routines must be assembled before the main program, because they contain the subroutine start address labels referred to in the main program, so they are entered first in the source code. The last instruction in the ISR must be RETFIE. This instruction pulls the interrupt return address from the stack and places it back in the program counter, where it was stored at the time of the interrupt call.

To illustrate context saving, the state of the LEDs is saved in register 'tempb' at the beginning of the interrupt, because Port B is going to be overwritten with 'FF' to switch on all the LEDs. Port B is then restored after the program has been restarted. Note that writing a '1' to the input bit has no effect. During the ISR execution, the stack will hold both the ISR return address and the subroutine return address.

9.3.4 More Interrupts

In larger PIC chips, additional interrupt sources are typically present, such as analogue inputs, serial ports and additional timers. These all have to be setup and controlled via additional

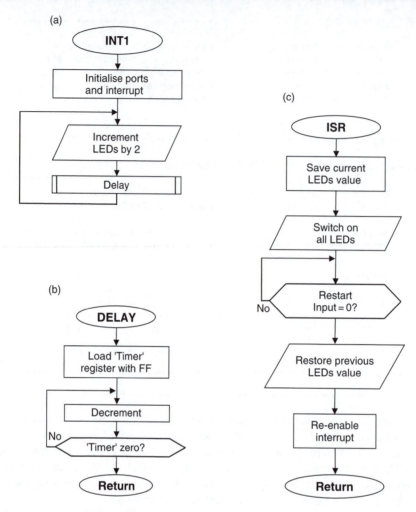

Figure 9.4   INT1 interrupt program flowcharts. (a) Main sequence; (b) Delay routine; (c) Interrupt service routine.

special function registers. As an example, the 16F877 has 14 interrupt sources, but still has only one interrupt vector address, 0004, to handle them. The interrupt bits must be checked in software to see which is active before calling the appropriate ISR. Also, the stack can still only hold eight return addresses, despite the program memory being 8k. The limit of eight levels of subroutine or interrupt can easily be exceeded if the program is too highly structured, so this must be borne in mind when planning the program implementation.

9.4   More Register Operations

The functions of the most commonly used registers are described in Chapter 8, and further operations using the PIC registers are outlined in this section.

9.4.1 Data Destination W

The default destination for operations which generate a result is the file register specified in the instruction. For example:

```
INCF     spare
```

increments the register labelled 'spare', with the result being left in the register. The above syntax generates a message when the program is assembled to remind the user that the 'default' destination is being used. This is because the full syntax is

```
INCF     spare,1
```

where '1' indicates the file register itself as the destination. If the result of the operation were required in the working register W, it could be moved using a second instruction

```
MOVF     spare,W
```

However, the whole operation can be done in one instruction by specifying the destination as W as follows:

```
INCF     spare,0
```

or

```
INCF     spare,W
```

The label W is automatically given the value 0 by the assembler. The result of the operation is stored in W, while the original value is left unchanged in the file register. All the register arithmetic and logical byte operations have this option, except CLRF (Clear File Register) and CLRW (Clear Working Register) which are by definition register specific, MOVWF and NOP (No operation). This option offers significant savings in execution time and memory requirements, which in PIC applications may be quite significant, and compensates for the lack of instructions to make direct moves between file registers.

9.4.2 Register Bank Select

The 16F84 file register set (Fig. 8.2) is organised in two banks, with the most commonly used registers in the default bank 0. Some of the control registers, such as the port data direction registers, TRISA and TRISB, and the OPTION register, are mapped into bank 1. Many of the SFRs can be accessed in either bank. Others have special access instructions, namely TRIS to write the Port A and B data direction registers, and OPTION which is used to set up the real time clock counter.

The manufacturer recommends using bank selection to access all these registers, and the instruction set warns that the instructions TRIS and OPTION may not be supported by future assemblers. Bank 0 is enabled by default, and bank 1 registers OPTION, TRISA, TRISB, EECON1 and EECON2 can be selected by setting bit 5, RP0, in the STATUS register, prior to accessing the corresponding register number. The alternative method to set Port B to output is therefore as follows:

```
STATUS   EQU     03          ; label for status register
TRISB    EQU     86          ; label for data direction register
         BSF     STATUS,5    ; select bank 1
         CLRW                ; load W with data direction code
         MOVWF   TRISB       ; set Port B as outputs
         BCF     STATUS,5    ; re-select bank 0
```

It is a good idea to re-select bank 0 immediately, as this is the most commonly used. However, if further bank 1 access is required, leave this step until later. Once a bank has been selected, it remains accessible until de-selected. Larger PIC chips which have more special function registers and provide more data registers have four register banks, requiring two bits for bank selection, status bits 5 and 6.

An alternative is to use the pseudo-operation 'BANKSEL' as follows:

```
BANKSEL    TRISB     ; select bank containing TRISB, bank 1
CLRW                 ; load code for all outputs
MOVWF      TRISB     ; set Port B as outputs
BANKSEL    PORTB     ; re-select bank containing PORTB, bank 0
```

BANKSEL selects the bank that the specified register is in, so, to change banks, any register in the required bank will do. The temperature control program (Program 15.1) uses this technique, and it is recommended as the best option for accessing registers not in bank 0.

Pseudo-operations, or special instructions, are explained in Section 9.8.

9.4.3 File Register Indirect Addressing

File register 04 is the File Select Register (FSR). It is used for indirect or indexed addressing of the other file registers, particularly the GPRs. If a file register address (00–4F) is loaded into FSR, the contents of that file register can be read or written through file register 00, the Indirect File Register (INDF). This method can be used for accessing a set of data RAM locations, by reading or writing the data via INDF, and selecting the next file register by incrementing FSR (see Fig. 9.5). This indexed, indirect file register addressing is particularly useful for storing a set of data which has been read in at a port, in, for example, a data logging application. An output data table of predefined values, such as seven-segment display codes, can use the program data table method described in Section 9.6.

The demonstration Program 9.3 loads a set of file registers, 20–2F, with dummy data (AA), using FSR as the index register. Here, FSR operates as a pointer to a block of locations and is incremented between each read or write operation. Notice that the data actually has to be moved into INDF each time.

9.4.4 EEPROM Memory

Many PIC chips have a block of electrically erasable read only memory (EEPROM) which operates as non-volatile read and write memory; the data written to this block is retained when the power is off. This is useful, for example, for security applications such as an electronic lock, where the correct combination can be stored and changed as required. Access to EEPROM is illustrated in Fig. 9.6.

The four registers used to access the memory are EEDATA, EEADR, EECON1 and EECON2. The data to be stored is placed in EEDATA, and the address at which it is to be written (00–3F) in EEADR. Bank 1 must then be selected, and a read or write sequence included in the program as specified in the data sheet, Section 3. The complex write sequence is designed to reduce the possibility of an accidental write to EEPROM, whereby valuable data is lost. Reading the EEPROM is more straightforward. The LOCK application program in Appendix B includes examples of the code sequences required to read and write EEPROM. Other devices use a different technique to access the EEPROM; the 8-pin PIC 12CE518/9 devices use serial

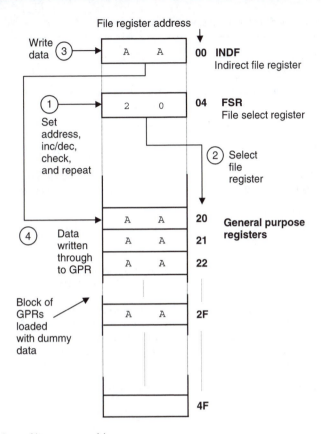

Figure 9.5 Indirect file register addressing.

access via the unused bits of the port register. The individual device data sheet must therefore be studied carefully to use this feature.

9.4.5 *Program Counter High Register, PCLATH*

The 16F84 has 1k of program memory (000–3FF), requiring a 10-bit address; the 8-bit PCL (program counter low byte) can only select one of 256 addresses. The 1k of program memory is therefore divided into four 256 word blocks (pages), one of which is selected with 2 extra bits in the PCLATH (program counter latch high) register. The PCL provides the address within each page of memory and is fully readable and writable. When a program jump is executed, PCL and PCLATH are modified automatically, that is, CALL and GOTO use a full 10-bit operand for jumps, so do not require any special manipulation of the address for jumping across page boundaries. However, if PCL is modified by a direct write under program control, PCLATH bits 0 and 1 may need to be manipulated to cross page boundaries successfully.

In other PIC devices, there may be other limitations to program branching operations. For example, CALL instructions in the 12C5XX group are limited to the first 256 locations of the program, even though the overall memory may be up to 1k. Check the data sheet carefully to avoid problems with this limitation.

Program 9.3 Indexed file register addressing

```
;   index.asm       M Bates        29-10-03
; ...................................................

;  Demonstrates indexed indirect addressing by
;  writing a dummy data table to GPRs 20 - 2F

; ...................................................

        PROCESSOR 16F84            ; select processor

FSR     EQU       04               ; File Select Register
INDF    EQU       00               ; Indirect File Register


        MOVLW     020              ; First GPR = 20h
        MOVWF     FSR              ; to FSR

        MOVLW     0AA              ; Dummy data
next    MOVWF     INDF             ; to INDF and GPRxx

        INCF      FSR              ; Increment GPR Pointer
        BTFSS     FSR,4            ; Test for GPR = 30h
        GOTO      next             ; Write next GPR

        SLEEP                      ; Stop when GPR = 30h

        END                        ; of source code
```

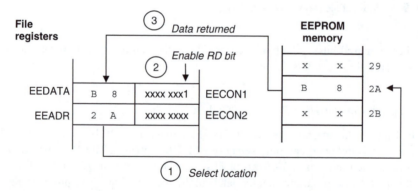

Figure 9.6 EEPROM read operation.

9.5 Special Features

PIC chips have a number of special features which enhance its flexibility and range of applications. Different oscillator types can be used, timers enabled to ensure reliable program start up and recovery, and in-circuit programming and code protection are available.

9.5.1 Oscillator Type

PIC chips can be operated with an external RC network, a crystal oscillator and an externally or internally generated clock signal. Typical oscillator circuits are illustrated in Fig. 9.7.

For applications where the precise timing of the program is not important, an inexpensive RC clock circuit (Fig. 9.7(a)) can be used. This requires only a resistor and capacitor connected as shown to the CLKIN pin of the chip. If a variable resistor is used, as in the BIN hardware, the clock rate can be adjusted, within limits, and therefore all output signal frequencies can be changed simultaneously (for example, the outputs from the program BIN1). The clock and output frequency can thus be 'trimmed' to a required value. On the other hand, the clock signal will not be very accurate or stable.

The crystal is slightly more expensive, but is far more precise than the RC clock. In the XT oscillator circuit (Fig. 9.7(b)) the crystal resonates at a fixed frequency, with an accuracy of around 50 ppm (parts per million), or 0.005%. This will allow the hardware timer to measure exact intervals and to generate accurate output signals. The overall execution time of the

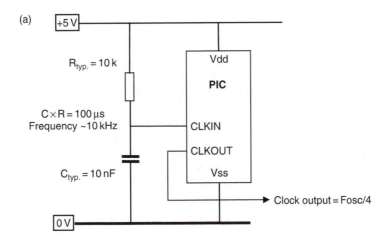

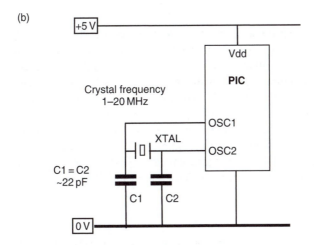

Figure 9.7 PIC clock circuits. (a) RC oscillator; (b) Crystal oscillator.

program blocks can also be predicted; this can be done by calculation, or, more readily, by use of the stopwatch in MPLAB.

If the PIC chip is part of a larger system, or one with more than one processor, a system clock signal generated by a master oscillator can be input at CLKIN. One of the crystal options must then be selected. The clock type must be selected when programming the chip, to match the target system hardware design. There are three types of crystal which can be used: standard (XT), low power (LS) or high speed (HS). XT mode should be used for clock speeds up to 4 MHz, and HS used up to 20 MHz.

In order to minimise the number of external components required, some PIC chips now have an on-board oscillator option, which provides a 4 MHz clock and 1µs instruction cycle. Because this is not a precise oscillator, it is tested in production and a calibration value supplied pre-programmed in the first program memory location. This value must then be loaded into the oscillator calibration register OSCCAL. Even so, the accuracy achieved is specified as only about 5% (3.8–4.2 MHz).

9.5.2   Power-on Timers

When a power supply is switched on, the voltage and current initially rise in an unpredictable way, depending on the design of the supply and the circuits connected to it. If the processor program tries to start immediately, before the supply had settled down, it may not start correctly. In a conventional microprocessor, an external circuit is typically connected to the CPU reset input, which provides a delay between the power being switched on and the processor starting.

The PIC has the required power-on timers built in to the chip. The reset input can therefore simply be connected to the positive supply (+5 V) for many applications, as is the case in the examples in this book. When the PIC is powered up, a power-on reset pulse is generated when the supply voltage detected at Vdd rises through about 1.5 V. This starts a power-up timer which times out after 72 ms, which in turn triggers an oscillator start-up timer, which delays for another 1024 clock cycles, to allow the internal clock to stabilise. An internal reset is then generated, and the program starts executing. The power-up timer should normally be enabled when programming the chip, as the resulting delay on start up will normally be insignificant.

9.5.3   Watchdog Timer (WDT)

This is an internal independent timer which, by default, forces the PIC to automatically restart after a fixed period (about 18 ms). The idea is to allow the processor to escape from an endless loop or other error condition, without having to be reset manually. This facility would be used by more advanced programs, so our main concern here is to prevent watchdog timeout occurring when not required, because it will disrupt the sequence and timing of our programs.

The WDT can be disabled by selecting the appropriate configuration setting during program downloading, and this is the usual option for simple programs. If the watchdog is to be employed, the WDT must be regularly reset within the program loop using the instruction CLRWDT. If this happens at least every, say, 1 ms (1000 instructions at 4 MHz), the WDT auto-reset can be prevented. If a program misbehaves in the simulator, check that WDT is disabled.

9.5.4   Sleep Mode

The instruction SLEEP causes normal operation to be suspended and the clock oscillator to be switched off. Power consumption is minimised in this state, which is useful for battery-powered applications. The PIC is woken up by a reset or interrupt; for example, when a key connected to Port B is pressed.

The SLEEP instruction is used (see Program 9.3) if the program is not required to loop continuously. If the program execution is allowed to run on into unprogrammed locations, there is a problem. The bits in the empty memory locations after the last instruction code default high. In the '84, this is in fact a valid PIC instruction, ADDLW FF, which means add literal 'FF' to W, so this instruction will be repeated throughout the unused locations. The program counter will roll over to zero after executing these meaningless instructions up to address 3FF, and the program at 000 will be restarted, so the program will loop by default. It is therefore a sensible precaution to terminate the program with a SLEEP instruction if does not run in continuous loop. If SLEEP is used to stop the program at the end, a power-on reset, external reset or an interrupt can then restart the processor.

9.5.5 In-Circuit Programming and Debugging

In-circuit programming allows the chip to be programmed without being removed from the circuit, which avoids possible mechanical (broken/bent legs) and electrical (static) damage. The programming module is connected to the serial port of the host PC and to the chip via two port pins (RB6 and RB7 in the 16XXX) via a suitable connector (Fig. 9.8). The program can then be downloaded in the usual way, in serial form. Note, however, that the circuit must be designed so that the normal operational connections to the port pins do not interfere with the downloading process. If in doubt, leave these pins exclusively for programming. When programming is complete, the connector can be removed, the board set to run and the port pins used for their normal function.

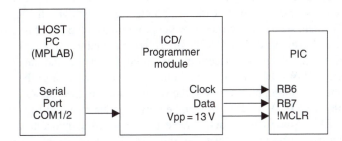

Figure 9.8 Serial programming and ICD connections.

PIC chips are now designed to allow this programming link to be used in a cheap but effective debugging system. If ICD is supported by the target hardware, an ICD module acts as programmer and debugging interface. The same MPLAB simulator tools can be used to test the program as it runs in the actual chip, rather than in the purely software model, with the real hardware acting as inputs and outputs. This allows the hardware to be verified, and timing critical operations to be tested easily and reliably. All the usual techniques are available: single stepping, breakpoints, register monitoring and so on. The finished program can be tested at full speed in the actual hardware, and any final bugs removed. Previously, an expensive in-circuit emulator would be needed for this type of testing (see Section 11.5.2). It is anticipated that this feature will be extended to more PIC devices, as it is a valuable low-cost tool for PIC program development.

By incorporating the programming interface into the target hardware, it is also possible for microcontrollers to be reprogrammed remotely after final installation. If a suitable communication link is available, a new control program can be downloaded while the target processor remains at its remote site. This is a great advantage in, for example, distributed sensing and monitoring applications, where a site visit would be expensive or time-consuming. Obviously, the new program would need to be fully tested on an identical local system before being downloaded to the remote system.

9.5.6 Code Protection

In commercial applications, the PIC program designer does not want the software supplied with a product to be copied by a market competitor. The 'Code Protect' fuse, selected during programming, is designed to prevent unauthorised copying. The chip can also be given a unique identification code during programming, if required. For our purposes, the code protection should not be enabled, as the program cannot then be read back for verification.

9.5.7 Configuration Word

The oscillator selection bits (2), watchdog timer, power-up timer and code protection are all selected by setting the bits of a configuration word, located at a special address which is only accessible when the chip is being programmed. These bits can be set via the programming dialogue in MPLAB, as described in Chapter 7. Alternatively, the configuration options can be set by including an assembler directive in the source code.

9.6 Program Data Table

A program may be required to output a set of pre-defined data bytes, for example, the codes to light up a seven-segment display with the correct pattern for each display digit, as in Program 12.2. The data set can be written into the program as a table within a subroutine, and the data list accessed using CALL and RETLW. To fetch the table value required, the position in the table is placed in W. '0' will access the first item, '1' the second and so on. At the top of the subroutine, ADDWF PCL is used to add the table pointer value to the program counter register so that the execution point jumps to the required item in the list. RETLW is then used to return the table value in W, and it can then be moved to the required file register.

Program 9.4, TAB1, shows how such a table may be used to generate a sequence at the LEDs in our BIN demonstration hardware. In this case, it is a bar graph display which lights the LEDs from one end, using the binary sequence 0, 1, 3, 7, 15, 31, 63, 127, 255.

Spare registers labelled 'timer' and 'point' are used. Port B is set as outputs and subroutines defined for a delay and to provide a table of output codes. In the main loop, the table pointer register 'point' is initially cleared, and will then be incremented from zero to 9 as each code is output. The value of the pointer is checked each time round the loop to see if it is 9 yet. When 9 is reached, the program jumps back to 'newbar', and the pointer reset to zero.

For each output, the pointer value (0–8) is placed in W and the 'table' subroutine called. The first instruction 'ADDWF PCL' adds the pointer value to the program counter. At the first call, this value is zero, so the next instruction 'RETLW 000' is executed. The program returns to the main loop with the value 00 in W. This is output to the LEDs, the delay run, and the pointer value incremented. The new value is tested to see if it is 9 yet, and if not, the call is made to the table with the next value, 1, and so on to 8. Each time the pointer value is added to PCL, so that the program jumps to the second, then third, then fourth code and so on, until finally the ninth code, which is 0FF, is returned to the main output loop for display. After this, the test of

Program 9.4 TAB1 table program

```
;****************************************************************
;        TAB1.ASM       M. Bates        13/6/99       Ver 1.3
;****************************************************************
;
;       Output binary sequence gives a demonstration of a
;       bar graph display, using a program data table...
;
;       Processor:         PIC 16F84
;
;       Hardware:          PIC Demo System
;       Clock:             CR ~10kHz (Cycle time ~0.7s)
;       Inputs:            none
;       Outputs:           LEDs (active high)
;
;       WDTimer:           Disable
;       PUTimer:           Enable
;       Code Protect:      Disable
;
;       Interrupts:        Disabled
;       Subroutines:       'delay' (no arguments)
;                          'table' (argument 'point')
;
; ****************************************************************
;   Register Label Equates.................................
;
PCL     EQU         02              ; Program Counter Low Register
PORTB   EQU         06              ; Port B Data Register
timer   EQU         0C              ; GPR1 used as delay counter
point   EQU         0D              ; GPR2 used as table pointer
;
; ****************************************************************
;
        ORG         000
        GOTO        start           ; Jump to start of main prog
;
;   Define DELAY subroutine.................................
;
delay   MOVLW       0xFF            ; Delay count literal
        MOVWF       timer           ; loaded into spare register
;
down    DECFSZ      timer           ; Decrement timer register
        GOTO        down            ; and repeat until zero
        RETURN                      ; then return to main program
;
;   Define Table of Output Codes ...........................
;
table   ADDWF       PCL             ; Add pointer to PCL
        RETLW       000             ; 0 LEDS on
        RETLW       001             ; 1 LEDS on
        RETLW       003             ; 2 LEDS on
        RETLW       007             ; 3 LEDS on
        RETLW       00F             ; 4 LEDS on
        RETLW       01F             ; 5 LEDS on
        RETLW       03F             ; 6 LEDS on
        RETLW       07F             ; 7 LEDS on
        RETLW       0FF             ; 8 LEDS on
```

continued...

```
;              Initialise Port B (Port A defaults to inputs).........

start      MOVLW   b'00000000'      ; Set Port B Data Direction Code
           TRIS    PORTB            ; and load into TRISB

; Main loop .............................................

newbar     CLRF    point            ; Reset pointer to start of table

nexton     MOVLW   009              ; Check if all outputs done yet
           SUBWF   point,W          ; (note: destination W)
           BTFSC   3,2              ; and start a new bar
           GOTO    newbar           ; if true...

           MOVF    point,W          ; Set pointer to
           CALL    table            ; access table...
           MOVWF   PORTB            ; and output to LEDs

           CALL    delay            ; wait a while...

           INCF    point            ; Point to next table value
           GOTO    nexton           ; and repeat...

; End of main loop ......................................

           END                      ; Terminate source code
```

the pointer being equal to 9 succeeds, the jump back to 'newbar' taken, and the process repeats. Note the use of 'W' as the destination for the result of the subtract (SUBWF) instruction. This is necessary to avoid the pointer value being overwritten with the result of the subtraction.

9.7   Assembler Directives

Assembler directives are commands inserted in PIC source code which control the operation of the assembler. They are not part of the program itself and are not converted into machine code. Many assembler directives will only be used when a good knowledge of the programming language has been achieved, so we will refer to a small number of selected examples at this stage. The use of some of these is illustrated in Program 9.5, ASD1. The assembler directives are placed in the second column, with the instruction mnemonics. We have already met some of the most commonly used directives, but END is the only one which is essential, all the others are simply available to make the programming process more efficient. For definitive information refer to the documentation and help files supplied with your current assembler version.

9.7.1   Control Directives

Processor

Specifies the PIC processor for which the program has been designed, and allows the assembler to check that the syntax is correct for that processor. The simulator also uses this to automatically select the right processor. The processor can be selected in the assembler command line, if so, this supersedes the source code directive.

Program 9.5 ASD1 assembler directives program

```
;****************************************************************
;  ASD1.ASM        M. Bates          17/12/03          Ver 1.1
;****************************************************************
;  Assembler directives, a macro and a pseudo-operation are
;  illustrated in this counting program ...
;  ************************************************************

;  Directive sets processor type:
   PROCESSOR 16F84

;  Set configuration fuses:
   __CONFIG B'1111111111110011'
;  Code protection off, PuT on, WDT off, RC clock

;  SFR equates are inserted from disk file:
   INCLUDE "C:\PIC\REG84.EQU"

;  Constant values can be predefined by directive:
   CONSTANT     maxdel = 0xFF, dircb = b'00000000'

timer    EQU        0C          ; delay counter register

; Define DELAY macro *****************************************

DELAY    MACRO

         MOVLW      maxdel      ; Delay count literal
         MOVWF      timer       ; loaded into spare register

down     DECF       timer       ; Decrement spare register
         BNZ        down        ; Pseudo-Operation:
                                ; Branch If Not Zero

         ENDM

;****************************************************************

;         Initialise Port B (Port A defaults to inputs)

         MOVLW      dircb       ; Port B Data Direction Code
         TRIS       PORTB       ; Load the DDR code into F86

;         Start main loop ...................................

         CLRF       PORTB       ; Clear Port B Data & restart
again    INCF       PORTB       ; Increment count at Port B
         DELAY                  ; Insert DELAY macro
         GOTO       again       ; Repeat main loop always

         END                    ; Terminate source code
```

__Config

The configuration directive allows the configuration bits to be specified in the source code, so that they do not have to be set up each time when downloading. This is obviously useful if the program has to be downloaded several times before completion of debugging. The significance of each bit is shown in the data sheet, Section 6.1. A 16-bit word is loaded into the configuration register by this directive. Bits 0 and 1 set the clock type ($11 = RC$, $01 = XT$), bit 2 disables the watchdog timer if cleared and bit 3 enables the power-up timer if cleared. All the other bits are set to 1 to disable code protection. The double underscore which starts the directive indicates an operation on the MCU registers.

Org

Sets the code 'origin', meaning the address which will be allocated to the first instruction following this directive. We have already seen (Program 9.2) how it is necessary to set the origin of the interrupt service routine as 004. The default origin is 000, the first program memory location, so if not specified, the program will be placed here. This is the reset address where the processor always starts on power up or reset. If using interrupts, an unconditional jump 'GOTO addlab' should be used as the first instruction at the reset address 000. This will jump over the ISR (or a jump to it) placed at address 004. The main program can then be placed at a higher address using the ORG directive.

End

Informs the assembler that the end of the source code has been reached. This is the one directive that must be present.

9.7.2   Conditional Directives

These directives allow selective assembly of source code blocks. That is, sections of code can be omitted during assembly, or repeated, by use of high level language type statements such as IF...ELSE...ENDIF. Assembler 'variables' are used to define the conditions for assembly.

9.7.3   Listing Directives

List

This directive has a number of options which allow the format and content of the list file to be modified, e.g. number of lines and columns per page, error levels reported, processor type and so on.

Page

Forces a page break when printing.

Title

Defines the program name printed in the list file header line, if you want it to be different from the source code file name (see also SUBTITL).

9.7.4   Data Directives

Equ

EQU is probably the second most commonly used directive, because it allows literal and register labels to be defined, and we have already used it routinely. It assigns a label to any numerical value (hex, binary, decimal or ASCII), and the assembler then replaces the label with the number. This allows recognisable labels to be used instead of numbers.

Include

This directs the assembler to include a block of source code from a named file on disk. If necessary, the full file path must be given. The text file is included as though it had been typed into the source code editor, so it must conform to the usual assembler syntax, but any program block, subroutine or macro could be included in the same way. This allows separate source code files to be included, and opens the way for the user to create libraries of reusable program modules. In the example ASD1, it is used to include a standard header file (REG84.EQU) which defines labels for all the special function registers in the PIC. Use of this option is recommended when the basics have been mastered; standard header files, which use labelling which is consistent with the SFR labels used in the register monitoring windows in MPLAB, are supplied with the development system files for all processors.

Data, Zero, Set, Res

Allow program constants and data blocks to be defined and memory allocated for specified purposes.

9.7.5   Macro Directives

Macro .... Endm

A macro is a block of source code which is inserted into the program when its name is used as an instruction. In ASD1, for example, DELAY is the name of the macro, and its insertion in the main program can be seen in the list file. Thus using a macro is equivalent to creating a new instruction from standard instructions, or an automatic copy and paste operation. The directive MACRO defines the start of the block (with a label), ENDM terminates it. It effectively allows you to create your own instruction mnemonics (see also LOCAL and EXITM).

9.8   Special Instructions

Special instructions are essentially macros which are pre-defined in the assembler. A typical example is shown in the program ASD1, 'BNZ down', which stands for 'Branch if Not Zero to label'. It is replaced by the assembler with the instruction sequence Bit Test and Skip and GOTO:

```
BNZ down = BTFSS 3,2
           GOTO  down
```

These two instructions are inserted into the program in place of the special instruction. The zero flag (bit 2) in the status register (register 3) is tested, and the GOTO skipped if it is set as a result of the previous operation being zero. If the result was not zero, the GOTO is executed, and the program jumps to the address label specified. Special instructions are designed to simplify operations using the carry or zero flag, and are equivalent to conditional branch instructions in complex instruction set processors. This type of instruction is included in the main instruction set of the more powerful 18XXX series of PICs.

9.9   Numerical Types

Literal values given in PIC source code can be written using different number systems. The default is hexadecimal, so if the type is not specified, the assembler will assume it is hex. However, it is very important to note that the assembler will still get confused between numbers and labels if the hex number starts with a letter (i.e. A, B, C, D, E or F). The literal must start with a number, so use a leading zero at all times. Then 8-bit literals will be written as three digits, with the first always zero (000–0FF).

The numerical types supported by the MPASM assembler are:

- hexadecimal
- decimal
- binary
- octal
- ASCII

To specify a type, the initial letter of the type can be used with quotes, such as:

```
H'3F'
D'47'
B'10010011'
A'K'
```

Binary is useful for specifying register values which are bit-oriented, as is the case for many SFRs; the state of each bit can be clearly seen. In particular, we have used binary to define port data direction codes in our demonstration programs.

If an ASCII character is specified, the corresponding 8-bit code in the range 00–7F will be loaded representing the code for each character in the set (see Table 9.2). This option is used

Table 9.2 ASCII character set

Low bits	High bits					
	0010	0011	0100	0101	0110	0111
0000	Space	0	@	P	`	p
0001	!	1	A	Q	a	q
0010	"	2	B	R	b	r
0011	#	3	C	S	c	s
0100	$	4	D	T	d	t
0101	%	5	E	U	e	u
0110	&	6	F	V	f	v
0111	'	7	G	W	g	w
1000	(	8	H	X	h	x
1001	)	9	I	Y	i	y
1010	*	:	J	Z	j	z
1011	+	;	K	[	k	{
1100	,	<	L	\	l		
1101	-	=	M	]	m	}
1110	.	>	N	^	n	~
1111	/	?	O	_	o	Del

in sending data to alphanumeric liquid crystal displays, for example. The character itself may then be used in the program, and the assembler does the code conversion:

```
MOVLW    'Y'      ; Converted to binary 01011001
MOVWF    PortB    ; send to display
```

Note that the A for ASCII can be left out, and the character still be correctly recognised by the assembler.

Summary

- Each PIC instruction takes four clock periods to execute (instruction cycle time). Jumps take two instruction cycles. Block execution times can therefore be calculated.

- The hardware counter/timers can be used to count inputs or time intervals. Programmable prescalers extend the range of the counter. Timer overflow sets the time-out flag, which can be used to trigger an interrupt.

- Interrupts allow an internal or external event to change the program sequence, and force the execution of an ISR. There are multiple interrupt sources, but no interrupt priority.

- Register and memory bank selection are sometimes necessary; EEPROM is available for non-volatile storage.

- The clock signal which drives the chip can be obtained from an RC or crystal circuit, or master system clock. Power-on timers, watchdog timer, sleep mode, in-circuit programming and code protection are available.

- Program data tables can be operated using CALL and RETLW, with an incrementing PCL offset.

- Assembler directives are instructions to the assembler which are not converted into machine code.

- Macros are user-defined instructions. Special instructions are pre-defined macros.

- Numerical types hex, decimal, binary, octal and ASCII character codes can be used in the source code.

Questions

1. State the number of clock cycles in a PIC instruction cycle and the number of instruction cycles taken to execute the instructions (a) CLRW (b) RETURN.

2. If the PIC clock input is 100 kHz, what is the instruction cycle time?

3. Calculate the pre-load value required in TMR0 to obtain a delay of 1 ms between the load operation and the T0IF going high, if the clock rate is 4 MHz and the prescale ratio selected is 4:1.

4. List the bits in the SFRs which have to be initialised to enable an RB7:RB3 interrupt.

5. Sketch the circuits for an RC and crystal clock, showing typical component values and chip connections. State one advantage of each type.

6. State the assembler directive that must be used in all PIC programs.

7. Explain the difference between a subroutine and a macro.

Answers

2. $40\,\mu s$

3. 6

4. TRISB, 3, 4, 5, 6 and INTCON bits 0, 3, 7

6. END

Activities

1. Calculate the time taken to execute one complete cycle of the output obtained from TAB1 with a clock rate of 10 kHz.

2. Modify the program TIM1 to use a timer interrupt rather than polling to control the delay.

3. Devise a program to measure the period of an input pulse waveform at RB0, which has a frequency range of 10–100 kHz. The input period should be stored in a GPR called 'period' as a value where $0A_{16} = 10\,\mu s$ and $64_{16} = 100\,\mu s$, with a resolution of 1 bit per μs. The clock uses a 4 MHz crystal. Estimate the accuracy of the frequency measurement at each end of the range.

Part C
Applications

Chapter 10
Application Design

This chapter will take you through the complete process of application design and development, based on a simple motor drive system. At each step, relevant design techniques will be explained and a suitable implementation developed, stage by stage.

Before designing hardware or writing a program, we have to describe as clearly as possible what an application is required to do. That means a specification is needed which defines the user's requirement. There are national and international standards which should be observed when designing commercial products to aid clear communication between the engineer, management and client; here we will simply establish some basic, 'common sense' rules.

Once the specification has been written, a useful starting point for hardware design is a block diagram. We have already seen numerous examples in previous chapters. It should represent the main parts of a system and the information flow between them, in a simplified form. This can later be converted to circuit diagrams and the hardware connections laid out and constructed on PCB. In a similar way, software can be designed using techniques which allow the application program to be outlined, and then the details progressively filled in. Flowcharts have been used already, and this chapter will explain in more detail the basic principles of using flowcharts to help with program design.

Pseudocode is another useful method for designing software. The program outline is entered directly into the source code text editor as a set of general statements which describe each major block, which would normally be defined as functions and procedures in a high level language, and subroutines and macros in a low level language. Detail is then added under each heading until the pseudocode is suitable for conversion into source code statements for the assembler or compiler for the target processor or programming language. The pseudocode versions can be described as level 1, 2, 3 and so on, as more detail is added.

At this stage, we will concentrate on flowcharts, because they are suitable for simple real-time applications, as their graphical nature makes them a good learning tool. The first step in the software design process is to establish a suitable algorithm for the program; that is, a processing method which will achieve the specification using the features of an available programming language. This obviously requires some knowledge of the range of languages which might be suitable, and experience in the selected language. Formal software design techniques cannot be properly applied until the software developer is fairly familiar with the relevant language syntax. However, when learning programming we have to develop both skills together, so some trial and error is unavoidable. When learning, it is useful to apply these design techniques retrospectively, that is, as an analytical tool or as part of the application documentation. For instance, a final version of a flowchart might be drawn after the program has been written and tested, when the suitability of the design algorithm has been proven.

In this chapter, a simple demonstration application will be used to illustrate the development process. Real software products will generally, of course, be far more complex, but the same basic design principles may be applied. If the design brief is not specific about the hardware, considerable experience and detailed knowledge of the options available are required to select the most appropriate hardware and software combination. The relative costs in the planning, development, implementation, testing, commissioning and support of the product should also be estimated to obtain the most cost-effective solution. Naturally, the example used here to illustrate the software development process has been chosen as suitable for PIC implementation.

The application program will be required to generate a pulse width modulated (PWM) output. This can also be generated by a specially designed interface in many microcontrollers, including PICs, as it is a common requirement. The software implementation will help us understand the operation of the hardware-based PWM interface, which will be described later.

10.1   Design Requirements

A system is required to provide a PWM drive signal for a small DC motor. Under PWM control, the motor runs at a speed which is determined by the average level of the signal, which in turn is dependent on the ratio of the on (mark) to off (space) time. This method provides an efficient method of using a single digital output to control output power from a motor, heater, lamp or similar power output transducer. PWM is also used to control small digital position servo units, as used in radio-controlled models, for example. The basic drive waveform is shown in Fig. 10.1.

A variable mark/space ratio (MSR) of 0–100%, with a resolution of 1%, is required. The frequency is not critical, but should be high enough to allow the motor to run without any significant speed variation over each cycle (>10 Hz). It is desirable to operate at a frequency above the audible range (>15 KHz) because some of the signal energy can radiate as sound from the windings of the motor, which can be quite irritating! However, a dedicated PWM interface is

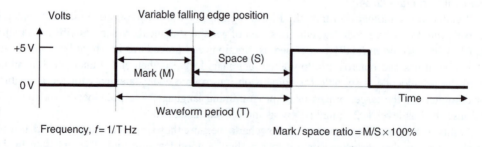

Figure 10.1 Pulse width modulated signal.

needed to achieve this. We will go for low-frequency operation just to demonstrate the principles involved. The hardware is also simplified; for example, instead of a single FET drive transistor as used here, a full bridge driver IC would typically be used to provide bidirectional motor control.

The motor speed will be controlled by two active low inputs which will increment or decrement the PWM output. An active low enable signal is also required to switch the drive on and off, while preserving the existing setting of the MSR. The system should start on reset or power up at 50% MSR, that is, with equal mark and space, and a reset input should be provided to return the output to the default 50% MSR at any time. The increment and decrement operations must stop at the maximum and minimum values; in particular 0% must not roll over to 100%, causing a zero to maximum motor speed transition in a single step. The inputs and outputs must be TTL compatible for interfacing purposes, allowing PWM control from another master controller for multiple motor control. A programmed device will be used so that it is possible to modify the control algorithm to suit different motors and to enable future enhancement of the controller options and performance. A logic table (Table 10.1) specifies the operation required.

In addition, a performance specification should be provided to quantify the performance criteria as far as possible:

PERFORMANCE SPECIFICATION
****
                Project: MOT1
                Variable Speed Controller for Small DC Motor

        1.   Maximum load: 500mA @ 5V (2.5W @ 100% MSR)
        2.   Manual or remotely controlled variable MSR:
                2.1 Start: at 50% MSR
                2.2 Reset: to 50% MSR
                2.3 Range: Min < 1%, Max > 99%
                2.4 Step Resolution: < 1%
                2.5 Manual Control:
                        2.5.1 Push Button Increment, Decrement
                        2.5.2 Hold MSR when inputs inactive
****

Table 10.1 MOT1 application control logic

Inputs				Operation	Output
!MCLR	!RUN	!UP	!DOWN	Description	dc motor
0	x	x	x	Initialise – set speed to 50%	Off
1	1	x	x	Disabled	Off
1	0	1	1	Run with MSR = 50% or run at current speed	Default speed or speed constant
1	0	0	1	Increment MSR (hold at max)	Speed increasing
1	0	1	0	Decrement MSR (hold at min)	Speed decreasing

10.2　Block Diagram

In the block diagram, the system inputs and outputs must be identified, and if necessary, a provisional arrangement of sub-systems worked out (see Fig. 10.2). The direction and type of information flow between the blocks should be identified clearly, using directed line segments (arrows). Small diagrams can be used to illustrate the waveform of a signal, if appropriate. Parallel data paths should be shown as broad arrows, or with suitable signal labelling.

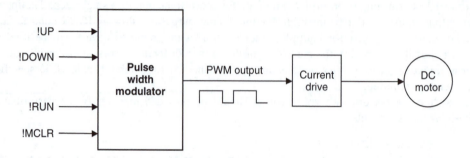

Figure 10.2　MOT1 system block diagram.

The block diagram can be drawn using the drawing tools in Word, other standard wordprocessor or DTP package, since it needs only basic shapes, arrows and text boxes. Experiment with your usual package.

In Word, the drawing toolbar may need to be enabled via the main menu 'view, toolbars, drawing'. The drawing can be embedded in a text file, but beware of interaction of drawing objects with the text cursor, which can disrupt the drawing. It is usually a good idea to move the text cursor below the drawing area. A drawing grid can be switched on to help line up the main drawing objects; from the 'draw' menu, select 'grid' and check the 'snap to grid' option. To make fine adjustments to drawing objects, the grid can later be switched off.

The main elements can be drawn using text boxes, and the same object used for labelling with 'no line' and 'no fill' options selected. Various line and arrow styles are available, and the 'freeform' line style in the 'autoshapes' menu is useful for multi-segment lines. This menu also provides various standard shapes for block diagrams and flowcharts.

When the drawing is finished, select all the drawing elements by looping with the 'select objects' tool and select 'draw, group'. This will create a single drawing object which will no longer be affected by the text cursor, and allows the whole drawing to be re-positioned on the page if necessary.

10.3　Hardware Design

Unless the program is being written for an existing hardware system, the general hardware configuration must be worked out as part of the design exercise. The nature and complexity of the software is an important consideration in the selection of a microprocessor or microcontroller, as is the number and type of inputs and outputs, data storage and interfacing.

The design requirements of MOT1 could be satisfied using relatively complex controller system, based on a conventional CISC processor, such as the 68000, and additional features could then easily be included. More inputs and outputs could allow control of several motors simultaneously, a standard serial interface to a host computer system would be available, and the larger memory could accommodate a more complex program, or a program written in a high level language such as 'C'. In the simple PWM example proposed here, however, the requirement is of minimal complexity with no special interfacing specified.

Alternatively, a purely hardware solution could be produced based around, for example, the 555 timer. However, this would not provide the push button digital control required as the 555 timing is controlled by an analogue input. Therefore a small microcontroller solution appears the most suitable.

A circuit derived from the block diagram is shown in Fig. 10.3. The motor is controlled by an FET, which acts as a current switch operated by the PIC TTL level output. The motor forms an inductive load, so a diode is connected to protect the FET from any back emf from the motor. The input control uses simple active low push buttons, with connections for remote system control.

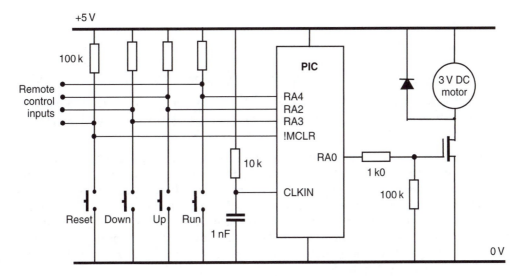

Figure 10.3 MOT1 circuit diagram.

The microcontroller only needs four I/O pins, so a 6 I/O 12XXX series device could be considered. However, an external reset is required, so our reference device 16F84 will be used. The controller I/O allocation can then be specified as shown in Table 10.2.

Table 10.2 MOT1 I/O allocation using PIC 16F84

Signal	Type	Pin	Description	Comment
Clock	System	CLKIN	RC clock	~100 kHz
Reset	System	!MCLR	Active low	Restart at default speed
PWM	Output	RA0	Pulse	FET drive
Run	Input	RA4	Active low	Enable motor
Up	Input	RA2	Active low	Increase speed
Down	Input	RA3	Active low	Decrease speed

The PIC provides motor speed control with a PWM output at RA0. The !RUN ('Not Run', active low) input has been allocated to RA4. This will be programmed to enable the PWM output to run the motor when low. When RA2 (!UP) is low, the MSR at RB0 should increase, and the motor speed up. When RA3 (!DOWN) is low, the MSR should be reduced, slowing the motor down. !MCLR (Master Clear) is the reset input to the PIC, which will restart the program when pulsed low, and hence reset the speed to the default value of 50% MSR.

We can now start work on the software using a flowchart to outline the program. A few simple symbols and rules will be used to help devise a working assembly code program. These are explained below.

10.4   Software Design

Computer and controller programs in general incorporate three main types of operation:

1.   Sequence (no jumps): A sequence of instructions is executed without branching. The program counter is not modified.

2.   Selection (conditional jump): A condition is tested for result true or false and a jump is made or not, depending on the result. In high level languages, these operations may be combined to provide a multiple choice branch depending on the value of the variable.

3.   Iteration (repeating loop): A conditional branch is used to jump back and repeat a sequence while a condition is true, or until a condition is met, or endlessly (this is the last option at the end of the program).

A program consists of a sequence of instructions in a low level language (LLL) such as PIC assembler, or statements in a high level languages (HLL) such as Basic, Pascal or 'C'. These instructions are executed in the order that they appear in the source code, unless there is an instruction or statement which causes a jump, or branch. Usually jumps are 'conditional', which means that some input or variable condition is tested and the jump made, or not, depending on the result. In PIC assembler, 'Bit Test and Skip if Set/Clear' and 'Decrement/Increment File Register and Skip if Zero' provide conditional branching when used with a 'GOTO label' or a 'CALL label' immediately following.

A loop can be created by jumping back at least once to a previous instruction. In our standard delay loop, for instance, the program keeps jumping back until a register which is decremented within the loop reaches zero. In high level languages, conditional operations are created using the IF (a condition is true) THEN (do a sequence), and loops created using the statements such as DO (a sequence) WHILE (a condition is true/not true).

10.4.1   MOT1 Outline Flowchart

Flowcharts illustrate the program sequence, selections and iterations in a pictorial way, using simple set of symbols. Some basic recommendations for laying out flowcharts will be made here which will help to ensure consistency in their use and will allow flowcharts to be used to create well-structured programs. An outline flowchart for the motor speed control program MOT1 is shown in Fig. 10.4.

The outline flowchart shows a sequence where the inputs (run, speed up and speed down) are checked and the delay count modified if either of the speed control inputs are active. The output is then set high and low for that cycle, using the calculated delays to give the mark/space ratio.

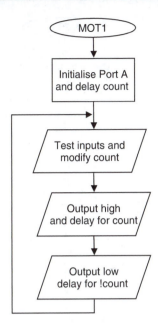

Figure 10.4 MOT1 outline flowchart.

The loop repeats endlessly, unless the reset is operated. The reset operation is not represented in the flowchart, because it is an interrupt, and therefore may occur at any time within the loop. The program name, MOT1, is placed in the start terminal symbol. Most programs need some form of initialisation process, such as setting up the ports at the beginning of the main program loop. This will normally only need to be executed once. Any assembler directives, such as label equates, should not be represented, as they are not part of the executable program itself.

In common with most so-called 'real-time' applications, the program loops continuously until reset or switched off. Therefore, there is an unconditional jump at the end of the program back to start, but missing out the initialisation sequence. Since no decision is made here, the jump back is simply represented by the arrow, and no process symbol is needed. It is suggested here that the loop back should be drawn on the left side of the chart, and any loop forward on the right, unless it spoils the symmetry of the chart or causes line segment crossovers (see below).

Note that when branching, the flow junctions must be *between* process boxes, to preserve a single input, single output rule for each process. Each process then always starts and ends at the same point.

10.4.2 MOT1 Detail Flowchart

The outline flowchart given in Fig. 10.4 may show enough information for an experienced programmer. If more detail is needed, boxes in the main program can be elaborated until there is enough detail for the less experienced programmer to translate the sequence into assembly code. A detail flowchart is shown in Fig. 10.5.

After the initialisation sequence, a set of conditional jumps is required to enable the motor, check the 'up' and 'down' inputs, and test for the maximum and minimum values of the value of 'Count' (FF and 01). Two different forms of the decision box have been used in this example, both of which may be seen in other references. The diamond-shaped decision symbol is used

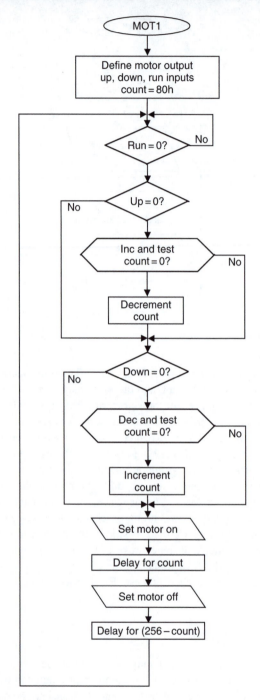

Figure 10.5 MOT1 detail flowchart.

here to represent a 'Bit Test and Skip If Zero/Not Zero' operation, while the elongated symbol represents an 'Increment/Decrement and Test for Zero' operation, which essentially combines two instructions in one. In either case, the decision box should contain a question with its outputs representing a 'Yes' or 'No' result of the test.

10.4.3 Program Structure

In the previous example program BIN4, a delay subroutine is used. Recall that this is a process defined as a separate block of code which can be used more than once. It is indicated in the main program flowchart (Fig. 7.3(a)) with the subroutine box with double sides. The delay routine sequence is then detailed in Fig. 7.3(b). It starts with a terminal symbol which contains the subroutine start label used in the program source code, and ends with 'Return'. All subroutines which are invoked with the CALL instruction must be terminated with a RETURN instruction. The CALL automatically pushes the return address onto the stack, and the RETURN pulls it back into the program counter. Failure to observe this rule will result in a stack error.

'Decrement and Skip if Zero' is used to create the standard software delay loop. The two delays required in BIN4 could be written separately, but the function is the same, with only the delay count differing; using the same delay block twice is more code efficient. Sometimes separate blocks might be used if the timing is critical. Alternatively, a macro could be defined for the delay which the assembler would insert twice (see Chapter 9) which would effectively create a 'delay' special instruction.

A subroutine will often use a value set up in the calling routine. In BIN4, the value for the delay time is placed in W for use by the delay routine; this is an example of 'parameter passing'. Other blocks in the main program can also be created as subroutines, but they are most useful when the routine is to be used more than once. The subroutine can then be copied for use in another program, or saved as a separate file for inclusion in new programs.

The application MOT1 does not require the use of subroutines.

10.4.4 Flowchart Symbols

A minimal set of flowchart symbols is shown in Fig. 10.6, most of which have already been used. For commercial applications, the relevant standards should be applied, and would override any recommendations made here.

Terminals

These symbols are used to start or end the main program or a subroutine (see Fig. 10.6(a)). The program name or routine start label used in the source code should also be used in the start box. If the program loops endlessly the END symbol is not needed, but RETURN must always be used to terminate a subroutine. In PIC programming, use the project name (MOT1) in the start symbol of the main program, and the subroutine start address label in subroutine start symbols.

Processes

The process box is a general purpose symbol which represents a sequence of instructions, possibly including loops inside it (see Fig. 10.6(b)). The top level flowchart of a complex program can be simple, with a lot of detail concealed in each box. A subroutine is a process

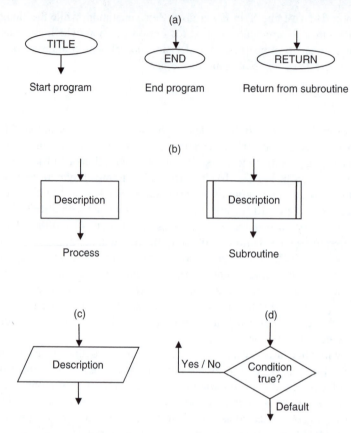

Figure 10.6 Flowchart symbols. (a) Terminals; (b) Processer; (c) Input/Output; (d) Decision.

which will be implemented in the source code as a separate block, and which may be used more than once within a program. It should be expanded into a separate subroutine flowchart, using the same name in the start symbol as that shown in the calling process. Subroutines can be created at several levels in a complex program.

Input/Output

This represents processes whose main function is input or output using a port data register in the microcontroller or microprocessor system (see Fig. 10.6(c)). Use a statement in the box which describes the general effect of the I/O operation, for example, 'Switch Motor On' rather than 'Set RA0'. This will make the flowchart easier to understand.

Decisions

The decision symbol contains a description of the selection as a question. There will be two alternate exit paths, for the answer 'yes' and 'no' (see Fig. 10.6(d). Only the arrow looping back or forward needs to be labelled 'yes' or 'no'; the default option, which continues the program flow down the centre of the chart, need not be labelled. In PIC assembly language, this symbol would refer to the 'Test and Skip' instructions. In the MOT1 detailed flowchart, an enlarged decision box is used to represent the 'Decrement/Increment and Skip if Zero' operation. This symbol allows more text inside, so is a useful alternative to the standard diamond shape.

10.4.5   Flowchart Structure

In order to obtain good program structure, there should be a single entry and exit point to and from all process blocks, as illustrated in the complete flowcharts. Loops should therefore rejoin the main flow between symbols, and *not* connect into the side of a process symbol, as is sometimes seen. Terminal symbols have a single entry or exit point. Decisions in assembler programs only have two outcomes, branch or not, giving two exits. Loops back should be drawn on the left of the main flow, and loops forward on the right of the main flow, if possible. For the main flow down the page, the arrowheads may be omitted as forward flow is clearly implied.

Connections between pages are sometimes used in flowcharts, shown by a circular labelled symbol. It is recommended here that such connections be avoided; it should be possible to represent a well-structured program with a set of separate flowcharts, each of which should fit on one page. An outline flowchart should be devised for the main sequence, and then each process detailed with a separate flowchart, so that each process can be ideally implemented as a subroutine or macro. In this case, the main program sequence should be as small as possible, consisting of subroutine calls and the main branching operations.

Therefore, the program should initially be represented as an outline flowchart on a single page, and each process expanded using subroutines or functions on separate pages. Keep expanding the detail until each block can be readily converted to source code statements. A well-structured program like this will be easier to debug and modify. Subroutines can be 'nested' to any required depth, depending on the stack size of the system. Smaller PICs tend to have an 8-level deep hardware stack, which means that only eight levels of subroutine are allowed.

10.4.6   Structure Chart

The structure chart is another method for representing complex programs (see Fig. 10.7). Each program block is placed in a hierarchical diagram to show how it relates to the rest of the

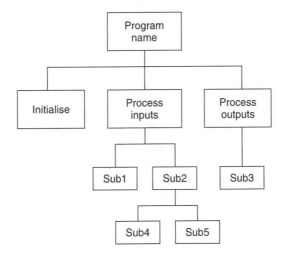

Figure 10.7   Structure chart.

program. This technique is more commonly used in data processing in business applications running on larger computer systems.

The program shown in the structure diagram has four levels. The main program calls subroutines to initialise, process inputs and process outputs. The input processing routine in turn calls Sub1 and Sub2 subroutines. Output processing only requires Sub3, but Sub2 calls Sub4 and Sub5 at the lowest level. At this level, three stack locations will be used up.

10.4.7 Pseudocode

Pseudocode is a text form of the program design. The main operations are written as descriptive statements which are arranged as functional blocks. The structure and sequence are represented by suitable indentation of the blocks, as is the convention for HLLs. A pseudocode version of MOT1 is shown in Fig. 10.8.

The pseudocode version of the program uses high level style syntax, such as IF ... THEN to describe the selections in the program. It has the advantage that no drawing is required, and the pseudocode can be entered directly into the text editor used for writing the source code. It can start as a brief outline, and be developed in stages until ready to be translated into assembler syntax. The pseudocode can be left in the source code as the basis of program comments, or replaced, whichever suits the programmer. Although used here to represent an assembler program, pseudocode is probably the more suitable for developing 'C' programs for applications for the more powerful PIC microcontrollers.

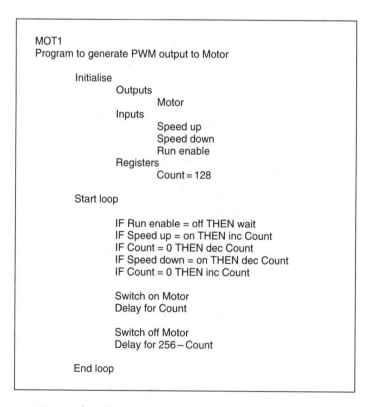

Figure 10.8 MOT1 pseudocode.

10.5   Program Implementation

When the program logic has been worked out using flowcharts, or otherwise, the source code can be entered using a text editor. Normally the program editor is part of an integrated development package such as MPLAB. Most programming languages are now supplied as part of an integrated edit and debug package.

10.5.1   Flowchart Conversion

The program design method should be applied so as to make the program as easy as possible to translate into source code. The PIC has a 'reduced' instruction set, meaning that the number of available instructions has deliberately kept to a minimum to increase the speed of execution and reduce the complexity of the chip. While this also means that there are fewer instructions to learn, the assembler syntax (the way the instructions are put together) can be a little more tricky to work out. For example, the program branch is achieved using the 'Bit Test and Skip' instruction. In CISC assembly code languages, branching and subroutine calls are implemented using single instructions. The PIC assembler requires two instructions. However, recall that 'special instructions' (essentially pre-defined macros) are available which combine 'test', 'skip' and 'goto' instructions to provide equivalents to conventional conditional branching instructions.

The representation of the program with different levels of detail is illustrated in Fig. 10.9. Figure 10.9(a) shows the process in detail so that each process box converts into only one or two lines of code. This may be necessary when learning the programming syntax. Later, when the programmer is more familiar with the language and the standard processes which tend to recur, such as simple loops, then a more condensed flowchart may be used, such as Fig. 10.9(b), where the loop is concealed within the 'delay' process. As we have seen above, this process can also be written as a separate, re-usable, software component, a subroutine. The corresponding source code fragment is shown in Table 10.3.

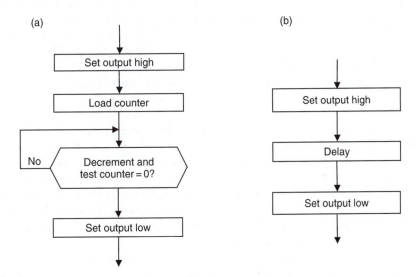

Figure 10.9   PIC program branch flowcharts. (a) Detail flowchart; (b) Outline flowchart.

Table 10.3 PIC program branch code fragment

```
;               Branch Program Fragment
                .
                .
                BSF         PortA,0         ; Set Output

                MOVLW       0FF             ; Set Count Value
                MOVWF       count1          ; Load Count
back1           DECFSZ      count1          ; Dec. Count & Skip if 0
                GOTO        back1           ; Jump Back

                BCF         PortA,0         ; Reset Output
                .
                .
```

Another limitation in PIC 16XXXX assembler is found when moving data between registers. It is not possible to copy data directly between file registers, it has to be moved into the working register, W, first, and then into the file register. This requires two instructions instead of the single instruction available in CISC processors. This problem is overcome to some extent by the availability of the destination register option with the byte processing operations. Nevertheless, the advantage of simplicity when learning PIC programming outweighs these limitations, especially when learning assembler programming for the first time!

We can see from the above examples that the software design techniques should be applied in a way which suits the application, the language and the level of expertise of the programmer.

10.5.2 MOT1 Source Code

The program source code for the MOT1 program is given in Program 10.1. The program produces the PWM output by toggling RA0 with a delay. A register labelled 'timer' holds the current value for the 'on' delay. The program does not use a subroutine for the delay, because the 'timer' value has to be modified for the 'off' delay. Note the use of the COMF instruction, which complements the contents of the timer register, which effectively subtracts the value from 256. The total PWM cycle time stays constant as a result. When incremented, the 'timer' value has to be checked to prevent it rolling over from FF to 00 , by decrementing it again if the new value is zero. The roll-under at the low end of the scale is prevented in a similar way.

The program source code has instruction mnemonics in upper case to match the instruction set in the data sheet. However, they are not case sensitive, so you will often see them in lower case. On the other hand, labels *are* case sensitive by default, so they must match exactly when declared and used. The label case sensitivity can be switched off as an assembler option if you wish. Upper case characters for the special function register names (PORTA) have been used to match the register names used in the data sheet, and lower case characters with the first letter capitalised used for general purpose registers (Timer, Count). The bit labels are lower case (motor, up, down, run), as are the address labels.

Using source code editing conventions like this is not obligatory, but consistent layout and presentation improves program readability and makes it easier to understand. Unfortunately, there is no generally accepted convention for assembler source code presentation.

Program 10.1   MOT1 source code

```
; **************************************************************
;      MOT1.ASM                      M. Bates              14/6/99
; **************************************************************
;
;      DC Motor Control using Pulse Width Modulation Motor (RA0) starts
;      with 50% MSR when enabled with RA4. Speed controlled with
;      RA2, RA3.
;
;      Hardware:        Simple Motor Circuit
;      Clock:           CR ~100kHz
;      Inputs:          Push Buttons (active low):
;                       RA2 = Speed Up
;                       RA3 = Slow Down
;                       RA4 = Run
;      Outputs:         RA0 (active high) = Motor
;
;      Chip Fuse Settings:
;      WDTimer:         Disable
;      PUTimer:         Enable
;      Interrupts:      Disable
;      Code Protect:    Disable
;
; **************************************************************
;
  PROCESSOR 16F84

; Register Label Equates.....................................

PORTA   EQU       05               ; Port A
Timer   EQU       0C               ; Delay Counter
Count   EQU       0D               ; Delay Count Pre-load

; Input Bit Label Equates ...................................

motor   EQU       0                ; Motor Output = RA0
up      EQU       2                ; Speed Up Input = RA2
down    EQU       3                ; Slow Down Input = RA3
run     EQU       4                ; Motor Enable Input = RA4

; **************************************************************

; Initialise ................................................

        MOVLW     b'11111110'      ; Port A bit direction code
        TRIS      PORTA            ; Set the bit direction
        MOVLW     080              ; Initial value
        MOVWF     Count            ; ...for delay

; Next Page .................................................

; Input Test ................................................

start   BTFSC     PORTA,run        ; Test Run input
        GOTO      start            ; & wait if HIGH
                                              continued...
```

```
            BTFSS       PORTA,up            ; Test Up input, if hi
            INCFSZ      Count               ; ...inc Count, test
            GOTO        test                ; and check down button
            DECF        Count               ; or dec Count again if 00
test        BTFSS       PORTA,down          ; Test Down input, if hi
            DECFSZ      Count               ; ...dec Count, test
            GOTO        cycle               ; and do an output cycle
            INCF        Count               ; or inc Count again if 00

; Output High and Delay......................................

cycle       BSF         PORTA,motor         ; Switch on motor

            MOVF        Count,W             ; Get delay count
            MOVWF       Timer               ; Load timer register
again1      DECFSZ      Timer               ; Decrement timer register
            GOTO        again1              ; & repeat until zero then

; Output Low and Delay.......................................

            BCF         PORTA,motor         ; Switch off motor

            MOVF        Count,W             ; Get delay count again
            MOVWF       Timer               ; Reload timer register
            COMF        Timer               ; Complement timer value
            INCF        Timer               ; Inc to avoid 00 value
again2      DECFSZ      Timer               ; Decrement timer register
            GOTO        again2              ; & repeat until zero then

; Repeat Endlessly...........................................

            GOTO        start               ; Restart main loop
            END                             ; Terminate source code
```

10.6   Source Code Documentation

Most programming languages allow comments to be included in the source code as a debugging aid for the programmer, and information for other software engineers who may need to fix the code at a later date. Comments in PIC source code are preceded by a semicolon; the assembler ignores any source code text following, until a line return is detected.

A header should always be created for the main program and the associated routines. It should contain relevant information for program downloading, debugging and maintenance. Examples have already been given. The layout should be standardised, especially in commercial products. The asterisk symbol (*) is often used to separate and decorate comments; rows of dots are also useful, but there is some scope here for individual touches!

The author's name, organisation, date, and a program description is essential. Hardware information on the processor or system type is important; for example, when a PIC program is assembled, the processor type must be specified, because there is some variation in the syntax required for each PIC chip type. The processor type may be specified in the header block as an assembler directive, or by selecting the processor in the MPLAB development system options.

Target hardware details such as input and output pin allocation is useful, and the design clock speed needs to be specified in programs where code execution timing is significant. Programmer settings which enable or disable hardware features such as the watchdog timer, power-up timer and code protection should also be listed.

The general layout of the source code should be designed to make the structure clear, with subroutines headed with their own brief functional description. Blank lines should separate the functional program blocks; that is, instructions which together carry out an identifiable operation. In this way, the source code can be presented in a way that makes it as easy to interpret as possible.

Summary

- The application requirements and target performance specification should be clearly stated as the first step in software design.

- A block diagram should be used to outline the hardware, components selected and a circuit designed.

- Programs consist of statements which allow sequence, selection and iteration.

- The software algorithm should be represented with a suitable software design aid, and elaborated until sufficiently detailed to translate into source code.

- Flowcharts should be structured, using separate charts to expand the processes in the higher level chart.

- Program flowcharts can be constructed from symbols representing terminals, processes, input/output, decisions and subroutines.

- Other methods for software design are pseudocode and structure charts.

- Source code should be fully commented for future reference, maintenance and modification.

Questions

1. Explain how PWM offers an effective way of controlling DC loads from a single digital output.

2. Explain briefly the role of the block diagram and flowchart in creating the final hardware and software design for an application.

3. State the three basic operations that make up a microcontroller program, how they are represented in a flowchart and give an example of how each is implemented in PIC assembler code.

4. Explain briefly the role of the subroutine in structured programming, and why it is generally desirable to use them.

1. Compare the source code for MOT1 with the flowchart in Fig. 10.5, and the pseudocode in Fig. 10.8, and check that they correspond.

2. (a)   Devise a block diagram for a motor control system which has a bidirectional drive and inputs which select the motor on/off and direction of rotation. Separate active high outputs will be used to enable the motor in each direction. Investigate and select a suitable component to provide the interface to a small DC motor.

    (b)   Construct a flowchart for a PIC program which will allow the user to turn the motor on and off from a single active low input, but only allow the direction to be changed when the motor is off. Produce a logic table, outline and detail flowcharts, and write the code observing the recommendations for source code documentation.

3. (a)   Devise a set of structured flowcharts for making a cup of tea (manually!).

    (b)   Draw a block diagram of a coffee machine, and devise a set of flowcharts for a control program. You may assume a PIC microcontroller will be used with suitable interfacing, sensors and actuators.

Chapter 11
Program Debugging

The design of a simple PIC motor control application MOT1 has been described in Chapter 10, and an assembler code program has been developed. In practice, it is unlikely that such a program will be written without any errors, especially when learning the language, so we now need to look further at the techniques and tools available for debugging (removing the errors from) PIC programs. We are going to continue with MOT1 as our example application, and will see how to resolve two main types of error.

Syntax errors. The syntax of a language refers to the way that the words are put together; any language, programming or spoken, must follow certain rules so that the meaning is clear. The rules in programming languages are very strict, because the source code must be converted into machine code without any ambiguity. Syntax errors are mistakes in the source code, such as mis-spelling an instruction mnemonic, or failure to declare a label before using it in the program. These errors are detected by the MPLAB assembler (MPASM), resulting in error massages being generated and displayed in a separate window. The source code is colour-coded in recent versions of MPLAB to indicate possible syntax errors instantly.

Logical errors. When a program is assembled without any syntax errors, it does not mean that it will function correctly when run in the hardware. Logical errors may well be present which prevent correct operation as originally specified. The software simulator (MPSIM) is used to detect and correct these errors prior to downloading to the chip. This allows the program to be run in a virtual processor, and logical errors detected by inspecting the program output, and comparing it with that which would be expected if the performance specification were correctly met. This normally requires inputs to be simulated as well.

If either type of error is detected, the program must be re-edited to remove the errors.

11.1   Syntax Errors

When the program source code for a PIC program has been created in the editor, it must be converted into machine code for downloading to the chip. This is carried out by the assembler program, which analyses the source code text line by line, and converts the instruction mnemonics into the corresponding binary codes for loading into the chip program memory, as PROGNAME.HEX. Only valid statements as defined in the PIC instruction set (see Appendix A) will be recognised and successfully converted. Assembler directives are also included, but these are not converted to machine code.

Before starting a project, create a folder to keep the project files in, named, for example, 'Motor'. In MPLAB 6, the source code is created in an edit window opened by hitting the 'new file' button. Type in the file name and save it immediately in the application folder as MOT1.ASM or any file name with an ASM extension. At the same time, create a backup version of the file on a different drive (floppy, network or removable drive).

When the program has been entered and saved, the project menu item 'Quickbuild' (or equivalent other MPLAB versions) will assemble a single file. If required, a project can be created, so that the MPLAB setup can be saved between sessions. In the Project menu, select 'New...' and call the project MOT1, or the same name as the source code. Now select 'Add Files to Project...' and select the source code created above. The program can now be assembled by selecting 'Build All', and the project saved.

In the source code file, numerical formatting, assembler directives and so on must all be used correctly. If they are not, error messages will be generated when the program is assembled. These describe the syntax errors which have been found. The error messages are saved in a text file PROGNAME.ERR, and displayed when the assembler is finished.

For demonstration purposes, deliberate errors were introduced into the example program MOT1.ASM (Program 10.1), and the error file MOT1.ERR generated by the assembler is listed in Table 11.1. There are three levels of error shown: 'Message', 'Warning' and 'Error'. The source code line number where the problem was found is indicated, and the type of problem that the assembler thinks is present. However, a word of warning – due to the presence of the error itself, the assembler may be misled as to the actual error. Consequently, the message generated is not always accurate. For example, the incorrect instruction mnemonic at line 58 caused the assembler to misinterpret 'Count' as an illegal op-code.

The PROCESSOR directive was misplaced, causing a non-fatal warning, which would not itself prevent successful assembly of the program. The TRIS instruction also caused a warning in the MPLAB assembler, because its use is not recommended, but will still be successfully assembled. It is used in our examples because the alternative method of port initialisation, using page selection, is more complicated to use.

Table 11.1 Selected error messages from assembly of MOT1

```
Deleting intermediary files... done.
Executing: "C:\Program Files\MPLAB IDE\MCHIP_ Tools\mpasmwin.exe"/q
/p16F84"MOT1.asm"
/l"MOT1.1st" /e"MOT1.err"
Warning[205] C:\MOT1.ASM 24 : Found directive in column 1.(PROCESSOR)
Warning[224] C:\MOT1.ASM 44 : Use of this instruction is not recommended.
Message[305] C:\MOT1.ASM 56 : Using default destination of 1 (file).
Warning[207] C:\MOT1.ASM 58 : Found label after column 1.(DEC)
Error[122]   C:\MOT1.ASM 58 : Illegal opcode (Count)
Error[113]   C:\MOT1.ASM 71 : Symbol not previously defined (Timer)
Error[113]   C:\MOT1.ASM 73 : Symbol not previously defined (again1)
Error[129]   C:\MOT1.ASM 92 : Expected (END)
Halting build on first failure as requested.
BUILD FAILED
```

The instruction mnemonic DECF was mis-spelt as DEC, causing the errors at line 58. This contributed to the register label count being misinterpreted. The register label 'Timer' was missed out of the EQU statements at the top of the program causing the error at line 71. The jump destination 'again1' has been incorrectly labelled 'again', causing the error at line 73. Finally, the END directive had been omitted at the end of the program, causing the message 'Expected (END)'.

The message 'Using default destination of 1 (file)' refers to the fact that the full syntax for MOVWF instruction has not been used. Using the full syntax, the destination for the result of the operation is specified as the file register or the working register, by placing a W (0) or F (1) after the destination register number or label. In the examples throughout this text, we take advantage of the assumption by the assembler that the destination is the file register if not specified in the instruction; this simplifies the source code. More advanced programmers will use this feature of the instruction set to save instructions when the data destination is the working register. When the error messages have been studied carefully, and printed out if necessary, the source code must be re-edited and re-assembled until it is correct.

11.2 Logical Errors

When all syntax errors have been eliminated, the program will assemble successfully, and the hex file created. However, this does not necessarily mean that it will function correctly when downloaded to the chip; in fact, it probably won't! Usually there will be logical errors, particularly when learning the programming method. Mistakes in the program functional sequence or syntax will prevent it operating as required. For instance, the wrong register may be operated on or a loop may execute correctly, but the wrong number of times. There may also be 'run-time' errors, that is, mistakes in the program logic which only show up when the program is actually executed. A typical run-time error is 'Stack Overflow', which is caused by CALLing a subroutine, but failing to use RETURN at the end of the process.

11.2.1 Simulation

Conventional microprocessor system hardware varies between each application because they are built from discrete chips. The configuration of the CPU, memory and I/O chips is designed to suit the application; therefore only the CPU itself can normally be simulated, not the whole system. Some logical errors can only be detected by running the program on the actual hardware, and testing for the correct outputs in response to the specified inputs. In some systems, it is possible to download the program to RAM for testing, so that it can be more easily corrected. In others the program can only be installed by programming an EPROM memory chip and actually fitting it in the hardware. This could mean erasing and reprogramming the EPROM repeatedly, in order to get the program right. This can obviously be quite time-consuming, and to be avoided if possible.

If the program is being tested in the hardware, it may also be difficult to work out exactly what the problem is. The program execution may stop, without any indication of the reason. If the system operating program has some debugging built in, and some means to display the errors, the program can be tested in the target hardware, but this adds complexity to the target system which may not be needed once the application program is up and running correctly. Alternatively, an in-circuit emulator (ICE) can be used for hardware testing (see Section 11.5.2), but these are relatively expensive.

The advantage of the microcontroller is that the design of the chip is fixed, so a full simulation model can be provided for each. This allows the program to be tested for logical errors before downloading. The simulation package (MPSIM) allows windows to be opened to show the source code, machine code, registers, simulated input, timing checks and so on. The program can then be run, stopped and stepped through one instruction at a time. Source level debugging shows the current execution point on the source code itself. Breakpoints allow the code to be stopped at any point, so that the registers and outputs can be inspected.

The software simulator is therefore a very useful tool for program testing. One of the advantages of the PIC range is that the development system, including the assembler and simulator, has always been provided free by Microchip to encourage the market for its chips.

11.2.2   Program Testing

The simulator must model the operation of the selected microcontroller as accurately as possible. The user must be able to provide the inputs which would occur in the actual system, and be able to monitor the effect on relevant registers. The program will need to be started, stopped at critical points and single stepped to check the sequence of operations, and timing measured. All possible input events and sequences must be anticipated and tested, to ensure that no unforeseen problems arise when the application is in use.

Inputs

The simplest method of simulating inputs is the asynchronous stimulus. Single bit inputs are changed via on-screen buttons at the required time while the program is executed in single-step mode. The other method is to use a stimulus file, which automates this process, allowing the same test sequence to be input every time the program is run in the simulator. This will save time while debugging more complex applications.

Outputs

For monitoring the outputs, various options are available. All the file registers may be displayed simply as hex numbers in a table. The SFRs can be displayed separately, in a choice of formats; hex, binary and decimal. Selected registers can be viewed in a 'watch' window, using the register labels from the source code.

Timing

Program timing can be checked using the stopwatch. First, the clock rate must be entered so that the actual timing can be predicted. The time taken for sequences to execute can then be checked. For example, the period of the waveform output by MOT1.

11.2.3   Testing in MPLAB 6

To test the program, the source code must be assembled or the project built. Ensure that the correct processor is selected, via the 'Configure', 'Select Device' dialogue. Select the 16F84 for the MOT1 project. At the same time, it is advisable to set the chip fuses via 'Configure',

'Configuration Bits', selecting oscillator = RC, watchdog timer off, power-up timer on and code protect on.

Now select 'Debugger', 'Select Tool' then 'MPLAB SIM'. The debugging tools should then appear in the toolbar:

RUN   Executes the program.
HALT   Stops the program with the current execution point indicated.
STEP INTO Executes program one instruction at a time.
STEP OVER Single step current routine, but execute subroutines at full speed.
RESET   Start again at the top of the program.

The program can now be run and stopped to make sure the simulator is working. When halted, the current execution point is indicated. We now need to set up the simulator so that the relevant information is displayed so that correct program function can be confirmed, or logical errors corrected. A typical display for MOT1 is shown in Fig. 11.1.

11.2.4   Setting up MPLAB 6

Simply running the program in the simulator does not generally provide enough information to confirm correct operation. Single stepping allows the program to be executed one instruction at time; the registers can then be checked for the correct results.

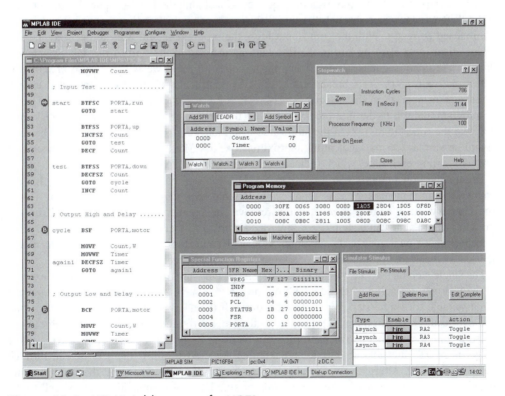

Figure 11.1   MPLAB 6 debug screen for MOT1.

View Outputs

The file registers can be displayed in simple hex by selecting 'View', 'File Registers' but this is of limited use. Instead, select 'View', 'Special Function Registers'; Port A is displayed in binary. Also select 'View', 'Watch' and display the labelled Count and Timer registers using the 'Add Symbol' dialogue.

Assign Inputs

We now need to set up the simulator inputs. Select 'Debugger', 'Stimulus' and the 'Pin Stimulus' tab in the Simulator Stimulus dialogue. 'Add Row' and click on the 'Pin' cell; select 'RA4'. Click on the 'Action' cell and select 'Toggle' mode. Type 'Run' in the Comment column. Repeat this for RA2 (Up) and RA3 (Down). Hide the blank columns. When the 'Fire' button is pressed, the input will change over, but only after the next step in the simulation.

Set up Stopwatch

Select 'Debugger', 'Settings' and the 'Clock' tab. Set the oscillator frequency to 100 kHz. Select 'Debugger', 'Stopwatch' to display the stopwatch. Arrange the windows for best visibility.

11.2.5   Testing MOT1

Set all the inputs high by clicking on each fire button and stepping once. Reset the program and single step through the initialisation sequence. Check that the Port A input bits are set high in the SFR window. Step through the initialisation sequence to 'start' and check that 'Count' is initialised correctly. With the 'run' input high, the program should wait at the 'start' label.

Now clear input RA4 by hitting the Fire button, and step to the start of the delay sequence. Once in the delay sequence, single stepping is not helpful, so we will now use breakpoints to test the main loop. Set a breakpoint at the BSF instruction by right-clicking on that line. A red marker appears in the margin. Set a break point at the BCF instruction as well. Reset the program and run to the first breakpoint. Zero the stopwatch and run to the next breakpoint and note the time (15 ms). Zero the stopwatch again and run to the first breakpoint. The time should be similar (16 ms), indicating approximately 50% MSR. At the same time, check that RA0 is toggling each time, which indicates that the output PWM signal is present.

The up and down control can now be tested. Hit the 'Up' button (RA2) and check that the Count value increments for each output cycle. Disable 'Up' (high) and enable 'Down' (low). The Count value should decrement. Now remove the breakpoints (right click) and run the program. The Count value should go down to 01 and stop there. Now enable the 'Up' input and check that Count goes to its maximum value (FF) and stops there. The stopwatch can now be used to check the minimum and maximum MSR values meet the specification.

The single step facility has two options, 'step into' and 'step over'. 'Step into' means execute all the instructions including those in subroutines. As we have seen above, the delay sequence is not suitable for single stepping as it is repetitive. If the delay is in a subroutine, as in BIN4, step over could be used; this will step through the current routine, but will run any subroutines called at full speed. This allows each program block to be tested separately. However, as the delays in MOT1 are not subroutines, breakpoints are used to allow them to be run through then at full speed. Alternatively, the process can be shortened by changing the count register value to a low value, either in the source code or simulator, or commenting out the call to the subroutine.

11.2.6 Stimulus File

Testing the program in the simulator can be automated by use of a stimulus file. This allows the state of an input or file register to be changed at particular step in the program. The same test sequence can then be applied each time the simulation is run, making the testing quicker and easier in more complex applications. In MPLAB 6, select 'Debugger', 'Stimulus' and the 'File Stimulus' tab and follow the help guide to create the necessary files.

11.2.7 Tracing

Tracing is another way of checking the program sequence. As each instruction is executed, the changes in relevant registers are logged to a file. In MPLAB 6, run the program, and select 'View', 'Simulator Trace' to see the trace record up to that point. Scroll down the bottom of the list to see the most recent events. Note that the 'probe' columns are not used in the software simulation, and should be deselected via the simulator settings.

MOT1 is now tested. Save the program if it has been corrected, and save the project.

11.3 MPLAB Tools

The most commonly used windows in the simulator are described below.

11.3.1 Edit Window

The Edit window is used for creating and editing the program. When one of the assembly options is selected, the source code in the current edit window is converted to machine code (PROGNAME.HEX). If syntax errors are then detected, the error file is displayed automatically, and the number of the incorrect line in the source code is given. When the program is stopped, the current instruction is indicated in the source code window. Breakpoints can be inserted by right-clicking on the line required; they are then indicated in the left margin.

11.3.2 Special Function Register Window

The SFRs are displayed in a number of different formats. It is particularly useful to see the contents in binary, as many of the SFRs are bit-oriented. The most commonly needed are:

	WREG	Working register (not an SFR).
	WREG	Working register (not an SFR).
01	TMR0	Timer zero hardware counter.
02	PCL	Program counter.
03	STATUS	Flag register, zero flag = bit 2.
05	PORTA	I/O port.
06	PORTB	I/O port.
85	TRISA	Data direction register for PORTA.
86	TRISB	Data direction register for PORTB.

11.3.3 Watch Window

A watch window allows selected registers to be displayed, and only those of interest in a particular application. In MPLAB 6, the 'Add SFR' button is used to display the SFRs, and 'Add Symbol' to display the registers identified by user label. The numerical format, among other characteristics, can be changed by right-clicking on the window and selecting 'Properties ...'.

11.3.4   Simulator Stimulus

To test the MOT1 program, the inputs which enable the motor and change the speed must be simulated. The simplest method is to use the 'Pin Stimulus' option in the simulator stimulus window. Each input is assigned a row in the table, and normally set to 'toggle' mode. When the table is complete, it can be saved as a separate file.

11.3.5   Stopwatch

The stopwatch window allows the simulated chip clock to be set to the frequency that will be used in the actual hardware. For simulating MOT1, the value 100 kHz is entered. The total number of instructions executed and the corresponding elapsed time are then correctly displayed. These can be reset to zero at any time. When testing MOT1, the time taken for the delays and the overall output period can be checked. This type of check can also be carried out using a break point.

11.3.6   Program Memory

If the 'Program Memory' is selected from the 'View' menu, the program can be viewed in different forms. The 'Opcode Hex' option shows the raw hex code. 'Machine' displays the disassembled program. This means, the simulator has taken the machine code program and converted it back into instruction mnemonics, to confirm that it is correct. The hex machine code (Opcode) for each instruction is shown against each instruction alongside the program memory location (Address) at which it is stored. Selecting the 'Symbolic' display restores the original labels used. When the simulation is run, the execution point is also shown in this window.

11.4   Test Schedule

Using the simulator, the program function can be tested against the specification. The specification therefore should be converted into a test procedure which will test all its functions, and all possible incorrect input sequences, especially where these are generated by an operator. A test procedure for MOT1 is suggested in Table 11.2.

The test procedure looks very detailed when written down, but in practice it does no more than test all the features of MOT1. The software product needs to meet the specification only once, in a prototype hardware. Once the software is proved, the hardware in the production units can be tested in conjunction with software which is known to be correct.

11.4.1   Typical Logical Errors

It is difficult to anticipate exactly what kinds of logical errors will arise, as they are usually the result of inexperience, but the following types of errors are typical.

Port Initialisation

If a port does not appear to respond to output operations, check that the initialisation is correct.

Table 11.2 Simulation test schedule for MOT1

Project:	MOT1	Simulator:	MPLAB 6.1

Setup: *Source code: MOT1.ASM*
Watch registers: PORTA, Timer, Count
Simulator Stimulus: RA2 = up, RA3 = down, RA4 = run (toggle mode)
Stopwatch: Clock frequency 100.00 Hz
Optional: Program memory

	Test	Action	Required Performance	✓/X	Fault/Comment
1	Initialise	RA2, RA3, RA4 = 1	Check watch window, PORTA		All inputs inactive
2	Start	Step over	Count = 80 Waits in start loop		Waiting for run enable
3	Enable run	Input: RA4 = 0 Step over	RA0 = 1 Runs into high delay Timer decrements to 0 RA0 = 0 Runs into low delay Timer decrements to 0 Repeats		One cycle of output at default MSR 50% stopwatch: Output period ≈33 ms
4	Disable run	Input: RA4 = 1 Step over	Returns to start loop		Waiting for run enable
5	Select increment	Input: RA2 = 0 RA4 = 0 Step over	Count increments to 81, 82 after next cycle, etc.		Count increments MSR increasing
6	Test for no roll-over	Run at full speed and stop	Maximum count = FF Count NOT to 00		Roll-over prevented
7	Select decrement	Input: RA3 = 0 RA4 = 0 Step over	Count decrements to FE, FD after next cycle, etc.		Count decrements MSR decreasing
8	Test for no roll-under	Run at full speed and stop	Minimum count = 01 Count NOT to 00		Roll-under prevented
9	Restart	All inputs = 1 run and stop	Returns to start loop		Restart correct
10	Program reset	Reset	Execution reset to first instruction		Reset correct
11	Recheck default output	RA4 = 0	Count = 80		Output toggles MSR ~50%

Tested: MBates   **Date: 4/11/03**

Register Operations

If a register is not responding as it should, ensure that the correct register is being modified, and the address label is correct.

Bit Test and Skip

Obtaining the correct sequence of operations in the program depends on these instructions. Make sure the skip condition is correct, clear or set, as this is easy to get wrong.

Jump Destinations

Make sure that the destination specified is correct, and the loop sequence includes all the necessary steps.

Program Structure

If the program gets lost during subroutine execution, check that call address labels are correct, and all subroutines are terminated with 'return' instructions. Note that the stack overflow/underflow warning which indicates that CALL and RETURN are not matched correctly is disabled by default in MPLAB 6. Stack error detection can be enabled via the 'Debugger', 'Settings', 'Break Options'.

11.4.2 *Limitations of Software Simulation*

The simulator allows the program logic to be tested before running the program in the actual hardware to ensure that it is functionally correct. However, the simulation cannot be 100% realistic, and its limitations need to be taken into account in testing the real system. The following is an example of the kind of error that might easily be missed, but seriously affects the operation of the application, and would compromise safe operation of the real system if a more powerful motor were used.

The simulation showed that after initialising Port A, with bit zero (RA0) set as output, this data bit goes high by default (all port bits default to logic '1'). The motor will therefore come on, even if the enable input has not yet been operated. This is obviously a major flaw, which can be fixed by following the port initialisation instructions with one to clear the motor output bit. This would still result in a very short pulse to the motor at the start of the program. If this caused a problem to the motor drive, an alternative fix would be needed; for example, an external circuit which ensured that the motor was not powered up until the controller had been started.

11.5   Hardware Testing

When the program has been fully debugged in the simulator, it can be downloaded to the chip, which is then placed in the target system. However, the target hardware layout and connections should be checked and tested separately before inserting the chip. The board should be carefully inspected for correct assembly; solder bridges between tracks and dry joints are common faults. The connections can be buzzed out with a continuity tester and checked against the circuit diagram.

Before fitting the chip, it is a good idea to apply power and check the rest of the circuit, but make sure the components connected to the chip outputs can be safely powered up with an open circuit input. For example, in the MOT1 circuit, the FET gate input should not be allowed to float, so there is a pull-down resistor fitted. The supply current should not be excessive, and components should be checked for overheating. The voltages at the chip power supply pins (Vdd and Vss) of the chip should also be checked, as incorrect connection of the supply is likely to damage most ICs.

If all is well, switch off the power and fit the chip using a suitable tool. Anti-static precautions should be observed, but PIC chips have not been found to be particularly sensitive in practice. Make sure it is the right way round! Pin 1 should be marked on the board. Switch on and check that the chip is not overheating or drawing excessive current. If left to overheat for more than a few seconds, the chip will probably be destroyed, but you may be able to rescue it if you switch off before the smoke appears!

11.5.1 In-Circuit Program Testing

Connect an oscilloscope to the output. On power up, there should be no output from MOT1. When the 'Run' button is pressed, the default output waveform with a 50% mark/space ratio should be observed, running at a frequency of about 30 Hz. The speed 'Up' and 'Down' buttons should be operated to ensure that the speed control stops at the minimum and maximum value, and does not 'roll over' from zero to full speed in one step. Note that the program algorithm does not give an MSR of 100% or 0%, but stops one step short of the maximum and minimum. Since there are 255 steps altogether, the step size is less than 1%.

The circuit should also be tested for 'fail-safe' operation, that is, no unplanned or potentially dangerous output is caused by an incorrect input operating sequence. In this case, operating both the 'Up' and 'Down' buttons together would be an erroneous input combination, which should result in no change in speed, because the increment and decrement operations cancel out.

Other examples of potential problems which would need to be considered are input switch bounce, variation in component performance (check specifications), dynamic operation of motor, minimum MSR required to make the motor run, and so on. More complex applications are likely to have more potentially incorrect input conditions and component-related problems, but the test schedule should ideally anticipate all possible fault modes (not easy!). If the circuit is being produced on a commercial basis, a formal test schedule would be needed, and the performance certificated as correct to the product specification.

A basic test schedule for the MOT1 program running in a PIC 16F84 is given in Table 11.3. Additional documentation should be prepared according to circumstances (education, commercial, research) to provide the application user or product customer with the relevant information on using the system.

11.5.2 In-Circuit Emulator

The simulator allows the software to be tested without the target hardware; however, the execution speed may be lower than in the real hardware, and input conditions due to the real components such as non-ideal voltage levels and transient signals may not be simulated. This limits the accuracy of the test.

On the other hand, an ICE allows the hardware to be tested without the microcontroller (or microprocessor). It allows a host computer with a special emulation pod attached to replace the

Table 11.3 Test schedule for MOT1 application board

TEST SCHEDULE	Project Name: MOT1		
Batch No:		Specification Number:	
Test	*Required Result*	*Tick*	*Comment*
Inspected	No faults observed		
Power Up	Motor off		Output low
!RUN operated	Motor on, 50% MSR		Frequency ~30 Hz
!UP operated	MSR increases >99%		Step Resolution <1%
!UP released	MSR constant		Speed held
!DOWN operated	MSR decreases <1%		Step Resolution < 1%
!DOWN released	MSR constant		Speed held
!MCLR pulsed	Speed reset 50% MSR		Restarts program
!RUN released	Motor Off		Output low
Tested by:			
Signed:			Date:

target processor (see Fig. 11.2). A header connector with the same pin out as the processor is connected to its socket on the board. The pod provides the specific processor emulation hardware, which is able to represent that particular processor at full speed. The program can then be tested with the actual hardware, giving more realistic results.

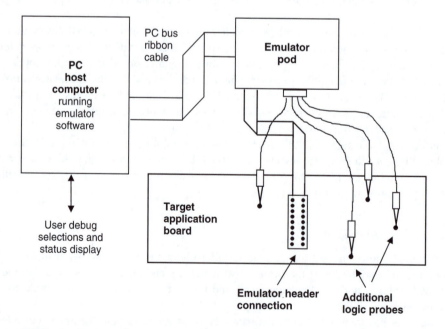

Figure 11.2 In-circuit emulator set-up.

Professional development systems use this technique as it allows the complete application hardware system to be tested as it interacts with a virtual processor. This type of equipment is usually regarded essential in the commercial development environment, but it is relatively expensive. Individuals, educational users and smaller companies, with limited development budgets, will find the low-cost ICD (see Section 7.7.2) option a very useful alternative.

Summary

- There are two main types of error that can occur in source code, syntax and logical errors.

- Syntax errors are invalid statements which are detected by the assembler; error messages are generated to assist debugging.

- Logical errors are mistakes in the program design or implementation, which can be detected by using the simulator to test the program operation.

- MPLAB is an IDE, featuring an editor, assembler, simulator, emulator support and downloading software.

- A typical MPLAB simulation setup would use the source code, program memory, SFR, watch, trace, stopwatch and stimulus windows.

- Logical errors can be identified using step into, step over and breakpoints functions with register monitoring.

- The ICE allows the software to be tested at full speed on the target hardware.

Questions

1. Explain briefly the difference between syntax and logical program errors, and how they are detected.

2. Explain the difference between 'simulation' and 'emulation' in the PIC system. Why does emulation need some extra hardware, and simulation does not?

3. How are the following used in program debugging: single stepping, breakpoint, pin stimulus, watch window?

4. An instruction in the program memory listing appears as follows:
   ```
   0005   1A05   start   btfsc   0x5,0x4
   ```
   Explain the meaning of each of the six elements in the line.

Activities

1. (a) In MPLAB, open a source file edit window, enter the source code for MOT1 and assemble it. Note any error messages generated. If the program assembles correctly first time, put some deliberate errors in the source code and inspect the error messages.

(b)   In MPLAB, create a project MOT1, assign MOT1.ASM to the project and test the program using the setup suggested in Section 11.4. Use the test schedule in Table 11.2 to check the program operation and confirm that it meets the specification.

2.   Modify the application by eliminating the inputs and generating a PWM output whose MSR increments automatically, once per output cycle. Allow the incremented count register to roll over from 100% MSR to restart at 0% MSR. Devise a test schedule to confirm correct operation. Download to a PIC chip and run the program in the hardware, monitoring the output with an oscilloscope. What output should you see? Predict the time taken (to the nearest half second) for a complete cycle from 0 to 100% MSR.

(Answer: 8.5 s)

Chapter 12
Prototype Hardware

12.1   **Hardware Design**

12.2   **Hardware Construction**

12.3   **Demo Board**

12.4   **Demo Board Applications**

We now come to the stage where we need to look at the techniques available for designing and building our PIC circuits. Circuit design, simulation and layout software has developed to the point where powerful packages are now available at a reasonable cost. Current software allows the circuit to be drawn, tested by simulation, and the circuit netlist (list of components and connections) produced. This is then imported into a PCB design package where the circuit is laid out on screen; this can be printed onto a masking sheet, or a file generated which can be used to automatically produce a PCB. Currently, a popular choice in the UK is Proteus™ from Labcenter Electronics. This consists of two main parts, ISIS™ for circuit design and schematic capture, and ARES™ for PCB layout. However, the design, layout and construction of PCBs will not be discussed in detail, because it involves learning to use this specialist PCB design software, which is beyond the scope of this book.

12.1   Hardware Design

Traditionally, circuits have been designed as sketches on paper and a final version produced by a draughtsman. This relied heavily on the experience of the electronics engineer to be able to predict the circuit performance from theoretical knowledge and practical experience. Numerous prototypes would typically be needed to arrive at a working solution.

This process has, since the development of increasingly powerful desktop computers, been radically improved. The designer still has to come up with the original ideas, but the proposed circuits can now be quickly drawn and tested on-screen, and a working design produced without a pencil touching paper or any component being inserted in a prototype board. The design cycle is much faster and the time taken from design concept to market is a major competitive factor in a rapid changing industry. Therefore, electronic computer aided design (ECAD) is now a vital tool for the electronics engineer, just as CAD has become for the mechanical engineer.

Thus, a circuit diagram can be drawn and converted into a PCB layout within a single software package. The circuit design can be tested for correct function by software simulation which incorporates mathematical models for the behaviour of each component, and their interaction. Libraries of microprocessor and microcontroller models, as well as interactive on-screen components, are now included; a circuit can be drawn on-screen, the application program attached to the microcontroller and the program tested by operating the on-screen inputs, such as switches and/or a keypad, with the mouse. The results are then displayed on simulated displays (LED and LCD) or operate animated output devices such as relays and motors.

As an example, a simple circuit design created in the schematic capture package ISIS is shown in Fig. 12.1. It is an electronic dice board with a push button, seven-segment display and buzzer controlled by a PIC 16F84. It can be programmed to display a random number between one and six when the button is pressed.

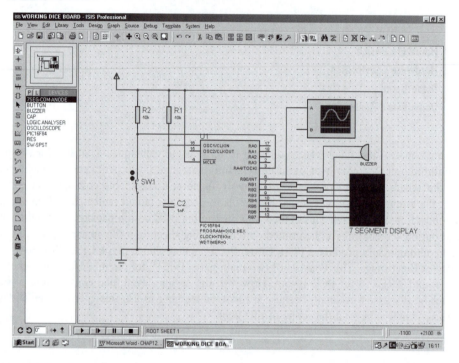

Figure 12.1 Circuit desgin for DICE board.

When a suitable program is assigned to the PIC in the simulation, the circuit becomes interactive on screen. When the switch is operated, the display will operate in the same way as the real device. If the chip is programmed to make a sound, the waveform can be displayed on a virtual oscilloscope, and even be reproduced from the PC audio output, if the CPU is fast enough.

12.2 Hardware Construction

First, we will look briefly at some traditional techniques suitable for building one-off boards and prototypes. A general purpose demonstration board will be then designed and laid out in prototype form, and some programs provided to demonstrate its features and the related programming principles. The DIZI board (display and buzzer with interrupt) has a seven-segment display and an audio output for some simple display and sound applications (see Appendix B).

12.2.1 Printed Circuit Board

The PCB is the standard method for making electronic circuits. In its basic form, it starts life as a sheet of insulating fibreglass board with a layer of copper on one side. The circuit connections are made by photographically transferring a pattern of conducting tracks and pads for the component connections onto the copper. For complex circuits, such as a PC motherboard, multilayer boards are used to accommodate the large number of parallel data connections.

The layout for a simple PIC circuit is shown in Fig. 12.2. It has a PIC 16F84, push button, seven-segment display, buzzer and associated components and can be programmed to operate as an electronic dice, generating a random number between one and six. The pattern of the copper tracks is shown, as well as the 'silk screen' printing which will be applied to the component side of the board to show where to place the components.

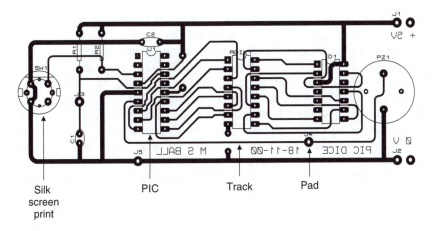

Figure 12.2 PIC dice board layout.

The layout is reversed as it will be printed onto a translucent mask, which is then used to create the pattern of connections on the copper side of the board. The copper layer is coated with a light-sensitive material, which is exposed to ultraviolet light through the mask. In the exposed areas of the board, the photo-sensitive material becomes soluble and is removed by a caustic solvent, exposing the copper below. This is then dissolved (etched) in an acid bath, leaving behind the copper layout where it was protected by the etch resisting layer. The components are then fitted to the silk screen side of the board, and the leads and pins soldered to the pads.

Once the layout has been designed, it can be used for batch production of the application hardware. Specialist companies are often used to manufacture the boards direct from the file output of the PCB design software. The final PCB-based product is shown in Fig. 12.3.

Even with the current ECAD packages, the PCB layout can take some time to create, and a considerable amount of skill is needed to use the software. Therefore, we will also look at how to prototype our hardware using traditional methods which do not require specialist software or PCB fabrication equipment.

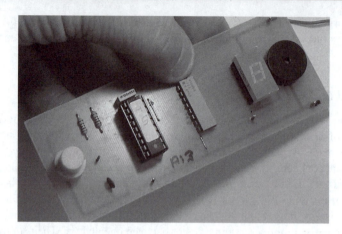

Figure 12.3 PIC 16F84 dice board.

12.2.2 Breadboard

A common method of constructing prototype circuits uses breadboard (plugblock) to wire up prototype circuits. The connecting wires are pushed into an array of interconnected sockets, allowing circuits to be quickly built and modified. A typical circuit, which will be discussed later, is illustrated in Fig. 12.9. The circuit for testing the BINx programs in Section B can also be built this way, as it is not too complicated, and may need modifying for experimental purposes.

The breadboard module has sets of terminals laid out on a 0.1" grid which will accept the manual insertion of component leads and insulated tinned copper wire (TCW) links. It has rows of contacts interconnected in groups placed either side of the centre line of the board, where the ICs are inserted, giving typically four contacts on each IC pin. At each side of the board, there are longitudinal rows of common contacts which are normally used for the power supplies. Some types of breadboard can be supplied in blocks that plug together to accommodate larger circuits, or are mounted on a base with built-in power supplies.

The layout for a simple circuit is shown in Fig. 12.4, with a PIC 16F84 driving an LED at RB0 via a current-limiting resistor. The only other components required are a capacitor and a resistor to form the clock circuit, but we must not forget to connect the !MCLR (master clear) pin to the positive supply, or the chip will not run. The chip could now be programmed to flash the output at a specified rate. To connect up the circuit, we will need to refer to the chip pinout, which is given in Fig. 12.5.

Breadboard circuits can be built quickly, with no special tools required. However, the connections are relatively unreliable, so bad connections are likely in more complicated circuits. Therefore, method of producing prototype circuits with more reliable soldered connections is useful.

12.2.3 Stripboard

Stripboard is a prototyping method which requires no special tools or chemical processing. The components are connected via copper tracks laid down in strips on a 0.1" grid of pin holes in an insulating board. The components are soldered in place and the circuit completed using wire links placed on the component side and soldered to the tracks on the copper side. The tracks must then be cut where the same strip is used for separate connections in the circuit. The components are generally placed across the tracks, so that each pin connects with a separate

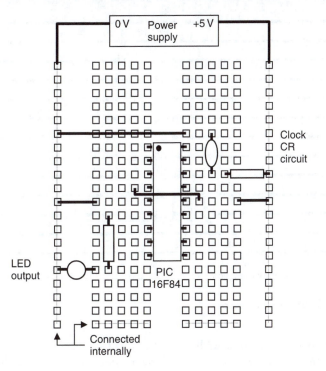

Figure 12.4 Breadboard layout.

Function		Label	Pin	Pin	Label	Function	
I/O	Port A bit 2	RA2	1	18	RA1	Port A bit 1	I/O
I/O	Port A bit 3	RA3	2	17	RA0	Port A bit 0	I/O
I/O	RA4 or timer input	RA4/T0CKI	3	16	OSC1/CLKIN	Crystal or RC oscillator	In
In	Master clear (Reset)	!MCLR	4	15	OSC2/CLKOUT	Crystal circuit (if used)	Out
In	Supply 0 V	Vss	5	14	VDD	Supply +5 V	In
I/O	Port B bit 0 + interrupt	RB0/INT	6	13	RB7	Port B bit 7 (+ interrupt)	I/O
I/O	Port B bit 1	RB1	7	12	RB6	Port B bit 6 (+ interrupt)	I/O
I/O	Port B bit 2	RB2	8	11	RB5	Port B bit 5 (+ interrupt)	I/O
I/O	Port B bit 3	RB3	9	10	RB4	Port B bit 4 (+ interrupt)	I/O

Figure 12.5 16F84 pinout.

track. The tracks must be cut between opposite DIL chip pins, and other required positions, using a hand drill. An example is shown in Fig. 12.11.

Care is required to avoid 'dry' joints (too little solder) or short circuits between tracks due to solder splashes and whiskers (too much solder!). A manual drawing may be used to draft the layout, if necessary, but a reasonably experienced constructor can build the circuit directly onto the board, with maybe some additional wastage of board area. It is also possible to work out the layout using a simple drawing package, or the drawing tools in a wordprocessor.

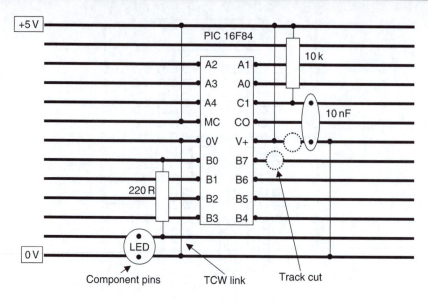

Figure 12.6 Stripboard connections.

Figure 12.6 shows how the simple PIC circuit can be laid out for construction on stripboard using general purpose drawing tools, such as those provided with Word. In the wordprocessor, the drawing toolbar needs to be switched on, and page layout view selected. In the 'Draw' menu, the grid should be switched on and set it to 0.1"; this allows layouts to be drawn actual size, since this is the spacing between standard in-line pins. The circuit can then be drawn using suitable line styles, text boxes and so on. When finished, use the 'Select Objects' tool to select the whole drawing and 'Group' it in the 'Draw' menu. This prevents text cursor movement from disrupting the drawing, and the whole diagram can be re-positioned on the page if required.

12.3   Demo Board

A circuit will be now designed, and a set of programs outlined, to illustrate the hardware design process and programming principles discussed in previous chapters. The DIZI board will allow the user to experiment with the various features of the PIC hardware and programming techniques within a single hardware module.

12.3.1   Hardware Specification

The microcontroller demonstration board will be suitable for demonstrating a wide range of processes incorporating display, audio, counting, timing and interrupt operations. The board will have a single digit seven-segment display for showing output data in hexadecimal and decimal form, and a low power audio transducer. Manually operated toggle switches will provide a 4-bit parallel input. Two input push buttons should be available; one to simulate input events to be counted, the other to simulate an external interrupt input. Timed events should be measured or generated with an accuracy of better than 1%. The circuit will be battery powered, with a push button power switch to ensure that the power cannot be left on, and a power 'on' indicator.

The board will be as small as possible, and the microcontroller must be easily reprogrammable, with flash memory.

12.3.2   Hardware Implementation

The seven-segment display will require seven outputs from the microcontroller. Active high operation can be provided by a common cathode LED display, and the display decimal point can be used as the power indicator. The audio transducer requires one output; a peizo buzzer was tested for suitability, since its power consumption is low. Although the device is specified to operate at a fixed frequency, it was found to be satisfactory in its frequency response. A miniature dip switch bank will be used for 4-bit input, and miniature push buttons used, to conserve space.

Fourteen I/O pins are required; the PIC 16F84A has only thirteen, so a chip with more I/O could be considered. However, the audio output and interrupt input could use the same I/O pin, because the high impedance of the buzzer will not interfere with input signals on the same pin. RB0 will be used as the dual function pin, since it is defined as the principal interrupt input, but can also be used as an output. The outputs can source up to 25 mA, but current-limiting resistors will restrict the current per display segment to 10–15 mA to control the maximum load on the port when all the segments are on.

The I/O allocation for the project is therefore as follows:

```
Seven-Segment display    Outputs    RB1-RB7
4-Bit switch bank        Inputs     RA0-RA3
Push button              Input      RA4
Push button interrupt    Input      RA0 (dual function)
Audio transducer         Output     RA0 (dual function)
```

A crystal clock of 4 MHz will be used to obtain the required timing precision, and the convenience of a 1 μs instruction cycle. The 16LF84A-04 (LF = low voltage) can operate from a supply of between 2.0 and 5.5 V, so the circuit will be powered from 2×1.5 V dry cells, giving a 3.0 V supply. The '04' suffix indicates that a maximum 4 MHz clock frequency can be used.

A block diagram of the proposed system is shown in Fig. 12.7. The inputs and outputs are given the labels which will be assigned in the application programs.

12.3.3   Implementation

A circuit for the DIZI board is shown in Fig. 12.8. The PIC 16LF84A drives an active high (common cathode) low current seven-segment LED display at Port B, RB1–RB7, via a block of 270 R current-limiting resistors. RB0 drives an audio sounder when set as an output, but can also be used to detect the 'Interrupt' push button when set as an input and the chip is initialised for this option. To prevent RB0 being shorted to ground if set as an output, the spare 270 R resistor is connected between the push button and RB0. This does not affect the operation of the sounder, which has a relatively high resistance. The flying lead is suggested because this would allow the all output pins to be monitored for audio output if, for example, BIN2 were run on this hardware.

A 4-bit DIP switch input is connected to Port A, RA0–RA3, with a push button connected to RA4, which can be used as an external pulse input to the counter/timer register RTCC. These operate as active low inputs with 10k pull-up resistors, as does the interrupt push button.

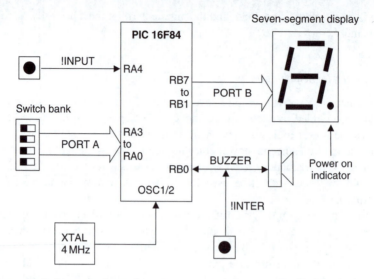

Figure 12.7   Block diagram of DIZI demonstration board.

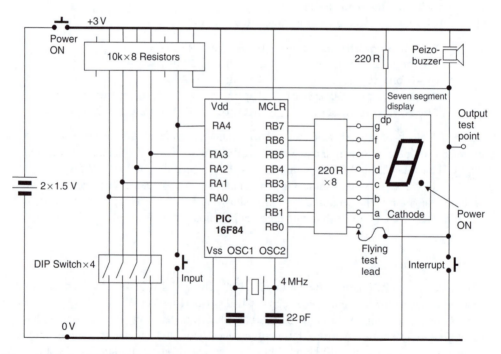

Figure 12.8   DIZI board circuit diagram.

A breadboard version of the circuit is shown in Fig. 12.9. A stripboard layout for the DIZI board is shown in Fig. 12.10. The detail of the component pin connections has been omitted due to the reduced scale of the illustration, but this information can be obtained from the component pin out data, when selecting particular components.

The finished stripboard circuit is shown in Fig. 12.11.

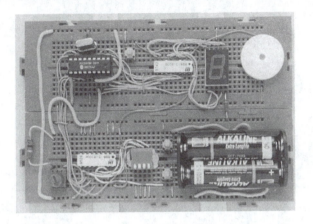

Figure 12.9 DIZI breadboard prototype circuit.

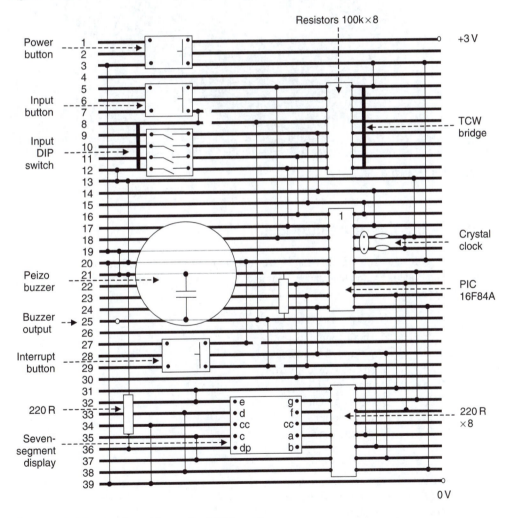

Figure 12.10 Stripboard layout for DIZI board.

Figure 12.11 DIZI stripboard circuit.

12.4 Demo Board Applications

A set of programs to run on this hardware is listed below. Selected applications will be developed and coded (*), and the reader is invited to investigate the others, using the techniques covered thus far.

Display

	FLASH1	Flash all segments
	STEP1	Step through segments
	HEX1	Binary to Hex converter
	MESS1	Message display
	SEC1	One-Second Clock
	REACT1	Reaction timer
*	DICE1	Electronic dice

Sound

	BUZZ1	Output single tone
	SWEEP1	Sweep tone frequency
*	TONE1	Switch tone on/off
	SEL1	Select tone on switches

	GEN1	Audio frequency generator
	MET1	Metronome
	GIT1	Guitar tuner
*	SCALE1	Musical scale
	BELL1	Doorbell tune

Interrupts

	STEP1	Step through scale
	STEP2	Step scale and display note
	BUZZ2	Output tone using TMR0
	REACT2	Reaction timer using TMR0
	SEC2	One-second clock using TMR0
	MET2	Metronome using TMR0

EEPROM

	STORE1	Store a display sequence in EEPROM
	STORE2	Store a tone sequence in EEPROM
	LOCK1	Store a code and buzz if matched

12.4.1 Program BUZZ1

A flowchart for the program BUZZ1 is shown in Fig. 12.12. It will generate a single tone at the buzzer when the input button is operated, by toggling the output to the buzzer, with a delay between each change of output state. If a count of 255 is used with a $1\,\mu s$ instruction cycle time, we have seen that the loop itself will take:

$$255 \times 3 \times 1 = 765\,\mu s$$

which will give a frequency of approximately:

$$1000000/765 \times 2 = 650\,Hz$$

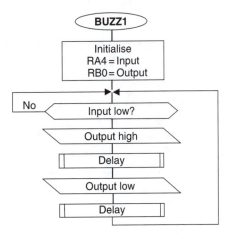

Figure 12.12 BUZZ1 flowchart.

The frequency is not critical, so we will ignore the additional loop instructions, because they will only make a small difference. The result is well within the audio range, and therefore is suitable. It can be adjusted by simply reducing the count value in the delay loop; 650 Hz is the minimum frequency available. A more precise calculation of the delay loop can be used to obtain a more exact frequency, or the hardware timer can be used. In either case, the period can be checked using the stopwatch in the simulator before downloading (see Program 12.1).

Program 12.1 BUZZ1 source code

```
;
;         BUZZ1.ASM               M. Bates            6/4/99
; ************************************************************
;
;         Generates an audio tone at Buzzer when the
:         Input button is operated..
;
;         Hardware:        PIC 16F84 DIZI Demo Board
;         Clock:           XTAL 4MHz
;         Inputs:          RA4      : Input (Active Low)
;         Outputs:         RB0      : Buzzer
;         MCLR:            Enabled
;
;         PIC Configuration Settings:
;         WDTimer:         Disable
;         PUTimer:         Enable
;         Interrupts:      Disable
;         Code Protect:    Disable
;
          PROCESSOR 16F84     ; Declare PIC device
; Register Label Equates...................................
PORTA     EQU          05     ; Port A
PORTB     EQU          06     ; Port B
Count     EQU          0C     ; Delay Counter

; Register Bit Label Equates ..............................
Input     EQU          4      ; Push Button Input RA4
Buzzer    EQU          0      ; Buzzer Output RB0

; Start Program ******************************************
; Initialise (Default = Input) ...........................
          MOVLW        b'00000000'  ; Define Port B outputs
          TRIS         PORTB        ; and set bit direction
          GOTO         check

; Delay Subroutine .......................................
delay     MOVLW        0FF          ; Standard Routine
          MOVWF        Count
down      DECFSZ       Count
          GOTO         down
          RETURN
```

```
; Main Loop ..........................................

check   BTFSC   PORTA,Input    ; Check Input Button
        GOTO    check          ; and wait if not 'on'

        BSF     PORTB,Buzzer   ; Output High
        CALL    delay          ; run delay subroutine
        BCF     PORTB,Buzzer   ; Output Low
        CALL    delay          ; run delay subroutine
        GOTO    check          ; repeat always

        END                    ; Terminate source code
```

12.4.2 Program DICE1

This program will generate a 'random' number at the display between 1 and 6 when the input button is pressed. A continuous loop will increment a register from 1 to 6, and back to 1. The loop is stopped when the button is pressed and the number displayed. The display is retained when the button is released. A table is required to work out the display digit codes.

First, the allocation of the segments to the pins on the display chip must be established. The segments of the display are labelled from a–g, as shown in Fig. 12.13. They must be lit in the appropriate combinations to give the required digit display; for instance, segments 'b' and 'c' must be lit for the digit '1' to be displayed. A table is useful here to work out the codes required for output to the display (Table 12.1).

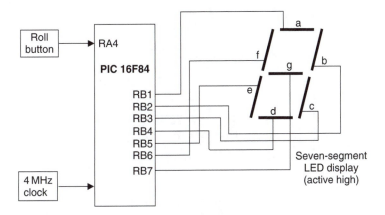

Figure 12.13 Block diagram for DICE1 system.

The display is 'active high' in operation. This means a '1' at the pin will light that segment. This arrangement is also described as 'common cathode', as all the LED cathodes are connected together at the common terminal. A 'common anode' display will therefore operate 'active low'. The binary or hexadecimal code for each digit will be included in the program in the form of a program data table.

The program represented in the flowchart, Fig. 12.14, uses a spare register as a counter which is continuously decremented from 6 to 0. When the button is pressed, the current number is

Table 12.1   DICE1 display encoding table

| Displayed digit | Segment code (1 = Segment on) | | | | | | | |
	g RB7	f RB6	e RB5	d RB4	c RB3	b RB2	a RB1	Hex (RB0 = 0)
1	0	0	0	0	1	1	0	0C
2	1	0	1	1	0	1	1	B6
3	1	0	0	1	1	1	1	9E
4	1	1	0	0	1	1	0	CC
5	1	1	0	1	1	0	1	DA
6	1	1	1	1	1	0	1	FA

(a)

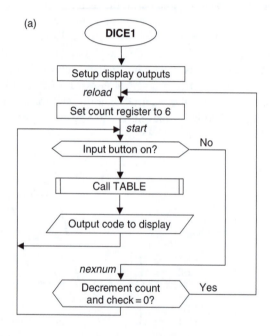

(b)

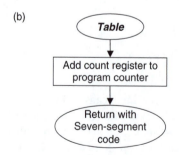

Figure 12.14   DICE1 program flowcharts.

used to select from the table of codes using the method described in Program 9.4. This results in the pseudo-random number code being displayed, and remaining visible until the button is pressed again. Because the number is selected by manually stopping a fast loop, the number cannot be predicted. In the flowchart, the jump destinations have been labelled, and these labels will be used in the program source code. The table subroutine is also named 'table' to match the source code subroutine start label (see Program 12.2).

Program 12.2 DICE1 source code

```
;         DICE1.ASM            M. Bates              6/4/99
;**************************************************************
;
;        Displays pseudo-random numbers between 1 and 6 when
;        a push button is operated.
;
;        Hardware:             PIC 16F84 DIZI Demo Board
;        Clock:                XTAL 4MHz
;        Inputs:               RA4       : Roll (Active Low)
;        Outputs:              RB1-RB7   : 7-Segment LEDs (AH)
;        MCLR:                 Enabled
;
;        PIC Configuration Settings:
;        WDTimer:              Disable
;        PUTimer:              Enable
;        Interrupts:           Disable
;        Code Protect:         Disable
;

; Set Processor Options.......................................

        PROCESSOR 16F84         ; Declare PIC device

; Register Label Equates......................................

PCL     EQU     02              ; Program Counter
PORTA   EQU     05              ; Port A
PORTB   EQU     06              ; Port B
Count   EQU     0C              ; Counter (1-6)

; Register Bit Label Equates..................................

Roll    EQU     4               ; Push Button Input

; Start Program **********************************************

; Initialise (Default = Input)

        MOVLW   b'00000001'     ; Define RB1-7 outputs
        TRIS    PORTB           ; and set bit direction

        GOTO    reload          ; Jump to main program

                                            continued ...
```

```
; Table subroutine ........................................

table       MOVF      count,W           ; Put Count in W
            ADDWF     PCL               ; Add to Program Counter
            NOP                         ; Skip this location
            RETLW     00C               ; Display Code for '1'
            RETLW     0B6               ; Display Code for '2'
            RETLW     09E               ; Display Code for '3'
            RETLW     0CC               ; Display Code for '4'
            RETLW     0DA               ; Display Code for '5'
            RETLW     0FA               ; Display Code for '6'

; Main Loop ...............................................

reload      MOVLW     06                ; Reset Counter
            MOVLF     Count             ; to 6

start       BTFSC     PORTA,Roll        ; Test Button
            GOTO      nexnum            ; Jump if not pressed
            CALL      table             ; Get Display Code
            MOVWF     PORTB             ; Output Display Code
            GOTO      start             ; start again

nexnum      DECFSZ    Count             ; Dec & Test Count=0?
            GOTO      start             ; Start again
            GOTO      reload            ; Restart count if zero

            END                         ; Terminate source code
```

12.4.3   Program SCALE1

This program will output a musical scale of eight tones. The frequencies for a musical scale from middle C upwards are:

$$262, 294, 330, 349, 392, 440, 494, 523\,(\mathrm{Hz})$$

These can be translated into a table of delay counts which gives the required tone period, since:

$$\mathtt{Period,\ T = 1/f\,(s)}$$
$$\mathtt{where\ f = frequency\,(Hz)}$$

The buzzer on the DIZI board is driven from RB0, so this needs to be toggled at a rate determined by the frequency of each tone. We therefore need to use a counter register or the hardware timer to provide a delay corresponding to half the period of each tone. We have seen in Section 9.1.2 how to calculate the delay time for a loop. Using a formula for the count value derived from this analysis, figures were calculated for a half cycle of each tone, which were then placed in the data table in SCALE1.ASM. To keep the program simple, each tone will be output for 255 cycles, so we will use another register to count the number of cycles competed during each tone. The scale will then be played over a period of about 5 s. The table of values can later be modified to play a tune in the doorbell program (see Program 12.3).

Instead of a flowchart, the SCALE1 program source code listing has been annotated with arrows to show the execution sequence. This informal method of analysis can be used to check the program logic prior to simulation. The eight tone frequencies are controlled by the value of 'HalfT', obtained from the program data table at 'getdel'. 'HalfT' is a counter value which will give a delay corresponding to a half cycle of the frequency required when the chip is clocked at 4 MHz. The eight tones are selected in turn by the value of 'TonNum', which is initialised to 8. This is used as the program counter offset in the data table fetch operation. It is decremented in the main loop after each tone has finished to select the next. The 'HalfT' values are thus selected from the bottom of the table upwards.

The tone is generated in the routine 'note', where RB0 is set high, the delay using 'HalfT' runs, RB0 is cleared, and the second half cycle delay executed. No operation instructions (NOP) have been inserted to equalise the duration of each half cycle. RB0 is toggled 255 times using the 'Count' register, which gives duration of around half a second, depending on which tone is being generated (the lower frequencies are output for longer). The main loop thus selects each of the eight values of 'HalfT' in turn, and outputs 255 cycles of each tone. The program is terminated with the SLEEP instruction to stop program execution running into the unused locations following the program.

Program 12.3 SCALE1 source code

```
; SCALE1.ASM              M.Bates              6/4/99
; ****************************************************
; Outputs a scale of 8 tones, 255 cycles per tone,
; tone duration of between a half and one second.
; Hardware: PIC 16F84
; XTAL 4MHz, !MCLR to start
; Audio Output: RB0


; Assign Registers ********************************

PCL     EQU     02              ; Program Counter
PORTB   EQU     06              ; Port B for Output
HalfT   EQU     0C              ; Half Period of Tone
Timer   EQU     0D              ; Delay Time Counter
Count   EQU     0E              ; Cycle Count
TonNum  EQU     0F              ; Tone Number (1-8)


; Initialise Registers

        MOVLW   B'11111110'     ; RB0 set..
        TRIS    PORTB           ; as output
        MOVLW   08              ; Set intial value of..
        MOVWF   TonNum          ; Tone Number
        GOTO    start           ; Jump to main program

                                        continued...
```

```
; Tone Period Table (HalfT)

getdel        ADDWF      PCL
              NOP
              RETLW      D'156'
              RETLW      D'139'
              RETLW      D'124'
              RETLW      D'117'
              RETLW      D'104'
              RETLW      D'92'
              RETLW      D'82'
              RETLW      D'77'

; Delay for half tone cycle

delay         MOVF       HalfT,W
              MOVWF      Timer
again         DECFSZ     Timer
              GOTO       again
              RETURN

; Output 255 cycles of tone

note          MOVLW      D'255'
              MOVWF      Count

cycle         BSF        PORTB,0
              CALL       delay
              NOP
              NOP
              NOP

              BCF        PORTB,0
              CALL       delay
              DECFSZ     Count
              GOTO       cycle

              RETURN
```

; **Main Loop Outputs 8 Tones**

```
start         MOVF       TonNum,W
              CALL       getdel
              MOVWF      HalfT
              CALL       note
              DECFSZ     TonNum
              GOTO       start
              SLEEP
              END        ; of source code
```

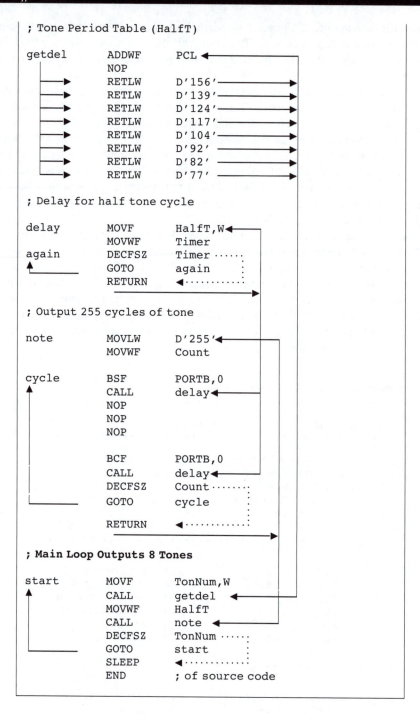

12.4.4   DIZI Application Outlines

Some further applications are outlined below, for the reader to develop for the DIZI hardware.

HEX1 Hex Converter

The hexadecimal number corresponding to the binary setting of the DIP switch inputs is displayed. The input switches select from a table of 16 seven-segment codes which light up the segments in the required pattern for each hex digit display:

$$0, 1, 2, 3, 4, 5, 6, 7, 8, 9, A, b, C, d, E$$

Note that numbers B and D are displayed in lower case on a seven-segment display so that they can be distinguished from 8 and 0 respectively. Use the switch input number to select one of the 16 codes from a seven-segment table.

MESS1 Message Display

A sequence of characters is displayed for about 0.5 s each. Most letters of the alphabet can be obtained on the seven-segment display in either upper or lower case, for instance 'HI tHErE'. Output a character code table with delay. The number of characters must be set in a counter, or a termination character used.

SEC1 One-Second Timer

An output is displayed which increments exactly once per second, from 0 to 9, and then repeats. A table of display codes is required as in the 'Hex Converter'. A one-second time delay can be achieved using the hardware timer (see Chapter 9) and spare register. A 'tick' could be produced at the audio output by pulsing the speaker at each step.

REACT1 Reaction Timer

The user's reaction time is tested by generating a random delay of between 1 and 10 s, outputting a sound, and timing the delay before the input button is pressed. A number representing the time between the sound and the input, in multiples of 100 ms, is then displayed as a number 0–9, giving a maximum reaction time of 900 ms.

GEN1 AF Generator

An audio frequency generator outputs frequencies in the range 20 Hz–20 kHz. The sounder output is toggled with a delay between each operation determined by the frequency required, as in the BUZZ program. For example, for a frequency of 1 KHz, a delay of 1 ms is required, which is 1000 instruction cycles at a cycle time of 1 µs. The information on program timing must be studied in Chapter 10. The delay time, and hence the frequency, can then be incremented using the input button, and range selection with the input switches might be incorporated, as there are only 255 steps available when using an 8-bit register as the period counter.

MET1 Metronome

An audible pulse is output at a rate set by the DIP switches or input buttons. The output tick can be adjustable from, say, 1 up to 4 beats per second, using the interrupt button to step the speed up and down, and the input button to select up or down. A software loop or the TMR0 register can be used to provide the necessary time delays.

BELL1 Doorbell

A tune is played when the input button is pressed, using a program look-up table for the tone frequency and duration. Each tone must be played for a suitable time, or number of cycles, as required by the tune. The program can be elaborated by selecting a tune using the DIP switches, and displaying the number of the tune selected.

GIT1 Guitar Tuner

The program will allow the user to step through the frequencies for tuning the strings of a guitar, or other musical instruments using the input button, or selecting the tone at the DIP switches. The program could be enhanced by displaying the string number to be tuned. The tone frequencies will be generated as for the doorbell application. The digit display codes would also be required in a table.

Summary

- Methods of circuit construction available include PCB, breadboard and stripboard.

- Design software is available to draw the application circuit, test it by interactive simulation and create a board layout, but it requires time to learn to use the software, and etching equipment is needed to produce a PCB.

- Breadboard is re-usable and allows circuits to be prototyped quickly and easily, but is unreliable for complex circuits.

- Stripboard is more reliable, but not re-usable. Connections can be laid out on paper, using computer drawing tools or directly onto the board.

- The DIZI demonstration board has a seven-segment display and audio output, with push button input and interrupt and a 4-bit switched input, and can be used for demonstrating a range of simple applications.

Questions

1. For constructing a circuit, state one advantage and one disadvantage of (a) breadboard (b) stripboard (c) PCB design software.

2. State the maximum rated current output for port pins on the PIC 16F84A. How do these ratings simplify the interfacing to this chip compared with standard digital outputs?

3. Explain why a common cathode seven-segment display operates as an active high display. State an input code for (a) all segments off and (b) all segments on for this type of display.

4. Outline an algorithm for generating a fixed frequency output of approximately 1 kHz from the DIZI board using the hardware timer.

5. Describe a method of outputting a set of codes to drive a seven-segment display of alphanumeric characters at a PIC parallel output.

Activities

1. Confirm by calculation that the values used in the program data table in SCALE1.ASM will give the required delays.

2. Devise a breadboard layout for the BIN circuit in Fig. 6.3. Build the circuit and test the BINx programs.

3. Devise a stripboard layout for the MOT1 circuit in Fig. 10.3, using a VN66 FET and 2 V small DC motor. Build the circuit and test MOT1.

4. Build the DIZI circuit on breadboard, stripboard or PCB and test the programs BUZZ1, DICE1 and SCALE1.

5. Design and implement any of the programs outlined for the DIZI hardware:

   HEX1, MESS1, SEC1, REACT1, GEN1, MET1, BELL1, GIT1.

6. (a) Investigate how an input from a numeric keypad can be detected. The typical keypad, shown in Fig. 12.15, has 12 keys in four rows of three: 1, 2, 3; 4, 5, 6; 7, 8, 9; #, 0, *. These are connected to seven terminals, and can be scanned in rows and columns. A key press is detected as a connection between a row and a column. The pull-up resistors ensure

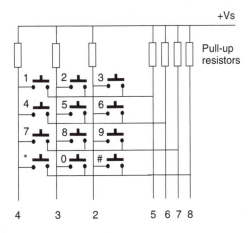

Figure 12.15 Keypad connections.

that all lines default to logic '1'. If a '0' is applied to one of the column terminals (2, 3, 4), and a key is pressed, this '0' can be detected at the row terminal (5, 6, 7, 8). If the keypad terminals are connected to a PIC port, and a '0' output in rotation to the three columns, a key can be detected as a combination of the column selected and the row detected. Terminals 2, 3, and 4 will be set as outputs, and 5, 6, 7 and 8 as inputs. Draw a flowchart to represent the process required.

A lock function may be implemented by matching an input sequence with a stored sequence of, say, four digits, and switching on an output to a door solenoid if a match is detected. Refer to Chapter 15, the temperature controller application, and Appendix B, the lock application, for further information. Produce structured flowcharts for a lock program.

(b) Design, build and test an electronic lock system using the keypad shown, a suitable PIC and an LED to indicate the state of the lock (ON = unlocked). Research the design for the interface to a solenoid-operated door lock.

Chapter 13
Motor Applications

13.1   Motor Control Methods
13.2   Motor Application Board
13.3   Control Methods
13.4   Position Control
13.5   Closed Loop Speed Control
13.6   Commercial Application

This chapter will focus on a PIC application in which program timing is critical. This kind of application is more demanding because the controlled device (motor) has its own dynamic characteristics which must be taken into account in the program design. The hardware design will be taken as given, so that we can concentrate on the software development for this application.

13.1   Motor Control Methods

There are two main types of control system, open loop and closed loop. An open loop system is essentially manually controlled. For example, a car requires the driver to monitor the direction of travel and correct the steering to follow the road. A closed loop system uses sensors to monitor the system outputs and control the process automatically, once the initial operating conditions have been set. A central heating system is a simple example; the thermostat monitors the temperature and switches the boiler on and off accordingly.

A small, inexpensive, DC motor will be used to demonstrate the use of the PIC microcontroller in a 'real-time' control application, allowing open and closed loop operation to be investigated. Real-time systems are those where the time factor is important, and where the dynamic response to control inputs and feedback from sensors is critical. Robot arm positioning is a good example of a closed loop motor control application.

Motor output is measured as the shaft speed or position. Open loop control of a motor would consist of simply switching it on and off for a fixed period to position it, or varying the speed, under manual control. There are obvious limitations to open loop control. A DC motor does not start until there is a reasonably large current, due to inertia, stiction and its electromagnetic characteristics. This makes its response non-linear, which means that the speed is not directly proportional to the current or voltage supplied. In addition, the speed cannot be accurately predicted for any given current, because the load on the shaft will affect it. The final position of the shaft when the motor stops cannot be precisely controlled either. Therefore, if the speed or position of a DC motor is to be controlled accurately, we need sensors to measure the output variables, and a control system for the motor drive.

A simple analogue potentiometer can measure position, by converting it to a voltage, or speed can be measured using a tachometer, which produces a voltage which is proportional to the motor speed. These 'transducers' have traditionally been used in analogue motor control systems, where all the signals are continuously variable currents and voltages. It is now more common to use a digital control method, and the microcontroller can be used as the basis of a programmable system in which the control algorithm can be designed to closely match the motor and load requirements. The dynamic (time) response can then also be adjusted in software.

The speed of a DC motor is controlled by the current in the armature, which interacts with the magnetic field produced by the field windings (or permanent magnets in small motors) to produce torque. An analogue control system gives continuous control over the motor current, and a digital to analogue drive converter can be used at the output if the feedback and control is digital. However, the control interface can be simplified if PWM is used, as described in Chapter 10. The PWM is a simple and efficient method of converting a digital signal to a proportional drive current. Many microcontrollers now provide dedicated PWM outputs, but we are going to generate the control signal in software.

Digital feedback can be obtained from a sensor which detects the shaft rotation. One way of doing this is to use a perforated or sectored disk attached to the shaft and an optical sensor to detect the slots or holes in the disk. The shaft position can be detected by counting pulses, and the speed by measuring their frequency. This signal can be fed directly to a microcontroller running a program which monitors the pulse input, and varies the output to control the speed and/or position of the motor.

13.2   Motor Application Board

The block diagram for a motor application (MOTA) board is shown in Fig. 13.1, and a circuit diagram in Fig. 13.2. A stripboard layout of the circuit is shown in Fig. 13.3 and a finished stripboard circuit is shown in Fig. 13.4.

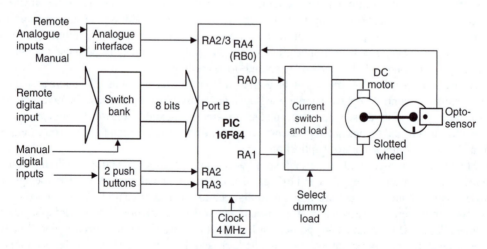

Figure 13.1   Block diagram of MOTA board.

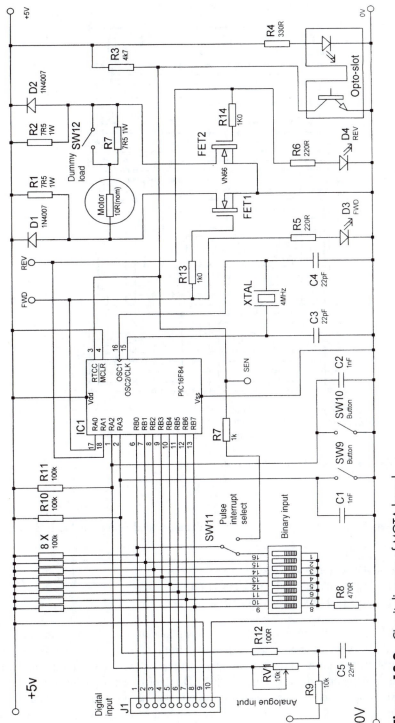

Figure 13.2 Circuit diagram of MOTA board.

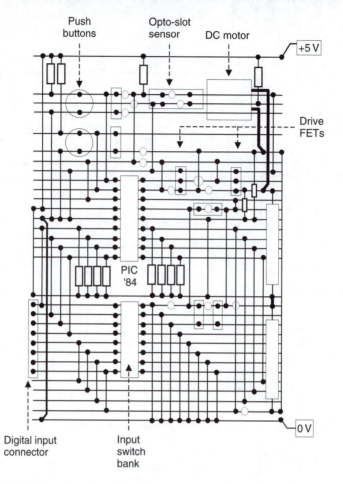

Figure 13.3 MOTA board stripboard layout.

A variety of motor control operations can be demonstrated using this hardware:

- MOTOR ON/OFF,
- MOTOR FORWARD/REVERSE,
- OPEN/CLOSED LOOP POSITION CONTROL,
- OPEN/CLOSED LOOP SPEED CONTROL.

The MOTA circuit is designed to demonstrate position and speed control with a PIC 16F84(A) microcontroller. Command inputs can be received from an 8-bit switch bank, a remote 8-bit master controller, two push buttons or an analogue input. The motor can be turned in either direction via bidirectional drive outputs, with LED motor direction indicators. The drive can provide position and speed control, and can be pulse-width-modulated to control the speed. The shaft speed and position are monitored by a slotted opto-sensor and disk with one slot, feeding back one pulse per revolution to the controller. The PIC is crystal clocked at 4 MHz, for precise feedback measurement.

Figure 13.4 MOTA stripboard circuit.

Motor Drive

The small, inexpensive 2 V permanent magnet motor is connected in a passive FET bridge with R1 and R2 as load resistors. The motor current direction, forward or reverse, is controlled by switching on one of two VN66 FETs from RA0 or RA1. A dummy load can be switched in series with the motor to allowing testing of closed loop control. Light emitting diodes indicate the motor direction. The use of a low-quality motor will emphasise the problems which will arise from imperfections in the motor itself.

Opto-sensor

This contains an LED and photodetector mounted either side of the slot in a plastic housing. The perforated disk attached to the motor shaft allows the light to pass through the holes and digital pulses are output from the sensor via a built-in amplifier, which allows the motor speed or position to be monitored by the controller. The sensor input is connected to the T0CKI (RA4) input of the PIC, so that the shaft revolutions can be counted in the counter/timer register. Alternatively, the slot pulse interval can be measured at RA4 using the timer. The pulse may also be used to trigger an interrupt at RB0 by setting the pulse interrupt select switch accordingly. This then means that only seven bits can be read from the binary input.

Switched Inputs

The control program can allow the push buttons connected to RA2 and RA3 to stop, start or change speed or direction. The binary input switches could then be used to select the speed or position. Alternatively, a remotely generated digital control code can be applied to the digital input connector pins from a master controller, which could be operating a number of motors in a system. In this case, part of the digital input would be a motor select code, and part would be a position or speed command. Serial commands could also be used, but a PIC with a dedicated serial port would be better for this (see Chapter 14).

If the parallel input is removed from the circuit, a smaller, cheaper PIC 12FXXX series device could be used instead. These have six I/O pins, so there would be three inputs available with which to control the motor speed, position and/or direction. Analogue inputs are also available, if the motor needs to be voltage controlled.

Analogue Input

Analogue input is possible via RA2 and RA3 using a software-based analogue conversion. However, it would be preferable to use a PIC with dedicated analogue inputs, such as the 16F818. This is a pin compatible with the 16F84, and could be used to measure analogue inputs at RA2 and RA3 directly. In this case, the relevant pins are initialised as analogue inputs by setting up the required registers during the initialisation phase of the program (see Chapter 14). The hardware should be modified accordingly.

13.3   Control Methods

The PIC 16F84 is crystal clocked at 4 MHz to give an instruction cycle time of 1 μs. No manual reset is required, but the power on timer should be enabled during programming to ensure a reliable start.

13.3.1   Open Loop Control

Open loop control of a DC motor has been described in Chapter 10 and a program developed which allows the speed to be controlled manually. In the MOTA circuit (Fig. 13.2), the motor can be driven in either direction by setting RA0 or RA1 high, with both set low to turn the motor off . Either can be pulsed for speed control, but they should not be high together; this would switch on both transistors, resulting in no current through the motor, and a lot of wasted power dissipated in the load resistors.

Open loop speed control can be implemented in various ways: programmed, push button manual input, manual/remote binary/analogue input. Sequence control can be incorporated in the program; for instance, the speed can be ramped up, held constant and then ramped down over a fixed period of time. The push button inputs could be programmed to run the motor in either direction or increment and decrement the speed in one direction by modifying the delay in a PWM program. The speed could be set at the binary inputs to Port B, either manually at the DIP switches, or with an 8-bit digital input code supplied from a master controller, and analogue control is possible from a manual input (RV1) or from a remote voltage source.

13.3.2   Closed Loop Control

The PIC motor control board has a slotted wheel and opto-sensor to monitor the rotation of the motor. A wheel with only one slot is used for the applications developed here to make the calculation of speed simpler. One pulse per revolution will also provide plenty of time between pulses for the control program to complete its processing tasks. The sensor is connected to RA4 of the PIC. As we know, the 16F84 contains an 8-bit counter/timer register, TMR0, which can be clocked from RA4 or the system clock, which we can use to measure the pulse number, frequency or period.

Closed loop position control involves counting the slots as the shaft turns. This sounds straightforward, but the dynamic characteristics of the motor have to be taken into account. For example, the motor can be switched on from the controller, and the pulses counted, and the motor turned off when a set number of pulses have been counted. However, the motor will probably overshoot the required position due to inertia of the rotor. A simple solution would be to keep counting the slots and turn the motor back by the requisite number of slots. This might have to be repeated several times.

With only one slot, the position can only be determined to the nearest whole revolution. This may be acceptable if a gearbox is fitted which reduces the angular rotation and speed. For instance, if the gearbox has a reduction ratio of 50:1, the output can be positioned within 1/50 of a revolution. In addition, a slotted wheel with more slots can be used, and a proportionate increase in accuracy obtained. With 100 slots, for example, and the gearbox, the accuracy will be $360/5000 = 0.072°$.

13.4   Position Control

A simple program, which moves the motor by a number of revolutions set on the DIP switches, is outlined in Fig. 13.5.

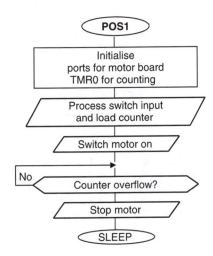

Figure 13.5   Flowchart for motor program POS1.

The hardware timer/counter (TMR0) is used to count the pulses from the sensor. The timer flag (T0IF) is set when the counter rolls over from 255 to 000, so the binary number at the switches has to be complemented (subtracted from 256) to set the count to a value whereby it reaches 256 (0) after the correct number of pulses. The option register has to be initialised to select the T0CKI (RA4) pin for input, and to allocate the prescaler to the watchdog timer, so that the count is not prescaled. The processor is put to sleep after the move has been completed, so it would have to be reset to repeat the operation (see Program 13.1).

As mentioned above, the position control achieved with the program may not be accurate, because, firstly, it can only count to the nearest whole revolution, and secondly, there is no provision for preventing overshoot. One way of achieving better accuracy is for the current

Program 13.1   POS1 basic motor position control

```
; *************************************************************
;          POS1.ASM              M. Bates              7/4/99
; *************************************************************
;
;          Basic position control program runs motor for
;          number of revs set on binary switch input.
;          Uses TMR0 to count revs, but does not correct
;          for overshoot...
;
;          Hardware:       PIC 16F84 Motor Board
;
;          Clock:          XTAL 4MHz
;          Inputs:         RB0-RB7  : DIP Switches (High)
;                          T0CKI    : Shaft Sensor (Low)
;          Outputs:        RA0      : Motor (High)

; Set Processor Options ...................................

  PROCESSOR 16F84          ; Declare PIC device

; Register Label Equates ..................................

PORTA    EQU    05         ; Port A
PORTB    EQU    06         ; Port B
TMR0     EQU    01         ; Counter/Timer
INTCON   EQU    0B         ; Interrupt Control

; Register Bit Label Equates ..............................

T0IF     EQU    2          ; Timer Overflow Flag = INTCON,2
motor    EQU    0          ; Motor Output = RA0

; Start Program  *****************************************

; Initialise .....................Port B defaults to input

         MOVLW    b'11111100'    ; Port A bit direction code
         TRIS     PORTA          ; Set the bit direction
         MOVLW    b'00100000'    ; Code for Option Register
         OPTION                  ; select T0CKI timer input
         BCF      INTCON,T0IF    ; Clear Timer Overflow Flag

; Main Loop ...............................................

         MOVF     PORTB,W        ; Read switches
         MOVWF    TMR0           ; into counter
         COMF     TMR0           ; and complement

         BSF      PORTA,motor    ; Motor ON
test     BTFSS    INTCON,T0IF    ; Test Overflow Flag
         GOTO     test           ; and wait until set
         BCF      PORTA,motor    ; Motor OFF

         SLEEP                   ; Suspend processor

         END                     ; Terminate source code
```

position to be continuously compared with the required position, and the motor driven at a speed proportional to the error. The motor will slow down as it approaches the target position. This type of process is referred to as PID control, where the response of the system can be tuned to give the best compromise between speed of response, accuracy and overshoot.

A simpler process called 'trapezoidal' control can also be used. This involves ramping the motor speed up and down at the ends of the move, with a constant speed period in the middle. Because the position control of a DC motor is not very practical without a gearbox, we will look at speed control in more detail. The speed can also be measured by using an oscilloscope to display the pulse period from the sensor, making it easier to monitor the motor behaviour.

13.5   Closed Loop Speed Control

A typical application may require the motor board to operate as a slave unit under digital control. A master controller would supply an 8-bit code to set the speed of the motor, with the local controller required to maintain it with a specified degree of precision. The MOTA board allows for such an input; it can also be simulated for test purposes using the switch bank. Suppose that the motor is to be controlled to a speed of exactly 50 revs per second (rps), which is about 40% of nominal full speed. This will produce 50 pulses per second (pps) at RA4 with a single slot in the wheel. The speed can be measured in one of two main ways:

1.   counting sensor pulses over a measured time period,

2.   measuring the period between sensor pulses.

13.5.1   Counting Pulses

The accuracy of the speed measurement using this method will depend on the number of slots counted, because the error is always +/–1 slot. Thus, if 100 slots are counted, the accuracy will be $1/100 = 1\%$. If the count were made over a period of 1 s at 50 pps, the precision would be $1/50 = 2\%$. Therefore, for 2% accuracy, each count would take 1 s, and speed could only be corrected once per second. This response time is too slow for most practical purposes, so this option will be rejected. It could however be used if there were more slots in the disk or the motor were running at high speed.

13.5.2   Measuring Pulse Period

At 50 pps, the target speed, the pulse period will be $1/50\,\mathrm{s} = 20\,\mathrm{ms}$. This can be measured by comparison with a 20 ms timer, which can be set up using the TMR0 hardware counter/timer (see Chapter 9).

The PIC instruction cycle time is 1 µs with a 4 MHz clock. The counter can therefore be clocked at a maximum 1 MHz (once per instruction cycle). The timer prescaler allows this to be divided by 2, 4, 8, 16, 32, 64, 128, or 256, by setting a 3-bit code in the option register. A possible prescale factor is calculated as follows:

Now $20\,\mathrm{ms} = 20000\,\mathrm{\mu s}$

$$= 20000 \text{ instruction cycles}$$

The timer can count 256 cycles, therefore the prescale multiplier required

$$= 20000/256$$

$$= 78.125$$

The nearest available prescale multiplier $= 128$
Using this prescale value, the number of counts for 20 ms

$$= 20000/128$$

$$= 156.25$$

$$= 156 \quad \text{(nearest whole number)}$$

The actual motor period will then be

$$= 156 \times 128 \times 1\,\mu s$$

$$= 19968\,\mu s$$

This count is accurate to within 1% of 20 ms. Remember, this value must be complemented before loading into the timer register, because it counts up to zero at rollover.

The binary switch can then be used to input the timer value, to provide a variable, but accurate speed control. The longest period then measurable will be

$$= 256 \times 128$$

$$= 32768 \; \mu s$$

Motor period $\qquad = 0.032768$ s/rev.

Motor speed $\qquad = 1/0.032768$

$$\sim 30 \text{ revs/s}$$

Therefore, this method should provide control from the motor maximum speed of about 125 rps down to about 30 rps. Lower speeds may be obtained by increasing the prescale factor in the timer, to give a longer time interval. This could be done 'on the fly' by adding code to allow the push buttons to select the operating speed range.

Alternative motor control algorithms will now be evaluated to illustrate the process whereby the most appropriate implementation is selected where more than one algorithm is possible.

13.5.3 *Synchronous Motor Control*

System signals which could provide closed loop control are illustrated in Fig. 13.6. The motor is driven with a PWM pulse whose length is controlled by a simple software loop. The shaft speed is measured using hardware timer, as outlined above. Initially, the motor mark (on) period will be set to the minimum value.

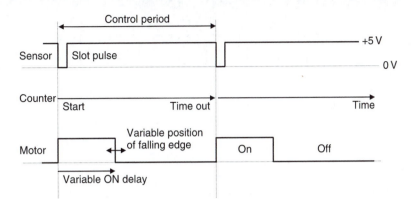

Figure 13.6 Timing diagram for closed loop speed control of motor.

At the start of the control cycle, the motor would be switched on and the timer TMR0 cleared and started. It will be then incremented once per 128 instruction cycles. The sensor and time out flag can be continuously checked to see whether the counter has finished, or the next pulse has arrived. If the timer times out before the rising edge of the next pulse, the motor is not going fast enough, so the speed must be increased by incrementing the ON time period of the motor. If the pulse arrives first, the motor is going too fast, so the speed must be reduced by decrementing the ON period. If the ON time starts at zero, it will be increased until the motor starts. The speed should then stabilise at the value corresponding to the switch input.

Using this method of control raises a few problems. Note that the drive signal and sensor signal are locked together, with the same period. This means that the drive frequency will be 50 Hz, which is too low. A higher frequency also prevents commutator switching in the motor from interfering with the drive switching. In addition, with this algorithm, the drive will be on for the same part of each revolution, potentially causing uneven wear on the commutator. We will therefore consider an alternative algorithm.

13.5.4 Asynchronous Motor Control

We would prefer the PWM drive to operate at a higher frequency, and asynchronously (not locked to the sensor cycle). The asynchronous algorithm still runs the motor using the software delay loop to generate a pulse-width-modulated drive signal to the motor, but checking the timer and sensor each time round the motor delay loop. The MSR is adjusted to control the speed, using the input code and its complement to set the delays for the mark and space.

The process is illustrated in the timing diagram (Fig. 13.7). The timing cycle starts at the falling edge of the negative going sensor pulse, where the timer is started. The program waits for the rising edge of the sensor pulse, then starts checking if the next pulse has arrived, or if the timer has timed out, once per motor cycle.

If the speed is too low, the timer times out first, before the pulse arrives. In this case the speed must be increased for the next timing cycle. On the other hand, if the slot arrives before the timer has timed out, it means that the motor is running too fast, so the speed must be decremented for the next cycle. When the speed is correct, the speed adjustment should alternate between

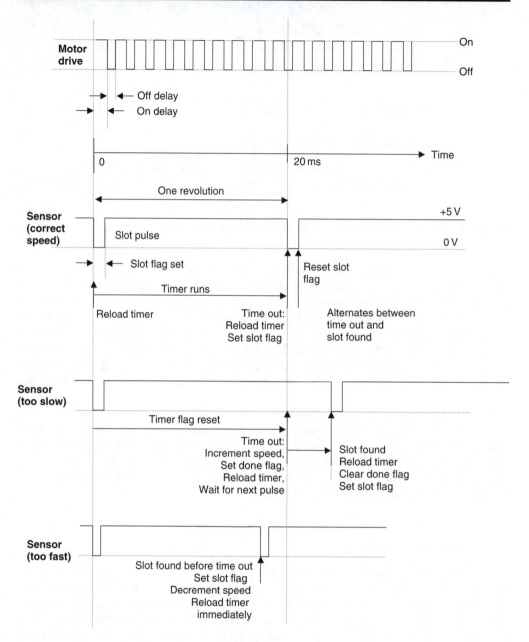

Figure 13.7 Timing diagram for high-frequency motor speed control.

incrementing and decrementing. This ideal performance may well be affected by imperfections in the motor. In practice, it has also been found that there was significant 'hunting' (variation) in the speed, because of the relatively low sampling rate.

Figure 13.8(a) shows the top level flow chart for the program. 'Speed' is a GPR which holds the value for the PWM ON time. The OFF time is derived by complementing this value. The total count for each motor drive cycle is then 256, which means the frequency will remain constant. The 'Reload Timer' (RELTIM) routine restarts the timer (Fig. 13.8c). TMR0 starts counting the internal instruction clock pulses as soon as it is reloaded. It is loaded with the complement of the input switch value because time out is detected when the TMR0 register rolls over from FF to 00. Therefore, the time interval is measured as the count from the loaded value up to 00.

The most complex part of the program is the TESTEM routine, which checks the inputs and modifies the value of 'Speed' accordingly (Fig. 13.8b). The requirements are:

1. To start the motor from rest, the MSR must be increased upwards when no slot is detected, so that the motor will eventually start. A fairly high MSR is required to get the motor moving initially.

2. When a slot is detected before a timeout, the timer must be restarted immediately.

3. When the timeout occurs first, the speed must be incremented and the timer must be restarted, to get the motor started initially. If a slot then arrives, the timer must be restarted again, so that it can correctly measure the time between the slot edges.

4. When the sensor is detected as low in the check cycle, it means that a falling edge has arrived, and the timer must be restarted. The program must then wait for the sensor to go high again before it starts looking for the next slot.

5. The speed must be stopped from rolling over from FF (maximum) to 00 (minimum), so the speed is checked after incrementing and decremented again if it is found to be equal to FE.

(a)

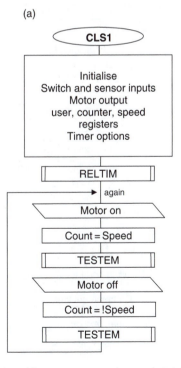

Figure 13.8 Flowcharts for closed loop motor speed control. (a) Main loop.

(b)

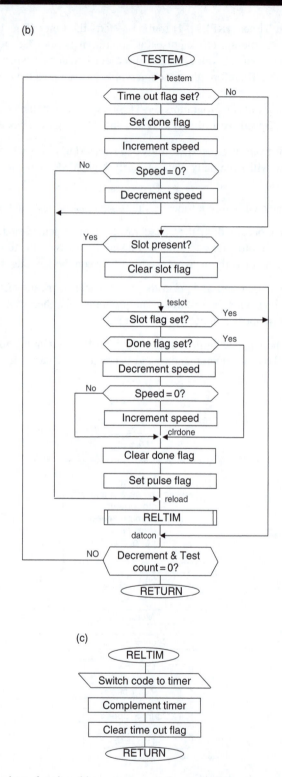

(c)

Figure 13.8 Flowcharts for closed loop motor speed control. (b) Input testing; (c) Timer reload.

To achieve these requirements, flags have been defined in a GPR to record the fact that the falling edge has been detected and acted upon (flag 'slot'), and another to record the fact that the timer has been reset, to make the program wait for the next slot to restart the timer.

13.5.5 Program Simulation

The source code for the closed loop speed control program is shown as Program 13.2. In order to avoid the need for an input stimulus file, this simulation version of CLS1 loads the timer with the literal value '156' in the subroutine RELTIM, rather than reading Port B switches. For running in the hardware, the comment delimiter on the switch input read to Port B must be removed, and the literal load operation commented out. The RA4 (sensor) input may be simulated using the asynchronous input window in MPLAB. The results shown in Table 13.1 should be obtained.

Program 13.2 Closed loop motor speed control source code

```
; ***************************************************************
;       CLS1.ASM              M. Bates              4/4/99
; ***************************************************************
;
;       Closed Loop DC Motor Speed Control using Pulse
;       Width Modulation (software loop) to control speed
;       and hardware timer to set reference time interval
;
;       Hardware:     PIC 16F84 Motor Board
;
;       Clock:        XTAL 4MHz
;       Inputs:       RB0-RB7   : DIP Switches (High)
;                     RA4       : Shaft Sensor (Low)
;       Outputs:      RA0       : Motor (High)
;
;       Configuration Settings:
;
;       WDTimer:      Disable
;       PUTimer:      Enable
;       Interrupts:   Disable
;       Code Protect: Disable
;

; Set Processor Options...................................

  PROCESSOR 16F84        ; Declare PIC device

; Register Label Equates.................................

PORTA     EQU     05          ; Port A
PORTB     EQU     06          ; Port B
TMR0      EQU     01          ; Counter/Timer
INTCON    EQU     0B          ; Interrupt Control

Speed     EQU     0C          ; Counter Pre-load Value
Count     EQU     0D          ; Delay Counter
Flags     EQU     0E          ; User Flags

                                        continued...
```

```
; Register Bit Label Equates...............................

timout  EQU     2               ; Time Out Flag = TMR0,2
motor   EQU     0               ; Motor Output = RA0
sensor  EQU     4               ; Shaft Opto-Sensor = RA4
slot    EQU     0               ; Slot Found Flag
done    EQU     1               ; Time Out Done Flag

; Start Program *******************************************

; Initialise ....................Port B defaults to input

        MOVLW   b'11111100'     ; Port A bit direction code
        TRIS    PORTA           ; Set the bit direction
        MOVLW   b'00000110'     ; Code for option Register
        OPTION                  ; Sets prescale 1:128
        MOVLW   080             ; Initial value for
        MOVWF   Speed           ; count pre-load value
        MOVWF   Count           ; and counter itself
        GOTO    start           ; Jump to main program

; RELTIM Routine  .........................................

; Reloads TMR0 timer/counter register with complement of
; switch input (or dummy value for simulation mode)

reltim  MOVLW   d'156'          ; Dummy value for timer (sim)
;       MOVF    PORTB,W         ; Input Switches (runtime)

        MOVWF   TMR0            ; Load Timer with input
        COMF    TMR0            ; Complement value
        BCF     INTCON,timout   ; Reset 'TimeOut' Flag
        RETURN

; TESTEM Routine  .........................................

; Increases speed if timeout detected or
; decreases speed if slot end detected...

testem  BTFSS   INTCON,timout   ; Time Out?
        GOTO    tessen          ; NO: Skip Speed Increment
        BSF     Flags,done      ; Set Time Out Done Flag
        INCFSZ  Speed           ; Test for maximum speed
        GOTO    reload          ; NO: jump to timer reload
        DECF    Speed           ; Decrement again
        GOTO    reload          ; & jump to timer reload

tessen  BTFSS   PORTA,sensor    ; Slot Present?
        GOTO    teslot          ; YES: jump to test slot
        BCF     Flags,slot      ; Reset 'Slot' Flag
        GOTO    datcon          ; & continue Count loop

teslot  BTFSC   Flags,slot      ; 'Slot' Flag Set?
        GOTO    datcon          ; YES: Skip speed decrement
        BTFSC   Flags,done      ; 'Done' Flag Set?
        GOTO    clrdone         ; YES: Skip speed decrement

        DECFSZ  Speed           ; Test for minimum speed
        GOTO    clrdone         ; NO: continue loop
        INCF    Speed           ; YES: increment again
```

```
clrdone    BCF      Flags,done       ; Clear 'Done' Flag
setslot    BSF      Flags,slot       ; Set 'Slot' Flag

reload     CALL     reltim           ; Reload timer

datcon     DECFSZ   Count            ; Decrement & Test Count
           GOTO     testem           ; Counter not zero yet
           RETURN                    ; End motor cycle if zero

; Main Loop ..................................................

start      CALL     reltim           ; reload timer to start

again      BSF      PORTA,motor      ; Motor ON
           MOVF     Speed,W          ; Put ON delay value
           MOVWF    Count            ; into Counter
           CALL     testem           ; Insert Delay Code

           BCF      PORTA,motor      ; Motor OFF
           MOVF     Speed,W          ; Put ON delay value
           MOVWF    Count            ; into Counter
           COMF     Count            ; and convert to OFF value
           CALL     testem

           GOTO     again            ; Insert Delay Code
           END                       ; Terminate source code
```

Table 13.1   Test table for motor control program CLS1

After...	Sensor input	Timout flag	Done flag	Slot flag	Speed register	Comment
Start	1	0	0	0	80	Motor stopped
Reload timer	1	0	0	0	80	
Timeout	1	1	1	0	81	Motor starting up
Reload	1	0	1	0	81	
Timeout	1	1	1	0	82	
Reload	1	0	1	0	82	
Slot 1 start and reload	0	0	0	1	82	First slot
Slot end	1	0	0	0	82	
Timeout	1	1	1	0	83	Inc. speed
Slot 2 start and reload	0	0	0	1	83	Second slot
Slot end	1	0	0	0	83	
Timeout	1	1	1	0	84	Inc. speed
Slot 3 start and reload	0	0	0	1	84	Third slot

continued...

Table 13.1 continued

After...	Sensor input	Timout flag	Done flag	Slot flag	Speed register	Comment
Slot end	1	0	0	0	84	
Timeout	1	1	1	0	85	
etc. etc. etc.						Repeats until up to
Slot start and reload	0	0	0	1	XX	Set speed
Slot end	1	0	0	0	XX	
Timeout	1	1	1	0	XX + 1	Inc. speed
Slot start and reload	0	0	0	1	XX + 1	
Slot end	1	0	0	0	XX + 1	
Slot start and reload	0	0	0	1	XX	Dec. speed
Slot end	1	0	0	0	XX	
Timeout	1	1	1	0	XX + 1	Inc. speed
Slot start and reload	0	0	0	1	XX + 1	
Slot end	1	0	0	0	XX + 1	
Slot start and reload	0	0	0	1	XX	Dec. speed
Slot end	1	0	0	0	XX	
etc.						Speed now stable – repeat inc. and dec.

13.5.6 Hardware Testing

The correct function of the closed loop control program can be tested in the target system by setting the binary input to 156 and checking the actual speed of the motor by measuring the period of the sensor pulse on an oscilloscope; it should be 20 ms. The binary input can then be varied, and the period should vary in proportion, within limits stated above. The transient and start up response can be examined by stalling the motor, and studying the motor response as it locks on to the target speed. A dummy load is also provided in series with the motor; when it is switched in, the additional series resistance will reduce the motor in the current, hence its speed. If the closed loop control is working, the drive should compensate and maintain the speed of the motor by increasing the drive MSR.

13.5.7 Evaluation of Algorithm

Ideally, the PWM speed control should operate at a frequency above about 15 kHz; the program CLM operates at only about 300 Hz, because of the time required to sample the timer status and sensor input, and to complete the software loop for one drive cycle, which uses a full 8-bit count (256). The frequency could be increased by reducing the total loop count to less than 256, but this will reduce the resolution of the control. Some compromise value could be arrived at by calculation and experimentation, based on the required resolution.

The MOTA hardware provides for alternative implementations of closed loop control. RB0 can be optionally connected to the output from the motor shaft sensor, so that an RB0 interrupt can be used to signal the arrival of a sensor pulse. Many PIC chips have more than one timer, which would be useful in this application. The PWM signal could be generated using a hardware timer, as well as the target sensor period. Some PICs have dedicated PWM outputs, which would simplify the software, but would require the PWM interface to be correctly initialised.

13.6 Commercial Application

Attempting closed loop dynamic control using the PIC 16F84, which has only one timer and one interrupt, illustrates its limitations, and why there is a range of more powerful processors available in the PIC family. A commercial design for a similar system uses a PIC 17C42 running at 16 MHz, with a dedicated PWM output. It is designed for use in printers and plotters for positioning the print or scan head quickly and precisely. A block diagram is shown in Fig. 13.9.

The motor is driven from a dedicated driver chip (LMD 18201) which requires only a single PWM input. A 50% MSR gives zero output which corresponds to motor being stationary. The motor can then be driven in either direction by varying the MSR above and below 50%. The driver circuit incorporates a full 'H-bridge' driver chip, which can supply motor current in either direction with minimal power consumption. The motor speed and position are monitored by a shaft encoder which produces two pulse trains. The relative position of the pulses indicates the direction of rotation of the motor. These are fed to a logic circuit which produces separate count up and count down pulses which are counted by 16-bit counters in the PIC to allow the current position of the motor shaft to be calculated. The control software is also far more complex than our demonstration system, with nearly 2000 instructions.

The PIC 17C42 has a serial port which allows commands to the motor to be sent via a single wire, and for the PIC to return information about the actual position, speed and so on, so that,

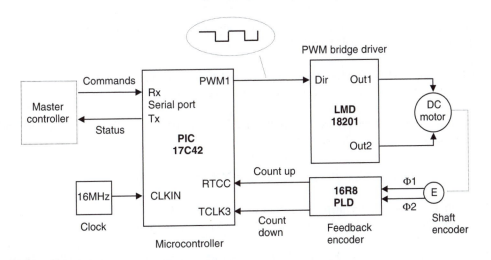

Figure 13.9 Block diagram of servo control unit.

Figure 13.10 PIC-based servo-control unit.

for instance, if the motor is stalled by a mechanical fault, the controller can detect the fault condition. The hardware is shown in Fig. 13.10.

Summary

- A small DC motor under PIC control can be used to demonstrate a range of real-time processes.

- The MOTA demonstration hardware allows motor control via a passive load bidirectional FET bridge driver, with analogue and digital inputs.

- Open loop control can be implemented relatively simply using pulse counting for position control and PWM for speed control.

- Closed loop control requires more complex algorithms which use pulse feedback to continuously modify the drive output for position and speed control.

- A practical servo typically uses additional hardware, a more powerful microcontroller and complex software for better performance.

Questions

1. Outline an open loop method of controlling the speed of a small DC motor, using a microcontroller. Identify the main hardware components required.

2. Explain how a slotted wheel can be used to provide speed and position feedback from a shaft to a microcontroller.

3. Explain why closed loop control is necessary for accurate speed control of a DC motor.

4. Calculate the positional accuracy in degrees of the output shaft of robot arm drive with a 90:1 gearbox, if a shaft encoder with 200 steps per revolution is attached to the motor shaft. (Answer: 0.2°).

5. Describe two alternative methods of measuring the speed of a motor shaft using an opto-slot detector with a crystal clocked microcontroller.

Activities

1. Construct the MOTA board using a method of your choice and devise a test schedule to confirm the correct operation of the hardware prior to fitting the PIC chip.

2. Devise a program to rotate the MOTA board output by exactly 100 revs from its start position. Evaluate the performance of the program in terms of speed of response, accuracy and reliability. What are the characteristics of the motor which affect the performance?

3. Investigate the performance of the program CLS1 in terms of reliability, response time, range of control (maximum and minimum speeds). Devise a method of loading the motor to test the performance of the controller with varying loads (the speed should be held constant within limits).

4. Modify CLS1 to read the input push buttons on the MOTA board to increase or decrease the set speed.

5. Modify the program for the MOTA board to use the timer interrupt to signal time out. Compare the performance of this alternative implementation with the program CLS1.

6. Modify the program for the MOTA board to use the RB0 interrupt to monitor the feedback from the motor. Compare the performance of this alternative implementation with the program CLS1.

7. Research and download the data sheet for a PIC chip which has analogue input/s, and redesign the MOTA board and control program so that the motor speed can be accurately controlled from an analogue input in the range 0–5 V.

8. Redesign the MOTA board circuit to use a full bridge motor driver IC instead of the dual FET and passive bridge drive circuit.

Part D
More Controllers

Chapter 14
More PIC Microcontrollers

14.1   **Common Features of PIC Microcontrollers**
14.2   **Selecting a PIC**
14.3   **Advanced PIC Features**
14.4   **Serial Communications**

The PIC 16F84 has been used as a reference device so far because its architecture and operation are relatively simple compared with other PIC microcontrollers. The range of flash memory PIC chips has now expanded such that alternative devices are now available which have more features at a lower unit cost. This chapter will review these features so that the most suitable device for a given application may be selected. Specifically, this means selecting the PIC chip which has the required number and type of inputs and outputs, program memory capacity sufficient for the application and so on, all at the minimum price.

We will continue to concentrate on PICs which have flash program memory, since these are the best choice for learning about application development, prototyping and producing one-off or low volume products. For larger production runs, OTP or mask programmed PICs are available. The OTP devices can be programmed by the user or supplied pre-programmed if the quantity justifies it, and the application program will not require any further modification. The masked ROM device has the program built in during production, and would only be used for high volume, mature products.

The main groups of PIC flash devices are shown in Table 14.1. They are divided into three groups, with a different prefix number: the 12XXXX series are 8-pin miniature PICs, the 16XXXX group might be described as the standard series and the 18XXXX devices as the high performance group. Their features are summarised in Table 14.1.

Full details of all these devices and the rest of the PIC range are provided at www.microchip.com, from where the individual data sheets can be downloaded as PDF files.

14.1   Common Features of PIC Microcontrollers

All PIC microcontrollers use the same basic architecture and instruction set to provide a product progression path from simple programs developed for the 12-series chip, through to the most complex applications for the 18-series device. Of course, the common architecture is used in the non-flash memory PICs as well. The architectural features may be compared by studying the

Table 14.1 PIC flash microcontrollers

12FXXX
- Low cost and small size
- 8-pin packages
- 6 I/O pins
- 33/35 x 12/14-bit instructions
- 1 k word program memory
- 20 MHz clock
- 4 MHz internal oscillator
- 8-bit and 16-bit timer
- Up to 4 analogue inputs*
- In-circuit programming and debugging*

16FXXX
- Mid-range cost and performance
- 14–40-pin packages
- 12–33 I/O pins
- 35 x 14-bit instructions
- 1–8 k word program memory
- 20 MHz clock
- 4/8 MHz internal oscillator*
- 2 x 8-bit and 1 x 16-bit timers*
- Up to 8 analogue inputs
- Serial communication ports, parallel slave port*
- 1/2 pulse width modulation outputs, capture & compare inputs*
- In-circuit programming and debugging*

18FXXX
- High performance
- 18–80-pin packages
- 13–68 I/O pins
- 58 x 16-bit instructions
- 2–64 k word program memory
- 40 MHz clock
- 8/10 MHz internal oscillator*
- 2–64 k program memory
- Up to 2 x 8-bit and 3 x 16-bit timers
- Up to 16 analogue inputs
- Serial communication ports, parallel slave port*
- Up to 14 pulse width modulation outputs, capture and compare inputs*
- CAN communication interface*
- In-circuit programming and debugging*

*Selected devices in the range

block diagram for each device found in its data sheet. The block diagram of the 16F84A can be seen in Appendix A, Figure 1-1.

The common features of the PIC architecture are:

- Harvard architecture
- RISC instruction set
- flash program ROM with ISP
- RAM block including SFRs
- EEPROM non-volatile data memory
- single working register
- dedicated, non-writable stack
- power-up and watchdog timers
- multiple interrupt sources
- hardware timers
- sleep mode
- serial in-circuit programming

Harvard Architecture

In conventional processor systems, the instruction codes and associated operands have to be transferred from memory using the same address and data bus as the system data, that is, the data read in via inputs or generated by the processor. The PIC architecture has separate paths for the instructions and the system data. The instruction fetch operation can therefore be carried out at the same time as the results from the previous operation are stored. As a result, the program executes faster at the same clock speed by carrying out these processes concurrently. The overlapping of instruction fetch and execution stages is described as pipelining.

Risc Instruction Set

The PIC has a small number of instructions compared with a conventional CISC processor. This has two main benefits – the instruction set is easier to learn and the code executes faster, because the instruction decoding hardware is less complicated. The down side is that more complex operations may have to be constructed from simpler ones, ending up taking longer to execute. Overall, the RISC performance is frequently better because, in a typical application, these complex instructions are not needed very often.

Flash Program Memory

Flash ROM is a great advance in memory technology which has developed over the last ten years. Writable, but non-volatile, memory is essential in embedded systems to store the control program. Previously, if erasable memory was required for development purposes, EPROM was used, but this must be removed from the system for erasing under ultraviolet light. Battery-backed RAM was an alternative, but, of course, batteries last only a limited time, and the program can be lost. Flash ROM can be easily re-programmed many times, and in-circuit serial programming (ISP) can be used to program the chip without the inconvenience and possible damage caused by removing it from the circuit.

RAM and SFRs

The individual bits in the SFRs need to be read and written when initialising the chip or during program operation. Because they are located in the same RAM block as the GPRs, they can be accessed using the same instructions. This means that special instructions for control register access are not needed, which helps to keep the instruction set small.

EEPROM Data Memory

This is very useful in applications where data read in at the ports or produced by the processor needs to be stored in non-volatile memory. For example, in a keypad-operated electronic lock, the lock code is entered by the user, and then must be retained to be checked against user keypad input to release the lock. Data logging applications, where sampled input data may need to be retained over a period of time, may also need to store the data while the power is off.

Working Register

Conventional processors tend to have a block of registers for storing current data. The Motorola 68000 processor, for example, has eight data registers. An architecture with only one working register, used in conjunction with the RAM register block, reduces the overall number and complexity of instructions required, as the options are reduced. This does mean, however, that loading a register with a literal takes two instructions, as it has to be loaded into W first, then into the register.

Stack

The stack size determines the number of subroutine or interrupt levels which can be used in the application program. The 12-series chips have only a 2-level stack, the 16-series 8 levels, and the 18-series 32 levels. This reflects the typical program complexity for each type. The application programmer needs to be aware of this limitation, and balance the advantages of a well-structured program using multiple subroutine levels, and the absolute limit imposed by the stack size. Unlike some processors, the stack cannot be overwritten by a program instruction, making it more secure.

System Timers

There is an array of features incorporated into the PIC microcontroller to ensure a smooth start-up to the application program when power is applied or a reset generated. A power-on reset is generated internally when the supply voltage has reached the required level. A power-on timer then provides a delay to allow the power supplies to stabilise, and an oscillator start-up timer provides a further delay to ensure that the clock is stable before program execution begins. The watchdog timer is another standard feature which allows the chip to reset itself automatically if the program execution fails to follow the normal sequence, thereby improving overall reliability. Brown-out protection allows the chip to reset in an orderly fashion if the power supply fails for a short period of time.

Interrupts

An interrupt is an internally or externally generated signal which forces the processor to suspend the current operation and execute an interrupt service routine. The ISR thus has a higher priority than the background process. PIC chips provide a variety of interrupt sources, for example, a change on a selected input, or a hardware timer time-out. There is an interrupt priority system available in the more advanced 18-series PICs; this allows the chip to be set up to ignore an interrupt source if a more important one is already active. In conventional microprocessors, such as the Motorola 68000, multiple interrupt vectors are available; that is, a different ISR address can be specified for up to eight interrupt sources. This means that a different ISR for each interrupt source can be specified. In the PIC, all interrupts have to be serviced via the single interrupt vector, at address 004 in program memory. Therefore, to differentiate between them, and to determine the action required, the ISR needs to check to relevant control register flags to find out which interrupt source is active, before branching to the required routine. As the number of peripheral devices increases, such as additional timers, serial ports and so on, the number of potential interrupt sources increases, making interrupt servicing via a single vector more complicated.

Hardware Timers

The number of hardware timers generally increases with the chip complexity. They are either 8-bit or 16-bit counters, with prescalers or post-scalers, which divide down the input or output of the counter to extend its range. If we take the motor program in Chapter 13 as an example, we can see how additional hardware timers would have been useful; the 20 ms time interval and the motor output cycle delay could both have been implemented as hardware operations, while the sensor pulse monitoring could have used RB0 interrupt. This would have simplified the control program significantly. Therefore a device with two hardware timers would have been a better choice for this application.

Sleep Mode

This is a very useful feature for power saving and also terminating programs which do not loop continuously. The processor shuts down when the instruction SLEEP is encountered, with current consumed dropping to around 1 μA. The device can then be woken up via an external signal when required; the advantage in battery-powered applications is obvious. SLEEP can also be used to terminate a program, so that program execution does not continue into unprogrammed locations. These locations default to all '1's, which generally corresponds to a valid instruction code (ADDLW in 14-bit code, NOP in 16-bit code). If a program is not terminated with a SLEEP or GOTO instruction, the program will carry on to the end of memory, the program counter will roll over to zero, and the program will restart unexpectedly.

In-Circuit Programming

The PIC microcontrollers use a common program downloading system, which consists of loading the program in serial form via one of the data pins when the chip is in programming mode. The chip can be placed in a programming unit for downloading the application code, and then transferred to the application board. Programming units usually have a ZIF socket which is large enough to accommodate a range of PIC chips.

Alternatively, the chip can be programmed in circuit, if the application hardware is designed for this option. This means that the chip can be left in circuit at all times, reducing the risk of damage, and can be programmed after the circuit has been manufactured, and reprogrammed at any time, via an on-board connector. This connector can be seen on the 16F877 temperature controller board circuit diagram and hardware in Chapter 15.

14.2   Selecting a PIC

Each type of PIC microcontroller provides a different combination of features, so that the most suitable can be selected for any given application. At the time of writing more than 140 are available, and increasing all the time. Some of the main selection criteria are:

- number of I/O pins available
- program memory size
- program memory type (ROM, EPROM, Flash)
- EEPROM data memory
- timers (8-bit or 16-bit), CCP
- interrupt sources
- analogue inputs (8-bit or 10-bit)
- serial communication interfaces (USART, SPI, I²C, CAN)
- internal oscillator
- in-circuit debugging
- maximum clock speed
- package/footprint (DIP, SOIC, PLCC, QFP)
- price

When developing an embedded application, the hardware will generally be specified and designed first. This will determine the number and type of inputs and outputs required. Simple switches will require single digital input, while a keypad will require several. A temperature sensor will need an analogue input, a motor will probably require a PWM output. Most systems use some kind of status or information display, and type of display (status LEDs, seven-segment LED or LCD) will determine the number of output pins needed to drive it. Serial communication will often be used if the PIC is part of a larger system or is connected to a master controller.

When the hardware requirements have been established, the program can be developed using MPLAB, and tested by simulation. The size of the program will then be known, so that chip memory size can be specified. In addition, the size of the stack in the selected device must be sufficient for the number of subroutine levels and interrupts; if not, the program can be restructured or a different chip used. MPLAB can also be used for provisional chip selection, since a specific device must be selected before the program is assembled.

We are assuming that flash memory PICs will be our first choice for experimental purposes, but for commercial production, the memory type must be selected on the basis of minimum cost for a given batch size. If overall size of the finished board is important, a surface mount implementation may be needed.

When the design parameters such as I/O requirements, program memory size and so on have been finally established, the most suitable device can be selected using the search facilities on the manufacturer's website. Summary information for selected PIC flash microcontrollers is provided in Table 14.2, as a guide to the features available.

14.2.1   I/O Pins

This is probably the most important criterion for an embedded controller. The number and type of inputs and outputs required should be clearly defined at an early stage in circuit design. The grouping of the pins may also be important, as they are generally arranged as 8-bit ports, with smaller chips having partial port implementations (e.g. Port A in the 16F84 = 5 bits). If analogue inputs are required, these must be deducted from the digital I/O total, since they share the same pins with digital I/O. The table shows that the number of I/O pins ranges from

Table 14.2 PIC flash microcontroller features

PIC device number	Total pins	I/O pins	Program ROM words	File RAM bytes	EEPROM bytes	Analogue inputs	Timers 8 bit / 16 bit	Max. clock (MHz)	Internal osc. (MHz)	In-circuit debug	CCP/PWM modules	Serial comms	Relative cost
12F629	8	6	1k	64	128	–	1 + 1	20	4	✓	–	–	1.02
12F675	8	6	1k	64	128	4 × 10-bit	1 + 1	20	4	✓	–	–	1.26
16F627A	18	16	1k	224	128	–	2 + 1	20	4	–	–	UART	1.49
16F628A	18	16	2k	224	128	–	2 + 1	20	4	–	1	UART	1.70
16F630	14	12	1k	64	128	–	1 + 1	20	4	✓	–	–	1.20
16F648A	18	16	4k	256	256	–	2 + 1	20	4	–	–	UART	1.83
16F676	14	12	1k	64	128	8 × 10-bit	1 + 1	20	4	✓	–	UART	1.38
16F72	28	22	2k	128	–	4 × 8-bit	2 + 1	20	–	–	1	–	2.10
16F73	28	22	4k	192	–	5 × 8-bit	2 + 1	20	–	–	2	All	3.27
16F74	40	33	4k	192	–	8 × 8-bit	2 + 1	20	–	–	2	All	3.97
16F76	28	22	8k	368	–	5 × 8-bit	2 + 1	20	–	–	2	All	4.10
16F77	40	33	8k	368	–	8 × 8-bit	2 + 1	20	–	–	2	All	4.58
16F818	18	16	1k	128	128	5 × 10-bit	2 + 1	20	8	✓	1	I²C, SPI	1.71
16F819	18	16	2k	256	256	5 × 10-bit	2 + 1	20	8	✓	1	I²C, SPI	1.71
16F84	18	13	1k	64	64	–	1	10	–	–	–	–	4.39
16F84A	18	13	1k	64	64	–	1	20	–	–	–	–	3.42
16F87	18	16	4k	398	256	–	2 + 1	20	8	✓	1	All	2.26
16F88	18	16	4k	368	256	7 × 10-bit	2 + 1	20	8	✓	1	All	2.41
16F873A	28	22	4k	192	128	5 × 10-bit	2 + 1	20	–	✓	2	All	3.98
16F874A	40	33	4k	192	128	8 × 10-bit	2 + 1	20	–	✓	2	All	4.35
16F876A	28	22	8k	256	368	5 × 10-bit	2 + 1	20	–	✓	2	All	4.28
16F877A	40	33	8k	256	368	8 × 10-bit	2 + 1	20	–	✓	2	All	4.68
18F1220	18	16	2k	256	256	7 × 10-bit	1 + 3	40	8	✓	1	UART	2.78
18F2320	28	25	4k	512	256	10 × 10-bit	1 + 3	40	8	✓	1	All	4.85
18F4320	40	36	4k	512	256	13 × 10-bit	1 + 3	40	8	✓	2	All	5.29
18F6520	64	52	16k	2048	1024	12 × 10-bit	1 + 3	40	–	✓	5	All	6.52
18F8621	80	68	32k	3840	1024	16 × 8-bit	1 + 3	40	10	✓	14	I²C, SPI	8.25
18F8720	80	68	64k	3840	1024	16 × 10-bit	1 + 3	40	–	✓	5	All	10.90

6 in the 8-pin chips to 68 in the 80-pin device. Most I/O pins have more than one function, one of which can be selected during initialisation by setting up the relevant control register. If no setup is performed for a particular pin, it will typically default to a digital input. Pins can be re-configured within the program sequence to have a different function at different times. If this is the case, the designer must ensure that the two functions do not interfere with each other, in terms of both the hardware and software.

14.2.2   Program Memory

The specification of memory size can only be finalised after the software has been developed, but an experienced application developer should be able to predict this requirement fairly early on. If the program is developed in 'C' language, the memory size will be greater, because each program statement can expand into several machine code instructions. In this case, an 18-series device is likely to be the best choice, as the memory available is larger. Microchip supply a 'C' compiler for the 18XXX chips, and third party compilers are also available. For prototyping and small production volumes, flash memory will be preferred. For intermediate volumes, where the design is proven, OTP memory will reduce costs per unit. For high volume production, a contract for mask programmed devices supplied by the manufacturer can be considered.

14.2.3   Data Memory

The file register RAM block tends to increase in size with the program memory size and chip complexity. The number of variables and temporary data storage blocks required should be totalled when the program has been developed, perhaps adding an allowance for future expansion or changes to the specification. If non-volatile data storage is needed, the EEPROM size must also be checked. As can be seen in the table, RAM ranges from 64 to 3840 bytes in the flash PICs.

14.3   Advanced PIC Features

14.3.1   Timers and CCP

The timer capacity of the flash PICs ranges from only one 8-bit timer in the 16F84 to five in some 18-series chips (2×8-bit plus 3×16-bit). Hardware timers should be used for most timing and counting operations, because the processor can then carry on with some other process while the timer process runs.

CCP stands for Capture/Compare/PWM. Capture mode provides input interval measurement (Fig. 14.1). The value in a timer register is captured (stored) when an input changes; the time between the timer start and input change can thus be measured. In the motor application, for

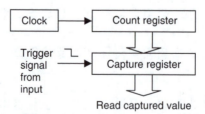

Figure 14.1   Timer capture operation.

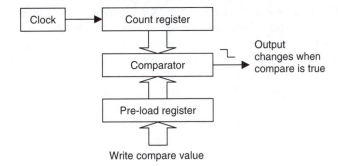

Figure 14.2 Timer compare operation.

example, the timer could be started when a pulse is received from the shaft sensor, and the time captured when the next pulse arrives, giving the period of the shaft sensor pulse. An interrupt can be enabled to signal this event.

Compare mode provides output interval generation (Fig. 14.2). A value is loaded into a register which is then continuously compared with a timer register as it runs. When the register values match, an output pin is toggled and an interrupt generated to signal the time-out event. This is a convenient way to generate a timed interval, so that, for example, an output can be switched a set time after an input has changed.

In PWM mode, preset values are loaded into two registers representing the mark and space period of the PWM output required (Fig. 14.3). The timer value is then compared with the mark register and the output toggled after the mark value is reached. The timer is then restarted and compared with the space value as it runs, and the output toggled when the space value matches. The process is repeated and the output from the flip-flop toggles after each mark and space interval to generate a PWM output.

14.3.2 *Analogue Inputs*

Many PIC chips incorporate analogue inputs so that they can be used in control systems with input sensors which produce a voltage, current or resistance change in response to an

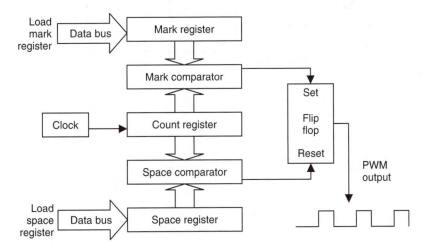

Figure 14.3 Pulse width modulation operation.

environmental variation. The temperature controller described in Chapter 15, for example, is designed to accept inputs from temperature sensors which give an output change of 10 mV per °C. The PIC then operates outputs which control the temperature in the target system.

Most PICs with this feature provide 10-bit conversion. This means that the input voltage is converted to a 10-bit number, giving a resolution (accuracy) of 1 in 1024, or better than 0.1%. This is good enough for all but the most demanding applications. If such resolution is not required, an 8-bit result can be obtained by ignoring the two extra bits. Multiple analogue inputs are usually available; the PIC 16F877 in the temperature controller has eight, although only four are used. Code for performing the analogue input conversion is given in Program 15.1. Refer to the PIC 16F87X data sheet for details of analogue input operation.

The analogue to digital conversion process is illustrated in Fig. 14.4. The port containing the ADC inputs can be set up with a combination of analogue and digital inputs, or all analogue. One of the analogue inputs is selected at a time for conversion, and the converter output is stored in an ADC result register. The maximum and minimum voltage levels to be converted can be set externally, or the internal supply voltages used. An external voltage reference of 2.56 V gives a convenient 0.01 V per bit conversion for an 8-bit result. A clock divider must be set up to allow the minimum specified conversion time (about 20 µs); for example, if the chip clock is 20 MHz, divide by 32 must be selected, and at 4 MHz, divide by 8. The GO/DONE bit in the control register is used to start a conversion; the same bit indicates when the conversion is finished. The ADC works by successive approximation, details of which can be found in standard electronics references.

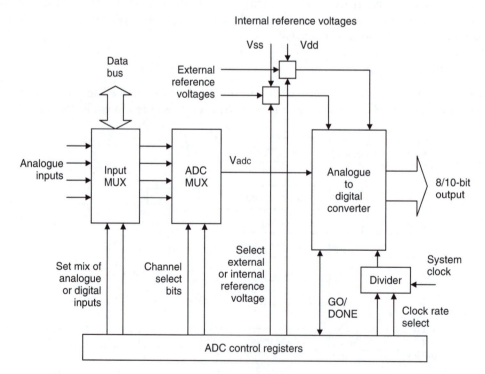

Figure 14.4   Analogue to digital conversion block diagram.

14.3.3 Parallel Slave Port

The microcontroller may need to operate as part of a larger system, exchanging data with a master controller, or other devices in the system, in the same way that a peripheral interface device operates as a slave to the microprocessor in a conventional system. An 8-bit port (Port D in the 16F877) can therefore act as a parallel slave port, allowing 8-bit data exchange with a master controller, which may be a conventional processor, a host computer or another microcontroller (see Fig. 14.5). Three bits of another port (E) provide a control signal interface, acting as active low read, write and chip select inputs to the slave PIC. In addition, the slave may need to output an interrupt signal, which can be generated in software at any convenient output pin.

14.3.4 Internal Oscillator

To save on external components, some PICs offer an internal oscillator option. It provides a fixed 4 MHz or 8 MHz clock based on an internal RC oscillator. It is therefore not very accurate, but its frequency can be trimmed using a calibration value pre-programmed into the chip by the manufacturer. Even so, the frequency is only accurate to about 5%, so an external crystal clock will still be needed to obtain accurate timing operations.

14.3.5 In-Circuit Debugging

By using an ICD hardware module instead of the usual programming interface, the serial programming system can double up as a debugging tool. When the application program has been downloaded, MPLAB can be used to monitor the program execution within the chip, with the real hardware. The program will respond to actual inputs, and output to the hardware devices. The program can be single stepped, breakpoints set and registers monitored, using the same techniques as used in the software simulation; it can be then checked for correct operation at full speed in the target hardware. When the program is fully debugged, the chip must be reset to operate in normal run mode. So ICD is a very useful addition to the PIC debugging toolset, and much cheaper than the alternative in-circuit emulation.

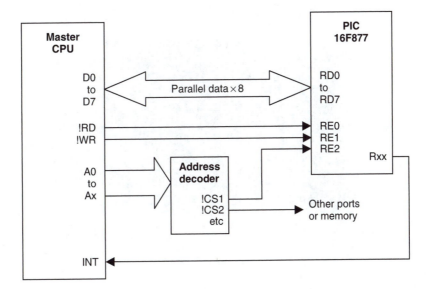

Figure 14.5 Parallel slave port connections.

The ICD system configuration is shown in Fig. 14.6(a). The ICD module sits in between the host PC running MPLAB IDE and the application board with the PIC chip fitted. There are two ways to connect to the chip, using a header and using an on-board ICP/D connector. If no provision has been made on the application board for direct in-circuit programming and debugging connections, an ICD header, which has the same pinout as the chip itself, can be fitted to the PIC socket. The header connector is shown with its demo board in Fig. 14.6(b). The PIC is then fitted into a corresponding socket on the header, which carries the ICP/D connections to Port B. When debugging is complete, the chip can be switched to run mode and plugged directly into the board.

A better option is to provide the ICP/D connection on the board, as this only requires a suitable 6-way connector, a resistor to protect the !MCLR input. The connections required are shown in Fig. 14.7. Vpp is the programming voltage supplied to the target chip by the ICD module, and the program is downloaded via PGD. When switched to ICD mode, the same connections allow the processor to be controlled from MPLAB, and register information to be relayed back to the host to assist with debugging. The ICD system is shown connected to the temperature controller board in Fig. 15.6.

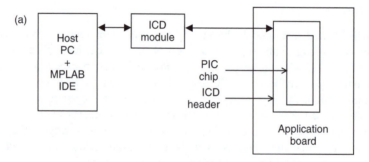

Figure 14.6   In-circuit debugging system via header connection. (a) Block diagram; (b) ICD header with demo board.

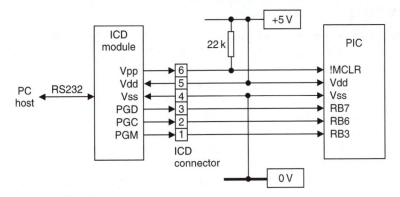

Figure 14.7 In-circuit programming and debugging on-board wiring.

14.3.6 Clock Speed

Clock speed is the main factor in the performance of any microprocessor system, and is critical in some applications. For example, in the motor control example, the higher the clock speed, the more precise the control can be, as the shaft speed can potentially be measured more accurately. Most of the flash PICs currently available operate at up to 20 MHz (12 and 16 series) or 40 MHz (18 series). This gives an instruction cycle time of 200 ns or 100 ns (nanoseconds) and an execution rate of 5 or 10 MIPs. All PICs use a fully static design which means that they can operate down to zero frequency. The clock rate is limited by the time taken by the internal signals to rise and fall, so correct performance is only guaranteed up to the maximum rated speed. The maximum speed is also limited by power consumption and its heating effect.

Power consumption is generally proportional to clock speed in CMOS devices, since most of the power is consumed when the component transistors switch on and off. This is illustrated by the current consumption curve for the 16F84A found in the data sheet under electrical characteristics, summarised in Fig. 14.8. Note that for operation at clock frequencies between 4 and 20 MHz, high speed (HS) mode must be selected when programming the chip, and below 4 MHz crystal (XT) mode. The power consumption at high speed, especially if operating in a small IC package, may necessitate additional cooling measures to keep a chip within its temperature limits. A heat sink or even a fan, as found on the processor in a typical PC motherboard design, can be fitted.

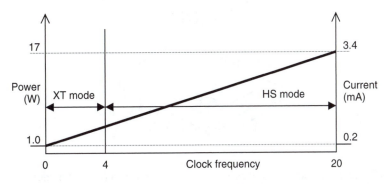

Figure 14.8 PIC 16F84A power consumption against clock rate.

14.3.7   Packaging

The standard package for integrated circuits is the plastic dual in-line (PDIP) chip, which has two lateral rows of pins spaced at 0.1" intervals (Fig. 14.9(a)). The maximum number of pins that can practically be accommodated in this type of package is 64, so other formats have been adopted for larger chips. The plastic leaded chip carrier (PLCC) package has the pins arranged around four sides of a square package, which is designed to fit in a recessed socket. The pin grid array (PGA) has pins arranged in a grid covering one side of the package, with a flat socket mounting.

Typically, the actual integrated circuit only occupies a small central portion of the DIP package, so miniaturised packages are easily designed. Surface-mount components are now increasingly used in commercial products, as chips become larger, circuits more complex, and products themselves miniaturised. The pins of the ICs are not fitted through holes on the board, but soldered onto the surface on flat pads. Surface-mount boards require very precise manufacturing techniques, generally being produced on automatic assembly machines.

The small outline integrated circuit (SOIC, Fig. 14.9(b)) is a surface-mount DIL package with a pin pitch of 0.05". The smaller shrink small outline plastic (SSOP, Fig. 14.9(c)) package has a pin pitch of 0.026". Quad flat pack (QFP) is a square-surface-mount package for larger chips, such as the 44-pin PIC 16F877, with pins on four sides.

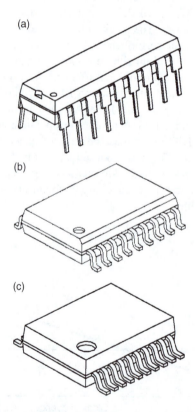

Figure 14.9   18-pin packages. (a) Plastic dual in-line package (PDIP); (b) Small outline integrated circuit (SOIC); (c) Shrink small outline package (SSOP).

14.3.8 Price

Price is determined by the complexity of the chip, but also by the volume of production. As the range is constantly updated, each design can be superseded by a chip with better features; as the volume builds up, the new device becomes cheaper due to economies of scale in production, and recovery of development costs. The older design may become relatively more expensive, as well as having less features, before slipping into obsolescence.

For example, at the current time the guide price quoted for the original 16F84 is $4.39, while the pin-compatible replacement, the 16F818, which has analogue inputs and other extra features, is only $1.71. Therefore, while the older chip has been used as an example, because it is less complicated, the reader should consider using the more recent chip in his or her own designs, even if the new features are not used to the full. It does mean, however, that the data sheet may be more difficult to understand, because it has to cover all the additional features.

The relative cost for each chip shown in Table 14.2 is based on the 'budgetary price' quoted by the manufacturer at the time of writing. The figures allow relative costs to be compared, while the actual price will obviously increase in time.

14.4 Serial Communications

Serial communication ports allow the PIC to communicate with other devices, or exchange data with a master controller, via a single connection. There are several protocols available:

- USART (universal synchronous/asynchronous receiver/transmitter)
- SPI (serial peripheral interface)
- I²C (inter-integrated circuit)
- CAN (controller area network)

14.4.1 USART

RS232 is a long established asynchronous communication protocol used at the serial (COM) port of the PC. It is low speed, but is easy to understand and has been used as the standard serial communication method between computers, terminals and other systems. It is used to download programs to the MPSTART programmer module. 'Asynchronous' means that no separate clock signal is provided with the data, so correct reception of data relies on the sender and receiver operating at the same speed. Serial data is sent and received as individual bytes using a pair of shift registers (Fig. 14.10(a)). After each bit of the data is shifted out of the send register onto the line, it must be shifted into the receiver register at the same time. In other words, the receiver must sample the line during the time that the bit is present. It must then take the next sample after a specified interval. The 'baud rate' sets this time interval; at 192000 bits per second, the time interval is about $50\,\mu$s. The sender and receiver must be set up to operate at the same baud rate; there are a set of standard rates of between 300 and 57600 bits per second. The signal is illustrated in Fig. 14.10(b). The line is high when inactive; the start of a byte is indicated by the falling edge of a start bit. The receiver then samples the line at the required interval to read each bit, and is re-triggered for each byte.

When the USART is operated in asynchronous mode (Fig. 14.11), there is a separate data path for send (TX) and receive (RX), which are symmetrical in operation. One byte of data is transmitted at a time down a serial line, with start, stop and optional error check bits. In synchronous mode, the TX pin is used instead to carry a clock (CK) signal, which is sent with

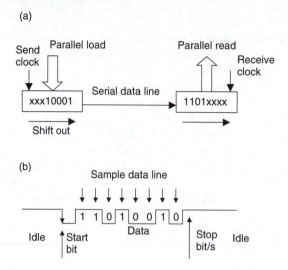

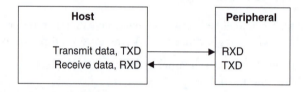

Figure 14.10 Serial data transfer. (a) Shift register operation; (b) Serial data signal.

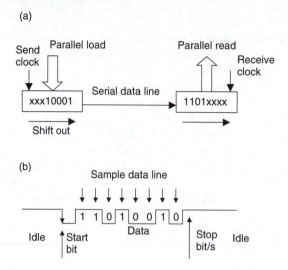

Figure 14.11 USART asynchronous mode connections.

the data to control the receiver, making the process more reliable. In this mode, each device can still send and receive, but only one at a time.

The RS232 type signal data signal produced by the PIC USART is output at TTL levels. Other terminals, such as the PC, will produce a signal which is transmitted at a voltage up to around $+/-24$ V to allow the signal to travel further on the line (up to about 100 m). If the PIC is to communicate with such a terminal, the signal must be passed through a line driver which will boost the voltage and also shift the level as required.

14.4.2 SPI

The SPI system uses three pins:

- serial data out (SDO)
- serial data in (SDI)
- serial clock (SCK)

It is a single master, multi-slave system using hardware slave selection (Fig. 14.12). To exchange data with a slave, the master selects it by taking the slave select input low (!SS). Synchronous 8-bit data is then exchanged via SDI or SDO as required, with a clock pulse to strobe in each bit to the destination register. Due to the hardware selection requirements, this system is most suitable for communication between devices on the same board. Data can be transmitted and received at the same time, at a clock rate of up to 5 MHz with a 20 MHz chip clock.

(a)

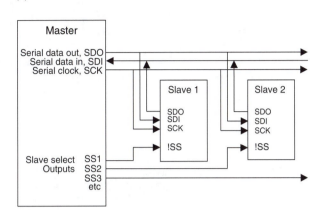

(b)

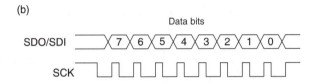

Figure 14.12 Serial peripheral interface (SPI). (a) SPI connections; (b) SPI signals.

14.4.3 I^2C

The I^2C system uses two pins:

- serial data (SDA)
- serial clock (SCL)

This system also uses synchronous master/slave communication, but with a software addressing system (Fig. 14.13). As in a network, the destination address for transmitted data in a multi-slave system is transmitted on the same line (SDA) prior to the data. A seven- or ten-bit address can be used (up to 1023 slaves), which must be programmed into an address register in each slave. The clock can operate at up to 1 MHz. This system is suitable for communication between separate microcontroller boards, since no slave selection hardware connections are needed. The hardware is simpler, but the software is more complex. Note that in the hardware diagram, the lines are pulled up to +5 V, giving active low, wired-OR operation on the serial bus and clock line.

14.4.4 CAN

The CAN system uses two pins:

- CANTX (Transmit)
- CANRX (Receive)

The CAN system (Fig. 14.14) is designed for transmitting signals in electrically noisy environments, such as motor vehicle control, using differential current drivers. It is only available in selected high performance PIC 18-series devices. A more complete description and

(a)

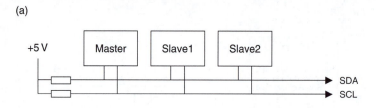

(b)

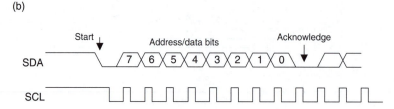

Figure 14.13   Inter-IC bus (I^2C). (a) I^2C connections; (b) I^2C signals.

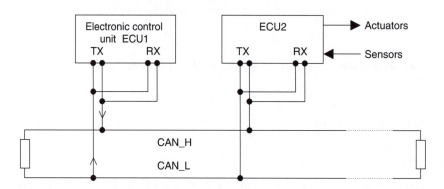

Figure 14.14   CAN bus system.

information for setting up these serial communication interfaces is provided in the PIC 16F87X, and other, PIC data sheets.

Summary

- The PIC family of microcontrollers share a common core architecture and instruction set. The 18XXXX group of high power MCUs have an enhanced architecture and instruction set.

- All PICs are designed with separate program and data busses, a reduced instruction set and two-stage pipelining to improve performance.

- To select a device for a given application, the number of I/O pins, the program memory type and size, and special I/O requirements, such as analogue input and PWM output, must be established.

- In-circuit programming is a common feature, with in-circuit debugging in selected devices.

- A range of serial bus communication interfaces is available in selected chips, as well as a parallel bus for operation within a master/slave system.

1. Summarise the differences between the PIC 12-, 16- and 18-series of microcontrollers.

2. Describe the features in the PIC chip which help to ensure that the program starts reliably.

3. State two advantages of in-circuit serial programming.

4. From the table of PIC flash microcontrollers, name (a) a device which has eight analogue inputs in a package with fewer than 18 pins, and (b) another device which could control 10 PWM motor outputs, runs at 40 MHz and can be programmed in 'C'.

5. What is the minimum time required to read four analogue inputs?

6. State two factors that limit the clock speed in a microcontroller.

7. Describe the essential difference between SPI and I^2C addressing. Which has more complex hardware requirements?

1. Download the data sheet and print the summary page for the PICs 12F675, 16F818 and 18F8720. Summarise the features and suggest a typical application, for each device.

2. A robot has five motors which are PWM driven with an analogue position sensor on each axis. Select the cheapest PIC chip from Table 14.2 which could be used as a controller for the robot positioning system. Download the data sheet and draw a block diagram for the system, identifying the pins which should be connected to the motors and sensors.

3. Sketch a block diagram for an alternative implementation of the robot controller in Activity 2 using a separate controller for each axis connected to a master SPI controller. Select the cheapest suitable chip for the slave controllers. If the pot on one axis rotates through 300 degrees, calculate the smallest movement which can be detected by the controller, using a 10-bit conversion. Assume the sensor pot is attached directly to the robot axis. Suggest an advantage of the master–slave system over the single controller solution proposed in Activity 2.

Question 5. $4 \times 20 = 80\,\mu s$

Activity 3. $300/1024 = 0.3°$

Chapter 15
More PIC Applications and Devices

15.1 **16F877 Application**
15.2 **16F818 Application**
15.3 **12F675 Application**
15.4 **18F452 Application**

In the previous chapter, the features of the three main groups of PIC flash microcontrollers have been outlined. The standard group (16FXXX) provides an intermediate range of features, with between 13 and 33 I/O pins and 1k–8k of program memory. The miniature group (12FXXX) of 8-pin chips have six I/O pins and 1k instructions. The high performance group (18FXXX) provides up to 64k program words and 68 I/O pins. In this chapter, an application for the 16F877 will be described in some detail, and how similar applications could be implemented using a range of other devices.

15.1 16F877 Application

The 16F877 is at the top of the range in this group, and will therefore be used to illustrate the range of features available within the 16 series, which have already been described in Chapter 14. Other chips in this group have different combinations of these features; the intention is to help the reader to make the best choice of chip for any given application.

The temperature controller described here uses most of the available I/O provided, including analogue inputs. The 8k memory should be sufficient for most application programs which might be developed for this hardware. A demonstration program is provided which will exercise the hardware for test purposes, but it will be left to the reader to develop a fully functional application.

15.1.1 Temperature Controller System

A temperature controller is required to control a system such as a greenhouse where the temperature must be kept within set limits ($0 - 50\,°C$) by a heating and ventilation system (Fig. 15.1).

The unit will be programmed to accept a maximum and minimum temperature, or a set temperature and operating range. The system operates on the average temperature reading from

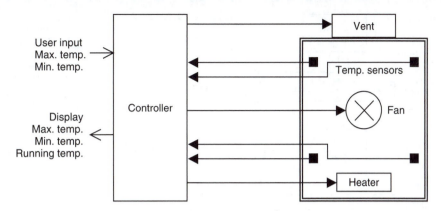

Figure 15.1 Temperature control system.

Table 15.1 Temperature controller function table

Measured temp.	Heater	Vent	Fan	Action
Temp. ≪ Min.	**ON**	OFF	**ON**	Forced heating
Temp. < Min.	**ON**	OFF	OFF	Heating
Min. < Temp. < Max.	OFF	OFF	OFF	Correct Temp.
Temp. > Max.	OFF	**ON**	OFF	Cooling
Temp. ≫ Max.	OFF	**ON**	**ON**	Forced cooling

four sensors to give a more accurate representation of the overall temperature in the enclosure. Using more than one temperature sensor also allows the system to tolerate a fault in one sensor, if the application software includes a check to see if one sensor is out of range. The temperature is maintained by a switched heater, switched vent and a fan which can be speed-controlled. The system should operate as specified in Table 15.1. The fan is fitted to the heater, so that it can be used for forced heating or cooling, depending on whether the heater is on. It needs to be connected to a PWM output on the controller if the rate of forced heating and cooling is to be varied. A demonstration system was constructed, where the heater was represented by a pair of filament lamps and the fan by a 5 V CPU fan. The temperature sensors were standard LM35 devices which have a built-in amplifier which outputs 10 mV per °C. Figure 15.2 shows the interfacing requirements for the application. The temperature sensor readings can be averaged, or processed with a weighting factor for each, to give a representative value for the measured temperature. A heater is then controlled via a suitable interface. A relay can be used if on/off control is sufficient. If proportional control is required, a PWM output would be required. In the hardware design provided here, the heater and vent interfaces are implemented as normally open switched relays, so that an external power supply can be used. The fan output demonstrates the alternative solid state interface, using a general purpose power FET. This would allow proportional control, but the external circuit must be operated at 5 V. For PWM control, the FET output would have to be re-allocated to one of the PWM outputs on the PIC 16F877.

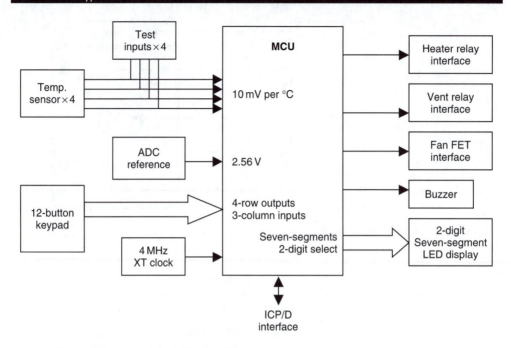

Figure 15.2   Temperature controller interfacing.

15.1.2   I/O Allocation

The I/O functions provided by the PIC 16F877 are detailed in Table 15.2. These were then mapped against the requirements of the application, and the most convenient grouping decided, giving the I/O allocation in Table 15.3.

Table 15.2   16F877 pin functions

Pin label	Function
	Port A (6 bits)
RA0/AN0	Digital I/O or analogue input 0
RA1/AN1	Digital I/O or analogue input 1
RA2/AN2/Vref–	Digital I/O or analogue input 2 or positive reference voltage for ADC
RA3/AN3/Vref+	Digital I/O or analogue input 0 or negative reference voltage for ADC
RA4/T0CKI	Digital I/O or input to timer 0
RA5/ AN4/SS	Digital I/O or analogue input 0 or slave select input (SPI mode)
	Port B (8 bits)
RB0/INT	Digital I/O or external interrupt input
RB1	Digital I/O
RB2	Digital I/O
RB3/PGM	Digital I/O or select serial programming mode
RB4	Digital I/O (interrupt on change)
RB5	Digital I/O (interrupt on change)

RB6/PGC	Digital I/O or in-circuit debugger or serial programming clock input (interrupt on change)
RB7/PGD	Digital I/O or in-circuit debugger or serial programming data input (interrupt on change)
	Port C (8 bits)
RC0/T1OSO/T1CKI	Digital I/O or Timer 1 oscillator output or Timer 1 clock input
RC1/T1OSI/CCP2	Digital I/O or Timer 1 oscillator input or Capture 2 input or Compare 2 output or PWM2 output
RC2/CCP1	Digital I/O or Capture 1 input or Compare 1 output or PWM1 output
RC3/SCK/SCL	Digital I/O or Synchronous serial clock input or output in SPI and I²C modes
RC4/SDI/SDA	Digital I/O or SPI data input or I²C data I/O
RC5/SDO	Digital I/O or SPI data output
RC6/TX/CK	Digital I/O or USART asynchronous transmit or USART synchronous clock I/O
RC7/RX/DT	Digital I/O or USART asynchronous receive or USART synchronous data I/O
	Port D (8 bits)
RD0/PSP0	Digital I/O or parallel slave port bit 0
RD1/PSP1	Digital I/O or parallel slave port bit 1
RD2/PSP2	Digital I/O or parallel slave port bit 2
RD3/PSP3	Digital I/O or parallel slave port bit 3
RD4/PSP4	Digital I/O or parallel slave port bit 4
RD5/PSP5	Digital I/O or parallel slave port bit 5
RD6/PSP6	Digital I/O or parallel slave port bit 6
RD7/PSP7	Digital I/O or parallel slave port bit 7
	Port E (3 bits)
RE0/RD/AN5	Digital I/O or PSP read select or analogue input 5
RE1/WR/AN6	Digital I/O or PSP write select or analogue input 6
RE2/CS/AN7	Digital I/O or PSP chip select or analogue input 7

The interfacing for this application is typical of that required for simple control systems using microcontrollers. Fortunately, this application does not need any analogue signal conditioning on the input side, as the temperature sensors can be connected directly to the PIC. Other references on interfacing will cover the range of design techniques needed for the most common sensors and output devices.

15.1.3   *Temperature Controller Circuit Description*

Figure 15.3 shows the circuit for the temperature controller. Each section of the external circuits can be described separately.

Table 15.3 Temperature controller I/O allocation

Device	Function	16F877 Pin	Initialisation
Temperature sensors	10 mV per °C 0–512 mV= 0–51.2 °C	RA0, RA1, RA2, RA5	AN0, AN1, AN2, AN4
ADC reference voltage	2.048 V	RA3	VREF+
Heater	Switched output	RE0	Digital output
Vent	Switched output	RE1	Digital output
Fan	Switched output	RE2	Digital output
4 X 3 keypad	Read column Scan row	RD3, RD4, RD5, RD6 RD0, RD1, RD2	Digital output Digital input
2 X 7-segment display	Segments Digit select	RC1–RC 7 RB1, RB2	Digital output Digital output
Buzzer	Audio alarm	RB0	Digital output
ICP/D interface	Program & debug	RB3, RB6, RB7	N/A

Analogue Inputs

The four temperature sensors are allocated to four of the eight analogue inputs available on the chip. In the demo system, standard sensors with an output of 10 mV per °C were used (0 °C = 0 mV). The controller is designed to operate at up to 50 °C, at which temperature the sensor output is 500 mV. This relatively low voltage is acceptable if the sensors are not connected on long leads, which could pick up electrical noise. For more remote operation, a DC amplifier should be used at the sensor end of the connection to increase the voltage to, say, 5.00 V at 50 °C. Alternatively, or in addition, screened leads could be used. The inputs are protected by a low pass RC filter; the input impedance at the ADC is high enough for this to have negligible effect on the input voltage.

The ADC normally operates at 10-bit resolution, giving output values 0–1024. It needs reference voltages to set the maximum and minimum values for the input conversion. These can be provided internally as Vdd and Vss (supply values), but Vdd does not give a convenient conversion factor. Therefore, an external reference value is provided from a 2.7 V zener diode and potential divider, giving Vref+ which is adjusted to 2.048 V. This then gives a conversion factor of 2048/1024 = 2 mV per bit. To simplify the software, and to cover the correct range, only the low eight bits of the ADC result will be used, with a maximum value of 255. At 50 °C, the input will be 500 mV/2 mV = 250, giving a resolution of 0.2 °C per bit. For test purposes, a set of four on-board pots are provided, so that input voltages in this range can be input manually, to check the operation of the software without having to heat and cool the target system. These can be switched in and out as required via a bank of DIP switches.

Outputs

Two types of output are provided, relay and FET. The relay gives a switched output that is isolated (electrically separated) from the controller. The external circuit operates with a separate

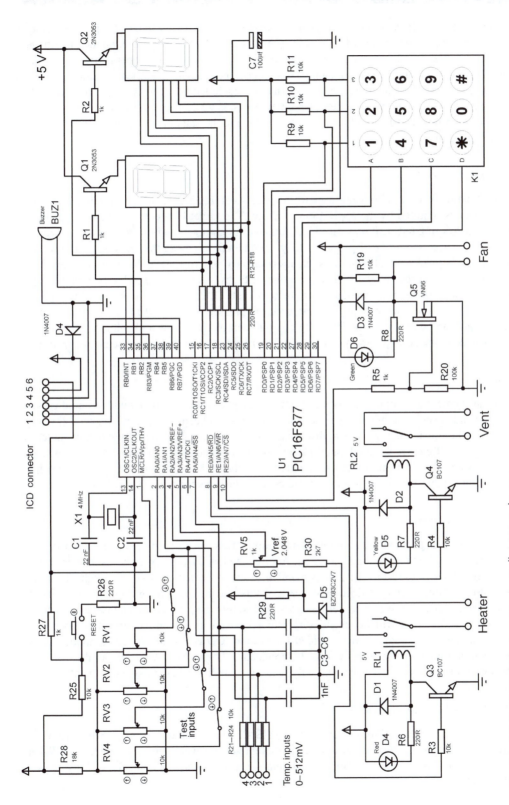

Figure 15.3 16F877 temperature controller circuit diagram.

supply, so the load (heater in this case) can be powered from a high-voltage supply if necessary. The relay also provides a high off resistance (an air gap). The FET interface, on the other hand, is more reliable, as it is solid state. The problem is that the load has to operate from the same supply as the FET, the 5 V board supply. It also does not provide electrical isolation between the controller and the load, unless an opto-isolator is included between the FET and the MCU. However, the FET can be switched at high frequency, while the relay cannot. The outputs include an on-board LED to indicate their status, in case the state of the outputs cannot be seen or are not connected.

Keypad

The 12-button keypad allows the user to input the required temperature and other operating parameters as required by the application program. The target temperature would typically be input as a 2-digit number. It may also be desirable to input upper and lower limits, alarm levels and so on. These should be displayed as they are entered, to ensure that the correct figures are stored. The keypad is simply a set of switches connected in a row and column arrangement (see Chapter 12) and accessed by a scanning routine. If the row inputs (A, B, C, D) are all set high initially, and no button is pressed, all the column outputs (1, 2, 3) will be high (pulled up to 5 V). If a '0' is output on each row in turn, and a button pressed, that '0' will appear at the column output, and can be read in to the PIC. The combination of active row and column identifies the key. The demonstration program (Program 15.1) includes a simple keypad scanning routine.

Display

A seven-segment display is used as it is relatively easy to drive, compared with a liquid crystal display, and is self-illuminating. The decoding process has been covered in Chapter 12, where a code to illuminate the segments to display digits 0–9 is looked up in a program data table. In this case, two digits are required, but they can both be operated from the same set of outputs by multiplexing. The digits are switched on alternately via Q1 and Q2; because they are switched too fast for the eye to perceive, they appear to be on at the same time, albeit at reduced brightness.

Other Circuit Elements

A buzzer is fitted to provide an audible alarm output. This can be used to signal system failure, temperature too low for too long and so on. Audible feedback from keystrokes is also desirable. A 4 MHz clock is used to give a convenient instruction execution time, although no timing critical operations are required. In the context of this circuit, cost saving on clock components would be insignificant. A manual reset is provided, so that the program can be restarted without powering down. This will be useful for testing as well as in normal operation. In-circuit programming and debugging are provided for via the ICD connector. The ICD module must be connected between the host PC and the application board. MPLAB IDE can then be used for testing the program in software initially, and then finally in ICD mode (see Chapter 14).

15.1.4   Hardware Development

The circuit was developed using Labcenter™ ISIS schematic capture software, which provides interactive components for circuit testing, and integrated software and hardware testing. When the circuit had been tested by interactive simulation, a stripboard implementation was devised (Fig. 15.4).

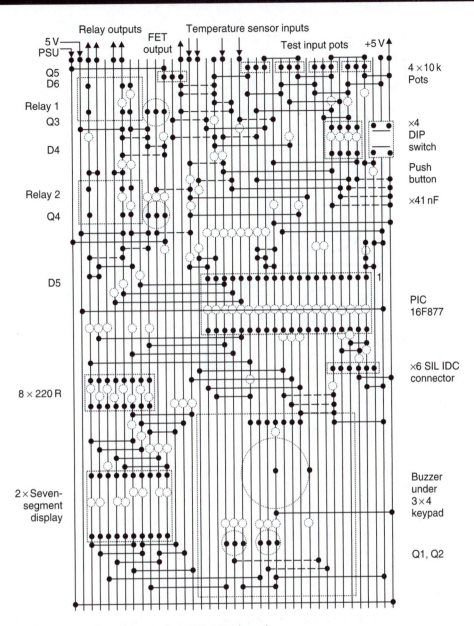

Figure 15.4 Stripboard layout for TEMPCON board.

A demo target system was then constructed, comprising two filament lamps as the heaters, operating from a high current 5 V supply, controlled by the relay output on the application board. A 5 V CPU fan was fitted as the cooling element, and the temperature sensors were arranged symmetrically inside the enclosure. The wiring of the target hardware is shown in Fig. 15.5. Note that there was a sensor output on the fan which could be used to monitor the actual fan speed, if a suitable interface were designed to convert the fan sensor pulse to TTL levels. The vent was not physically implemented at this stage.

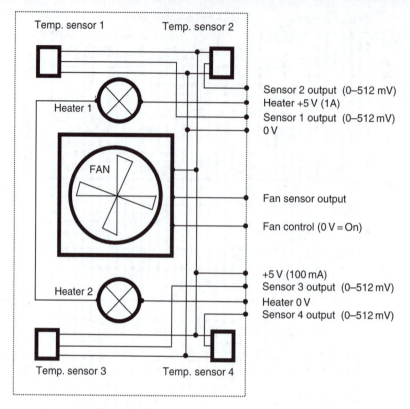

Figure 15.5 Greenhouse simulator wiring.

The photo of the prototype system (Fig. 15.6) shows the simulator at the right of the picture, with the ICD module (enclosed in ABS box), which is connected to ICD input of the TEMPCON board. 5V power supplies and a host PC would complete the system.

When final hardware testing was completed, an application board was created using Labcenter ARES™ PCB layout software, shown in Fig. 15.7. This incorporated an on-board +5 V supply for operation from a mains adapter.

15.1.5   *Temperature Controller Test Program*

Program 15.1 was written to exercise the hardware and to help get the reader started in developing applications for the TEMPCON hardware.

The program will read in the analogue inputs and display the raw data on the displays. An apparently random pattern results, which changes if the analogue inputs (test pots) are varied, indicating that the hardware input and display interfaces are working. Pressing a key on the keypad will select an analogue input for display; key '1' for input 1, and so on to '4', then repeating for keys 5–8. Key 9 will enable the buzzer test, while '*', '0' and '#' will operate the heater, fan and vent, respectively. A full header has been included with as much information as possible; details of target system, program description, register initialisation, port allocation and so on. The ports and analogue control registers have been initialised using bank selection, as recommended.

(a)

(b)

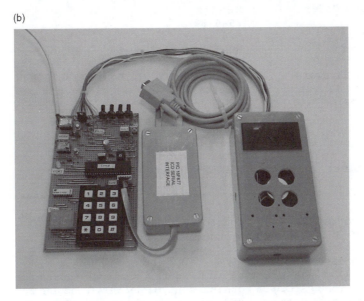

Figure 15.6 TEMPCON system. (a) Stripboard version of temperature controller board; (b) Temperature controller system with ICD module and dummy load.

Figure 15.7 Temperature controller board.

Program 15.1 Test program for TEMPCON board

```
;**********************************************************************
;       Source File:        CON0.ASM
;       Design & Code:      M Bates
;       Date:               29-6-02
;       Version:            1.0
;       Customer:           HCAT
;
;       Target Hardware:    16F877 Controller Board (TEMPCON Board)
;       Design & Layout:    M Bates
;       ISIS Design File:   CON0.DSN
;       Layout:             Stripboard prototype
;
;       Development System: MPLAB ICD Evaluation kit
;                           MPLAB IDE Ver 5.30.00
;       Assembler:          PICSTART Plus 2.40.00
;
;**********************************************************************
;
;       Test program for PIC877 Controller Board using ICD system
;       Use with ISIS CON0.DSN for simulation testing
;
;       Circuit description:
;       PIC 16F877 flash microcontroller gets 4 analogue inputs from temp
;       sensors (or test input pots) to control Heater, Vent and Fan in a
;       target system such as a greenhouse. Target temp will be set up
;       using keypad input and displayed on 2-digit multiplexed LED display.
;       Test program:
;       Checks all inputs and outputs for correct hardware operation.
;       - Press keypad buttons 1-4 to display raw data from input pots
```

```
;    - Buttons 5-8 ditto
;    - Button 9 to sound buzzer
;    - Button * to operate HEATER output
;    - Button 0 to operate VENT output
;    - Button # to operate FAN output
;
; PROGRAMMING OPTIONS ***********************************************
;
;    When programming PIC 16F877 in ICD mode, select:
;
;    XT clock mode (4MHz, 1us per instruction)
;    Power-up timer enabled
;    Watchdog timer disabled
;    Code Protection off
;
; I/O ALLOCATION ***************************************************
;
;    INPUTS.........................................................
;
;    Analogue temp sensors      AN0 - AN3 (0 - 5V)
;    Keypad column detect       RD0 - RD2 = 0
;
; OUTPUTS...........................................................
;
;    Buzzer                                          RB0 = toggle
;    Keypad row select                               RD3 - RD6 = 0
;    7-segment display      Select lo digit          RB1 = 1
;                           Select hi digit          RB2 = 1
;                           Segments                 RC1 - RC7 = 0
;    Relay interfaces Heater                         RE0 = 1
;                                                    Vent RE1 = 1
;    FET interface                                   Fan RE2 = 1
;
;    PORT DATA DIRECTION CODES REQUIRED ............................
;
;    TRISA = 11111111
;    TRISB = 11111000
;    TRISC = 00000000
;    TRISD = 00000111
;    TRISE = 00000000
;
;    RB3, RB6, RB7 reserved for ICD operation
;
; ADC SETUP *******************************************************
;
;    ADCON0   Bits 76          01 = A/Dclock = f/8)
;             Bits 543         Channel Select (AN0 - AN7)
;             Bit 2            Go = 1 / Done = 0
;             Bit 0            A/D module enable = 1
;
;    ADCON0 = 01xxx001 depending on channel required
;
;    ADCON1   Bit 7            0 = left justify result in ADRESH/ADRESL
;             Bits 3210        0010 = RA0-RA5 analogue, RE0-RE2 digital
;
;    ADCON1 = 00000010                                   continued ...
```

```
; ASSEMBLER DIRECTIVES *******************************************
;
;       Create list file and select processor:
        list p = 16f877
;
;       Include file containing register labels:
        include "p16f877.inc"
;
count   EQU     020                 ; assign GPR1 for counter
;
;       Set origin at address 000:
        org     0x000
;
; START PROGRAM **************************************************

        nop                         ; No op. required at 000 for ICD mode

; Initialise control registers ...................................

        banksel TRISA               ; Select DDR resgister bank 1
        movlw   b'11111000'         ; Setup buzzer and display digit
                                    ; select..
        movwf   TRISB               ; ..as outputs
        clrf    TRISC               ; Setup 7-segment driver port as outputs
        movlw   b'00000111'         ; Setup keyboard port for..
        movwf   TRISD               ; .. row outputs and column inputs
        clrf    TRISE               ; Setup relay port as outputs

; Setup ADC ......................................................

        banksel ADCON1              ; Select register bank 1
        movlw   b'00000010'         ; Set A/D mode left justify,4 channels
        movwf   ADCON1              ; and write A/D control word

        banksel ADCON0              ; Select register bank 0
        movlw   b'01000001'         ; Set A/D frequency Fosc/8, select AN0
        movwf   ADCON0              ; and write A/D control word

; Initialise outputs .............................................

        banksel PORTA               ; select port data register bank 0
        clrf    PORTE               ; switch off all outputs
        goto    start               ; jump over subroutines to main loop

; Subroutine to wait about 0.8 ms ................................

del8    clrf    count               ; Load time delay of 256 x 3=768 us
again   decfsz  count               ; Decrement and test counter
        goto    again               ; until zero
        return                      ;

; Subroutine to get analogue input ...............................
; Wait 20us ADC aquisition settling time ..

getAD   movlw   007                 ; Load time delay of 7 x 3=21 us
        movwf   count               ; Load counter
```

```
down      decfsz    count          ; Decrement and test counter
          goto      down           ; until zero (3us per loop)


; Get analogue input ..

          bsf       ADCON0, GO     ; Start A/D conversion
wait      btfsc     ADCON0,GO      ; Wait for conversion to complete
          goto      wait           ; by testing GO/DONE bit
          return                   ; from subroutine with result in
                                   ; ADRESH


; Subroutines to process keys ...............................

proc1     movlw     b'01000001'    ; Select analogue channel 1
          movwf     ADCON0         ; and
          call      getAD          ; and get analogue input
          return                   ; for next key

proc2     movlw     b'01001001'    ; Select analogue channel 2
          movwf     ADCON0         ; and
          call      getAD          ; and get analogue input
          return                   ; for next key

proc3     movlw     b'01010001'    ; Select analogue channel 3
          movwf     ADCON0         ; and
          call      getAD          ; and get analogue input
          return                   ; for next key

proc4     movlw     b'01011001'    ; Select analogue channel 4
          movwf     ADCON0         ; and
          call      getAD          ; and get analogue input
          return                   ; for next key

proc5     movlw     b'01000001'    ; Select analogue channel 1
          movwf     ADCON0         ; and
          call      getAD          ; and get analogue input
          return                   ; for next key

proc6     movlw     b'01001001'    ; Select analogue channel 2
          movwf     ADCON0         ; and
          call      getAD          ; and get analogue input
          return                   ; for next key

proc7     movlw     b'01010001'    ; Select analogue channel 3
          movwf     ADCON0         ; and
          call      getAD          ; and get analogue input
          return                   ; for next key

proc8     movlw     b'01011001'    ; Select analogue channel 4
          movwf     ADCON0         ; and
          call      getAD          ; and get analogue input
          return                   ; for next key

proc9     bsf       PORTB,0        ; Toggle buzzer on
          call      del8           ; delay about 0.8ms
          bcf       PORTB,0        ; Toggle buzzer off
          call      del8           ; delay about 0.8ms
          return                   ; for next key
```

continued ...

```
procs      bsf PORTE,0        ; switch on heater output
           return             ;

proc0      bsf PORTE,1        ; switch on vent output
           return             ;

proch      bsf PORTE,2        ; switch on fan output
           return             ;
```

; Routine to scan keyboard ....

```
scan       movlw   0FF        ; Deselect...
           movwf   PORTD      ; ...all rows on keypad
```

; scan row A of keypad ....

```
           bcf     PORTD,3    ; select row A of keypad

           btfsc   PORTD,0    ; test key 1
           goto    key2       ; next if not pressed
           call    proc1      ; process key 1

key2       btfsc   PORTD,1    ; test key 2
           goto    key3       ; next if not pressed
           call    proc2      ; process key 2

key3       btfsc   PORTD,2    ; test key 2
           goto    key4       ; next if not pressed
           call    proc3      ; process key 3
```

; scan row B of keypad ....

```
key4       bsf     PORTD,3    ; deselect row A
           bcf     PORTD,4    ; select row B

           btfsc   PORTD,0    ; test key 4
           goto    key5       ; next if not pressed
           call    proc4      ; process key 4

key5       btfsc   PORTD,1    ; test key 5
           goto    key6       ; next if not pressed
           call    proc5      ; process key 5

key6       btfsc   PORTD,2    ; test key 6
           goto    key7       ; next if not pressed
           call    proc6      ; process key 6
```

; scan row C of keypad ....

```
key7       bsf     PORTD,4    ; deselect row B
           bcf     PORTD,5    ; select row C

           btfsc   PORTD,0    ; test key 4
           goto    key8       ; next if not pressed
           call    proc7      ; process key 4
```

```
key8    btfsc   PORTD,1     ; test key 5
        goto    key9        ; next if not pressed
        call    proc8       ; process key 2

key9    btfsc   PORTD,2     ; test key 2
        goto    keys        ; next if not pressed
        call    proc9       ; process key 3

; scan row D of keypad  ......

keys    bsf     PORTD,5     ; deselect row C
        bcf     PORTD,6     ; select row D

        btfsc   PORTD,0     ; test key *
        goto    key0        ; next if not pressed
        call    procs       ; process key *

key0    btfsc   PORTD,1     ; test key 0
        goto    keyh        ; next if not pressed
        call    proc0       ; process key 0

keyh    btfsc   PORTD,2     ; test key #
        goto    done        ; next if not pressed
        call    proch       ; process key #

; all done .................

done    return              ; to main loop

; Main program  ***************************************************

start   bcf     PORTB,2     ; switch off high digit of display
        bsf     PORTB,1     ; and switch on low digit
        call    scan        ; and read keypad
        movf    ADRESH,W    ; move ADC result
        movwf   PORTC       ; to display

        bcf     PORTB,1     ; switch off low digit of display
        bsf     PORTB,2     ; and switch on high digit
        call    scan        ; and read keypad

        movf    ADRESH,W    ; move ADC result
        movwf   PORTC       ; to display

        goto    start       ; repeat main loop

        END                 ; of source code .................
```

The routine to read in an analogue input is based on the model routine provided in the data sheet, where a delay of about $20\,\mu s$ is included to ensure that the input has had time to settle, in case the input is changing rapidly. The conversion is then started by setting the GO bit in the ADC control register, and then waiting for it to be cleared by the ADC to indicate that the conversion is complete. In this program, only 8 of the 10 bits of the ADC result are used, so the result is 'left justified' to place the most significant 8 bits in the ADRESH register for output to the display. This allows the full range of the input (0–2.048 V) to be checked.

In a working program, the analogue input value would be converted into a 2-digit decimal value for the display. Using the conversion scaling calculated above, a temperature of 50 °C would give a result of 250 (in binary) in ADRESL, with the result right justified. Only a quarter of the full ADC range is then being used. This result can then be converted into the corresponding display digits '5' and '0', and so on down to zero. The keyboard-scanning routine uses a simple method to check if each key in each row has been pressed, calling the required action if it has. A more elegant and compact keyboard-scanning method is possible when reading in numerical values.

A full working program would allow the user to enter the maximum and minimum values for the target temperature, and then go into run mode, where the temperature would be controlled within the set range by operation of the heater, vent and fan. Pseudocode for the application software is shown in Program 15.2.

Program 15.2   Pseudocode for TEMPCON Control software

```
TEMPCON
        Initialise
                    Ports
                                Port A = Temp sensor inputs (4)
                                Port B = Display digit select (2), ICP/D (3)
                                Port C = Display segments (7)
                                Port D = Keypad (4 outputs, 3 inputs)
                                Port E = Heater, Vent, Fan outputs

                    ADC     Left justify, 4 channels
                            ADC frequency Fosc/8, select input AN0

        GetMaxMin
                    Scan keyboard
                    Store & display first digit of maxtemp
                    Scan keyboard
                    Store & display second digit of maxtemp
                    Convert to byte MaxTemp (0-200)

                    Scan keyboard
                    Store & display first digit of mintemp
                    Scan keyboard
                    Store & display second digit of mintemp
                    Convert to byte MinTemp

        Cycle
                    Read tempsensor1
                    Read tempsensor2
                    Read tempsensor3
                    Read tempsensor4

                    IF sensor out of range
                            replace with previous value
                    Calculate AverageTemp

                    Display AverageTemp
                            MSD = AverageTemp/10
                            Get 7-seg code & display MSD
```

```
                                         LSD = Remainder
                                         GEt 7-seg code & display LSD
                           IF AverageTemp > Mintemp
                                         switch heater OFF
                                         ELSE switch heater ON
                           IF AverageTemp > Maxtemp
                                         switch vent ON
                                         ELSE switch vent OFF
                           IF AverageTemp > Maxtemp + 4
                                         switch fan ON
                                         ELSE switch fan OFF

           GOTO Cycle
```

15.1.6 Application Modifications

There are some features of the microcontroller which are not yet utilised in this application, which could enhance it. As mentioned above, the PWM module could be used to control the speed of the fan. In addition, a serial communication port could send the temperature data to a master controller, and receive new operating parameters. If a PC were acting as the host, the USART could be used, and the PC could display the operating data, perhaps as in graphical form, or as a plot of temperature variation over time. This data could then be saved on disk, and sent via a network to a supervisory system.

When designing the application initially, a top of the range device such as the 16F877 is a good choice, as it has most of the available features. When the design has been finalised, it may turn out that some features are not required, some I/O is unused or the program could be fitted into a smaller memory. The designer can then review the alternative, cheaper or otherwise more suitable devices, and transfer the application to that device as long as the hardware redesign required is not excessive. The software reconfiguration should also not be too much of a problem within the same PIC group, since the chips are designed to be interchangeable. This idea is illustrated below, where the temperature controller is redesigned for other PIC chips.

15.2   16F818 Application

The PIC 16F818 is a replacement part for the 16F84. It has a compatible pin-out, and additional features at a lower cost. Sixteen I/O pins are available, including five analogue inputs. It has 1k words of program memory; if extra memory is needed, the 16F819 has the same features but 2k program memory.

As can be seen from the pin-out (Fig. 15.8), each pin has multiple functions, other than the two supply pins. Analogue inputs can be selected on RA0–RA4, or external reference voltages. There is a CCP module and a synchronous serial port offering SPI or I^2C modes. Other special features are a variety of power saving modes in addition to the usual 'sleep', an internal oscillator which obviates the need for external clock components, and in-circuit programming and dedugging.

Thus, many of the features of the more powerful 16F87X group are now available in the smaller 18-pin package. It is recommended that, when the user is familiar with all the options available on this chip, it can be used as a default choice when developing new PIC applications, if the number of I/O is sufficient.

RA2/AN2/Vref–	1	18	RA!/AN1
RA3/AN3/Vref+	2	17	RA0/AN0
RA4/AN4/T0CKI	3	16	RA7/OSC/CLKI
RA5/!MCLR/Vpp	4	15	RA6/OSC2/CLKO
Vss	5	14	Vdd
RB0/INT	6	13	RB7/T1OSI/PGD
RB1/SDO/CCP1	7	12	RB6/T1OSOT1CKI/PGC
RB2/SDO/CCP1	8	11	RB5/!SS
RB3/CCP1/PGM	9	10	RB4/SCK/SCL

Figure 15.8　The PIC 16F818 pin-out.

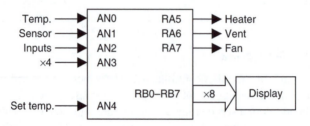

Figure 15.9　PIC 16F818 temperature control block diagram.

This chip could be used in the temperature controller if the keyboard were eliminated, and the set temperature input from a pot via one of the analogue inputs (see Fig. 15.9). A fixed control range might be necessary, as there would be no facility for entering maximum and minimum temperatures. The display digit selection can be reconfigured to use only one output. The application then only needs 16 I/O pins. Operational parameters could be transferred via the serial interface if the display were left out (RB1, RB2 and RB4).

15.3　12F675 Application

The 12-series of PIC mini-chips offers flash memory in 8-pin packages. At the current time there is limited choice, but no doubt the range will be expanded. The pin-out for the 12F675 illustrates the I/O features available (see Fig. 15.10).

The chip can be configured with six plain digital I/O pins, but also offers two timers, an analogue comparator or four analogue input channels. The 12F629 is the same, except that it does not include the ADC and is therefore a little cheaper. An internal oscillator and in-circuit programming are also featured.

A temperature controller could be implemented using this chip if only two analogue inputs are used (see Fig. 15.11). It could operate with a fixed set temperature, or another analogue input could be used as a set temperature input. With no display, a dial on the set temperature pot may be necessary.

Vdd	1	8	Vss
GP5/T1CKI/OSC1/CLKIN	2	7	GP0/AN0/Cin+
GP4/AN3/!T1G/OSC2/CLKOUT	3	6	GP1/AN1/Cin–/Vref
GP3/!MCLR/Vpp	4	5	GP2/AN2/T0CKI/INT/Cout

Figure 15.10　The PIC 12F675 pin-out.

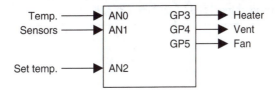

Figure 15.11 PIC 12F675 temperature controller.

15.4   18F452 Application

The 18-series of PIC microcontrollers is the most powerful in the flash family, ranging from the 16 I/O, 4k memory 18F1220 to the 68 I/O, 64k memory 18F8720 at the current time. The group offers different combinations of the advanced features, and the larger memory size means that 'C' can be used for application programming. The instruction set of 75 16-bit instructions is designed with this in mind.

A small selection of the available 18FXXXX devices are listed in Table 14.2. At the time of writing there are a total of 33 in production, with several others listed as future products. The architecture is somewhat more complex than the 14-bit devices, with extra blocks for multiplication, a hardware data table access, additional file select registers and other advanced features. The data bus is still 8-bit. Taking the 18F452 as an example, in terms of peripheral features it is comparable to the 16F877 described in Section 15.1, so a comparison of the two devices will be made to illustrate the differences and similarities of the two groups (Table 15.4).

As can be seen, the 18-series device has some advantages: 40MHz clock rate, 16k program memory and more data memory. However, bear in mind that a program written in 'C' will not be as code-efficient as an assembly language equivalent, so these advantages may or may not translate into extra performance, depending on the application and the way that it is structured. The main advantage is that more complex operations such as mathematical functions are easier to program in 'C'. The 18-series PIC has a richer instruction set, including instructions such as multiply, compare and skip, table read, conditional branch and move directly between registers, so still has advantages even when programmed in assembly language.

15.4.1   PIC 'C' Programming

For those readers unfamiliar with 'C' programming, a simple example is shown in Program 15.3. The program will give the same output as BIN1.ASM assembly language program. The program must be converted to PIC 16-bit machine code using a compiler such as MPLAB C18 Compiler, which is supplied as an add-on to the development system. This compiler recognises ANSI 'C', the standard syntax for microcontrollers (ANSI = American National Standards Institute). The 'C' compiler must be selected in the development mode dialogue when building the application.

The main elements of the program functioning are as follows.

/* comment */
Comments in 'C' source code are enclosed between /* and */; and can be run over several lines.

#include<p18F456.h>
This is a compiler directive which calls up a header file named 'p18F458.h'. This contains pre-defined register labels for that particular processor, such as TRISB and PORTB, and the corresponding addresses 06h and 86h.

Table 15.4 Comparison of the 16F877 and the 18F458

Feature	16F877	18F458
Total pins	40	40
Input/output pins	33	33
Ports	A, B, C, D, E,	A, B, C, D, E
Clock	20 MHz	40 MHz
Instruction bits	14	16
Program memory (instructions)	8k	16k
Instruction set size	35	75
Data memory (bytes)	368	1536
EEPROM (bytes)	256	256
Interrupt sources	14	21
Timers	3	4
Capture, compare, PWM modules	2	2
Serial communications	MSSP, USART	MSSP, USART, CAN
Parallel port	Yes	Yes
Analogue inputs	8 x 10 bits	8 x 10 bits
Resets	POR, BOR	POR, BOR, Stack, Programmed
In-circuit serial programming and debugging	Yes	Yes

int counter;

This assigns a label to a register and declares that it will store an integer, or whole number. A standard integer in 'C' is stored as a 16-bit number, requiring two data RAM (GPR) locations.

void main(void)

This rather peculiar syntax simply indicates, as far as we are concerned here, the start of the main program sequence. The following brace (curly bracket) encloses the main program with a matching brace at the end. These are lined up in the same column and the main program tabbed in between them, so that they can be matched up correctly.

counter = 1;

A value of 1 is initially placed in the variable location (low byte).

TRISB = 0;

A value 0 is loaded into the data direction register of Port B to initialise the port bits for output to the LEDs.

while(1)

This starts a loop which will run endlessly. A condition is placed in the brackets which controls the loop. For example, the statement could read 'while(count<256)', in which case the following group of statements within the curly brackets (braces) would execute 255 times, counting up

Program 15.3 A simple PIC 'C' program

```
/*   BIN1.C          M Bates         Version 1.0

     Program to output a binary count to Port B LEDs

***************************************************************/

#include <p18f458.h>              /* Include port labels for this chip */
#include <delays.h>

int counter                       /* Label a 16-bit variable location  */

void main(void)                   /* Start main program sequence       */
{
    counter = 0;                  /* Initialise variable value         */
    TRISB = 0;                    /* Configure Port B for output       */
    while (1)                     /* Start an endless loop             */
    {
            PORTB = counter;      /* Output value of the variable      */
                counter++;        /* Increment the variable value      */
            Delay10KTCY(100);     /* Wait for 100 x 10,000 cycles      */
    }
}                                 /* End of program                    */
```

to the maximum binary value and stopping. The value 1 means the condition is 'always true', so the loop is endless, until reset.

PORTB = counter;
The value in counter is copied to Port B data register for display on the LEDs

counter++;
The variable value is incremented each time the loop is executed. This causes the output to be incremented the next time.

Delay10KTCY(100);
This calls a pre-defined block of code which provides a delay, so that the LED output changes are visible. At a maximum clock rate, the processor instruction cycle time is 0.1 μs, so the delay works out to 0.1 s (10000 × 100 cycles). The overall count cycle will then take 25.6 s. This function will require initialisation and loading of hardware timers and associated operations which will clearly be quite complex in machine code.

The delay function is an example of a function call, which is one of the biggest advantages of 'C' – the collection of standard routines, which are automatically available, means that the programmer does not have to keep 're-inventing the wheel', or even invent it for the first time – it is ready-made.

The layout of the program, with tabs, is important for understanding the program and checking the syntax if there are logical errors. However, the layout does not affect the program function, only the sequence of characters. However, the statements must be all on one line; line returns are not allowed within a statement.

Each complete statement is terminated with a semicolon; note that some are not complete in themselves and do not have a semicolon. For example, 'while(1)' is not complete without the loop statements, or at least the pair of braces. The close brace terminates the 'while' statement. The whole of the main loop, and any functional sub-block, must be enclosed between braces.

15.4.2 Advantages of 'C' Programming

The 'C' compiler converts the program into PIC 16-bit machine code. Most of these 'C' statements translate into more than one machine code instruction. This can be confirmed by studying the list file which is produced by disassembling the machine code.

The pseudocode for the temperature controller above can probably be more easily translated into 'C' than assembly language. For example, the conditional control operations defined using IF...THEN statements will translate directly, whereas, in assembler, it has to be implemented by suitable combinations of 'Bit Test and Skip' with 'Goto' or 'Call'. In addition, the comparison of the 'average temperature' with the set values can be done in one statement in 'C', but needs a subtract or compare prior to a bit test, which is much more complicated. On the other hand, checking bit inputs is not so easy in 'C' as in assembler, as ANSI C contains no individual bit operations. Bit status in a register has to be checked by using a logical or numerical range check.

There are many references on 'C' programming. To program a microcontroller in 'C', only the basic set of statements and simple data structures will probably be needed, so if the reader has some knowledge of 'C' already, using it to develop PIC applications should not be too difficult. However, a full treatment will not be attempted here. We can now see the advantages of using 'C' with the 18-series PIC. The chips themselves have a good range of peripheral interfaces and other features, and can be programmed more easily using the high level language.

Summary

- The PIC 16F877 has a good range of peripheral interfaces, including analogue inputs, serial ports, CCP and PWM, and in-circuit debugging.

- The application designed around the PIC 16F877 operates as a temperature controller, with four sensors, three outputs, a keypad and 2-digit display.

- The temperature controller can be programmed to maintain the temperature in a heating/cooling system within a set range, display these parameters and operate alarms.

- A similar application can be implemented using the 16F818 without the keypad.

- A similar application can be implemented using the 12F675 without the display.

- The 18F458 has comparable I/O features to the 16F877, but can be programmed in 'C' and runs at twice the speed.

Questions

1. (a) What interfacing modifications are recommended for the LM35 temperature sensor if the connections are over 1 m long?

   (b) Calculate the output of the LM35 sensor at 25 °C, and the decimal value that would be found in the ADRESL on completion of an A/D conversion of this input, if the result is right justified.

2. State one advantage of (a) the relay output and (b) the FET output as configured in the temperature controller.

3. Describe briefly how a multiplexed seven-segment LED display works. Study the test program and explain how it ensures that each digit has the same brightness.

4. State two reasons why the PIC 16F818 should be preferred over the 16F84 for most applications.

5. Compare PIC assembler and ANSI 'C' programming, outlining the advantages of each.

Answer

1. (b) 125

Activities

1. Devise a more code-efficient keypad-scanning routine than that in Program 15.1 using the rotate instruction, such that the binary value for the keys 0–9 are stored in a suitable register. Assign keys '*' and '#' the hex values A and B, respectively.

2. Design and implement the fully functional program for the temperature controller based on the pseudocode provided in Program 15.2. The user will enter upper and lower temperature limits, set the controller to run mode where the outputs are operated to maintain the temperature between those limits. The system should tolerate a fault in one sensor, which puts its output outside the normal operating range. Devise a full design and performance specification for the controller.

3. Design a temperature-controlled enclosure with heaters, fan and vent which will allow a fully functioning temperature control program to be tested. Investigate the design of an interface for the fan sensor, so that the fan speed could be controlled by PWM with feedback. Investigate the set up required to use the PWM output of the 16F877, and redesign the hardware as necessary.

4. Implement the redesign of the temperature controller for the 16F818 or the 12F675.

5. Study relevant 'C' programming references, the Microchip manual *MPLAB C18 C Compiler, Getting Started* and modify the program BIN1.C such that the output can be stopped, started and reset by push button inputs at RA0 and RA1. Why is reading inputs more difficult in 'C'?

Chapter 16
More Control Systems

In this chapter, we will look at the range of other technologies available with which to build controller systems. There are numerous families of other microcontrollers which compete with the PIC range. Conventional microprocessor systems may be a better solution for larger microsystems, and the programmable logic controller provides a self-contained device which requires little additional interfacing. The PC itself can also be adapted as a controller host, with suitable additional hardware. The object is to help the reader to select the most appropriate solution for any given control problem.

16.1   Other Microcontrollers

There are several other families of microcontroller which are well established in the embedded systems market. This book has concentrated on the PIC because it is, arguably, the most suitable for learning about the principles of microcontroller application design and implementation. The skills and knowledge gained with the PIC can then be transferred to other types of microcontroller. Many of the alternative types of MCU are designed principally for complex embedded applications such as motor vehicle engine management, mass-produced consumer goods and telecommunications. Digital signal processors, which are devices which specialise in high-speed processing of analogue signals, are also becoming widely used in audio and communications systems. The main manufacturers at the current time include National Semiconductor, Texas Instruments, Analogue Devices and Hitachi. The product range from other selected manufacturers is outlined below.

16.1.1   Intel 8051

First introduced in 1980, the 8051 is the most well-established and widely used microcontroller. As can be seen in the block diagram (Fig. 16.1), the original design was a mid-range device,

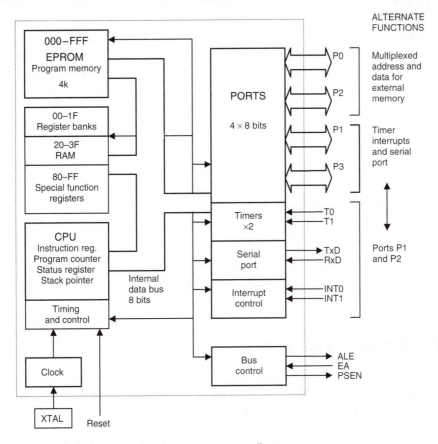

ALTERNATE
FUNCTIONS

Figure 16.1   Block diagram of Intel 8051 microcontroller.

with multiple parallel ports, timers and interrupts and a serial port. The 8051 can be used as conventional processor, as well as a microcontroller. It can access external memory using Port 0 and Port 2, which act as multiplexed data and address lines. Some of Port 1 and Port 3 pins also have a dual purpose, providing connections to the timers, serial port and interrupts. The 8031 was a version of the 8051 without internal ROM, the application program being stored in a separate EPROM. Many other microcontroller manufacturers now supply variants of the 8051.

16.1.2   *Motorola MC68HC11*

This series is based on the architecture and instruction set of the standard Motorola 68000 microprocessor, which is discussed below. The MC68HC11A1P is a typical member of the family, offering 8k program ROM, 512 bytes of data, 256 bytes of static RAM, 16-bit multifunction timer, synchronous and asynchronous serial communications, eight A/D channels and 38 I/O pins. It can also operate in microcontroller or microprocessor mode with external memory accesses via multiplexed address and data busses at the multipurpose port pins. The Motorola range is well established in motor vehicle and similar applications.

16.1.3 Zilog Z8

The Z80 microprocessor was for many years the standard 8-bit microprocessor for industrial applications, and the Z8 series is based on its architecture and instruction set. The Z8 Encore! flash MCU family currently ranges from a device with 4k program memory and 11 I/O (Z8F0441) to one with 64k memory and 60 I/O (Z8F6423) running at 20 MHz, most with 10-bit analogue inputs and the usual selection of peripheral interfaces.

16.1.4 Atmel AVR

Atmel also offers 8051-compatible devices as well as a further range of chips including the AT90 series of mid-range microcontrollers, the miniature/mid range 'tiny' series and the power 'mega' group. The I/O pins available in flash memory devices range from 3 to 54, with clock speeds from 4 to 16 MHz. A typical example is the ATtiny2313: an 8-bit MCU, which has 2k flash program memory, 120 instructions, 32 working registers, 18 I/O pins, on-chip debugging, serial ports, and two timer/counter/compare modules in a 20-pin package. The larger instruction and register set may give the Atmel chip the advantage in some applications, and the AVR series is currently a popular choice where this is the case.

16.2 Microprocessor System

The conventional microprocessor system was the forerunner of the microcontroller. That is, the elements of the microcontroller were originally developed as separate devices before being integrated into one chip to produce the microcontroller. The PC, as outlined in Chapter 1, is an example of a conventional system, where the individual CPU, memory and I/O devices are linked together by system address, data and control busses.

The advantage of the conventional microprocessor system is that it can be designed to suit the application; it will include only those peripherals that are actually needed, and memory chips as required. Obviously, the system is more complex to design and build, and so this type of system tends to be used for larger applications, where for example, large amounts of data storage is required.

16.2.1 M68000 Hardware

The current PC architecture is quite complicated, with a hierarchy of busses operating at different speeds. We will therefore look at a simpler example based on another commonly used processor type, the Motorola 68000 (M68k), which has been widely used in both home computers and industrial applications for many years. It is the CISC processor that has most often been used in education and training because of its relatively regular architecture.

The block diagram of a typical development and training system is shown in Fig. 16.2. The target board has a 68000 with CPU, EPROM, RAM and port chips on board which would typically measure about 150×200 mm (Fig. 16.3). It can be connected to an applications board which has a range of peripheral transducers, such as switches, LEDs and PWM-controlled motor and shaft opto-sensor. This is controlled by the 68000 CPU via a standard 68230 Parallel Interface/Timer (PI/T) which has three 8-bit ports, of which Port A provides data transfer and Port B the individual control and data lines.

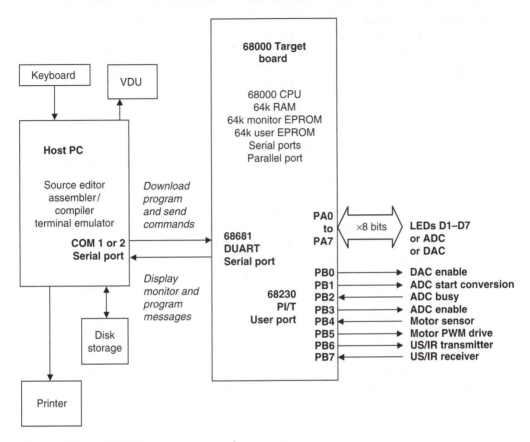

PA0
to
PA7

×8 bits

LEDs D1–D7
or ADC
or DAC

PB0 → DAC enable
PB1 → ADC start conversion
PB2 ← ADC busy
PB3 → ADC enable
PB4 ← Motor sensor
PB5 → Motor PWM drive
PB6 → US/IR transmitter
PB7 ← US/IR receiver

Figure 16.2 M68000 microprocessor demonstration system.

Figure 16.3 View of 68000 board.

The program for the 68000 is prepared on a host PC, in a similar way to the PIC programs. Assembly language source code is written using a text editor and converted to machine code by an assembler program. Alternatively, the source code can be written in the high level language, typically standard 'C'. A compiler then converts the source code initially into assembler code, which is then assembled. The machine code program created is downloaded via the PC serial port to the serial port of the target board and hence into its RAM block.

A 'terminal emulator' utility is used for downloading, which also allows the target board to use the PC screen and keyboard as a user interface for its monitor program, which functions as a minimal operating system. Simple monitor commands are used to run and debug the program. In single step or trace mode, the 68000 can display its register contents on the PC screen. The PC provides the keyboard, screen, disk storage and printer during program development. Once an application is up and running on the 68000 board, a user interface may no longer be required, or if it is, a simple keypad and display may be sufficient. At this stage, the program can be blown into EPROM to run independently and the PC disconnected.

The block diagram of the M68000 target board is shown in Fig. 16.4, so that it can be compared with the PIC 16F84 internal architecture. Note that in the PIC, the internal architecture of the processor is clearly illustrated in the manufacturer's block diagram, whereas in the 68000 system, it is concealed within the CPU. Therefore, to see all the details of the 68000 system, both the internal CPU architecture and the board circuit diagram must be studied.

16.2.2   M68000 program

A simple program for the 68000 system is shown in Program 16.1, so that the syntax for a CISC processor can be compared with the PIC assembly language. The program has a similar function to the PIC program BIN2, outputting a binary count to LEDs with a delay.

Comments

The comments are delimited with a star.

use tim.ini
This is equivalent to the include directive in the PIC; it incorporates a file 'tim.ini' which contains standard register labels, Port A and DircA. Port A is the 8-bit port data register and DircA the data direction register (DDR).

move.b #$ff,DircA
Move the literal FF into the DDR to set all bits as output. The '.b' means this is a byte operation (16- and 32-bit words can be moved in the 68000). '#' means this is a literal (immediate data in 68000 speak). '$' indicates a hex number. Note that in the 68000, a '1' in the DDR set that bit as output – this is the opposite to the PIC.

again move.b d0, PortA
The word 'again' is an address (line) label, 'd0' is the first data register in a set of eight (d0–d7) and PortA is the output register to which the LEDs are connected.

addq #1,d0
This means add 1 to (increment) d0. Surprisingly, the 68000 does not have an increment (or decrement) instruction. 'addq' means 'add quick' used for adding a small number to a register.

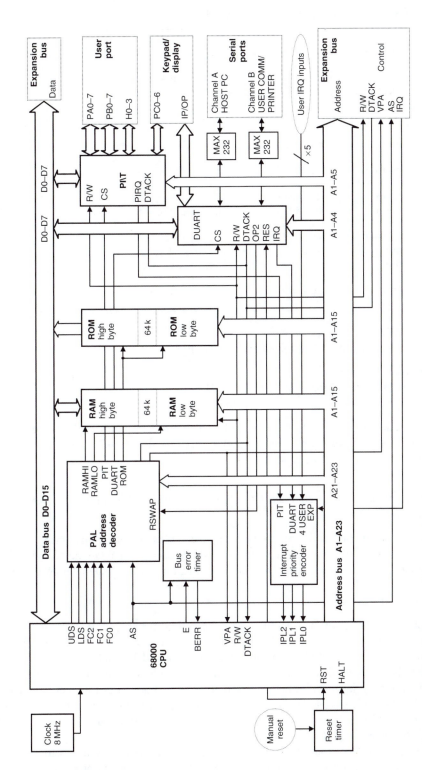

Figure 16.4 Block diagram of M68000 board.

Program 16.1   Simple program for 68000 board

```
*        OUT3.ASM          MPB     27/8/97
*
*        A demo program using general purpose system
*        initialisation file TIM.INI
*
* ------------------------------------------------------------

         use        tim.ini        Initialise system

         move.b     #$ff,DircA     Port A data direction code

again    move.b     d0,PortA       Output data to LEDS
         addq       #1,d0          Increment output value

         move.w     #$0fff,d1      Initialise delay count
delay    subq.w     #1,d1          Decrement and
         bne        delay          Loop until zero

         bra        again          Repeat forever..........
```

move.w #$0fff,d1
Move a 16-bit word (w) into d1 to initialise the delay loop.

delay subq.w #1,d1
Decrement (subtract 1 from) the counter register 'd1'.

bne delay
This means 'branch if not zero' to the label delay. The program jumps back and repeats the decrement until the result of the previous operation (decrement) was zero. This is available in the 14-bit PIC only as a pseudo-operation.

bra again
This is an unconditional jump equivalent to the GOTO label in PIC programs, to make the program repeat endlessly.

It can be seen that the 68000 syntax is a more complex because, first, there are more instructions and, second, there are more ways to use them, because there are more registers and addressing modes. This is a good reason to choose the PIC for learning assembler programming.

16.2.3 Program Execution

The M68000 system (Fig. 16.4) is controlled by the CPU driven at 8 MHz. The RAM holds the user data and program, and the EPROM stores the monitor and communications program which allows the program in RAM to be downloaded, run and debugged.

The system has a 16-bit data bus, and 24-bit addressing giving a maximum address space of 16 Mb (megabytes). In this minimal system, only a small fraction of this memory space is used; the EPROM and RAM blocks are all 64k bytes, installed as pairs of identical chips which store the upper and lower byte of the 16-bit data word. The 68000 instructions are of variable length, 2, 4 or 6 bytes long, stored in adjacent locations.

The lowest memory locations 0000–0FFF are used by the system control software in the 68000, so the user's machine code program is typically stored in RAM at a set of locations from address $1000 ($=hex). Assuming it has already been downloaded to RAM using the monitor commands in EPROM, the program is started by issuing a command from the terminal (PC), specifying the start address; for example, 'G $1000' to start at the default user program origin.

The program is executed by the CPU fetching each 16-bit code in turn from memory into the CPU via the data bus. The program codes are found in memory by the CPU sending out the addresses in sequence, starting with $1000, from its program counter register. The address is 'decoded' by the external logic which is contained partly in a PAL (programmable array logic) chip and partly in the memory chips themselves. In this way, each individual memory location (1 byte) can be accessed for reading and writing.

The address decoder chip generates the system chip select (CS) signals from the address lines A21, A22 and A23, which 'enable' the memory or I/O device which is to be accessed. The lower order address lines are fed directly to the memory chips to select an individual location within the memory array. The read/write (R/W) controls the data direction for the data transfer between the CPU and the other chips.

Note that in the conventional architecture, the program instructions and data transfers between memory, ports and CPU use the same data bus lines. This slows down the system; this is why, in the PIC design, the instruction and data busses are separated (Harvard architecture), allowing faster operation at the same clock speed.

16.2.4 Ports

The PI/T chip connects the 68000 board to the application board. In a 'real' application, this board would be replaced by the hardware required by that particular application, and a keypad and display connected, if required. The DUART serial port allows the program to be downloaded to RAM from the host PC. The MAX 232 chips are line drivers that boost the signal power on the serial connection between the host and the target board.

The ports can request service from the processor by using the interrupt signal (PIRQ, IRQ) so that more important data transfers can be completed quickly. The interrupt priority encoder converts the active interrupt line number into a 3-bit code which identifies the device requesting service to the CPU, which then runs a corresponding interrupt service routine. Here, the conventional CPU has the advantage – it provides an interrupt priority system which allows more important operations to take precedence over less urgent ones.

16.2.5 Bus Control

The 68000 has what is called an 'asynchronous bus'. This means that the memory or I/O data transfer is not completed unless the CPU receives a Data Acknowledge (DTACK) signal from the peripheral chip. The 68000 I/O chips (PI/T and DUART) are designed to provide DTACK, but memory chips are not processor specific in their design, so the DTACK for the memory access cycles must be generated by the PAL decoder. If not received within a preset time, the bus error timer generates an error signal to the CPU and the bus cycle is aborted, and an error reported.

The busses and most of the system signals are available at an expansion connector so that additional devices can be attached or data passed to another processor. The address decoder generates a VPA signal which allows external devices to be added to the decoding system.

16.2.6 M68k Application

Figure 16.5 shows a block diagram of the M68000 board used as a supervisory controller, making use of its multiple I/O facilities and the keyboard and display option. The motor control board is designed to operate six motors under closed loop speed control. Each motor has its own microcontroller, while the 68000 acts as a master controller. It could be used as a robot drive system, for example.

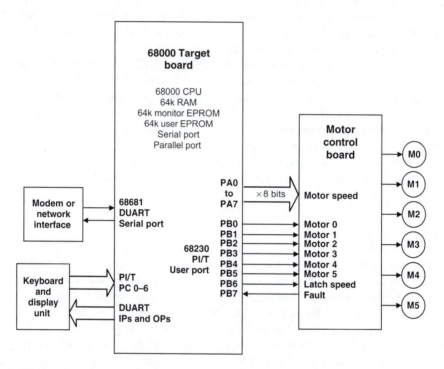

Figure 16.5 Block diagram of M68K motor control system.

In a robot controller, the command to move to a particular position, given as a set of three-dimensional co-ordinates, could be received via the serial port. The 68000 would work out the move required for each motor (quite complicated!) and then select each motor in turn and download data to move the motor through a calculated angle. The individual controllers could ensure that each joint speeds up and slows down gradually to achieve smooth motion. Motor operating sequences could be stored in the RAM block, while the communication and control firmware would be resident in ROM.

16.3 Control Technologies

Microprocessors and microcontrollers fit into a range of technologies which may be used to implement control systems, which includes:

- electromechanical relays
- programmable logic controllers

- microcontrollers
- standard microprocessor boards
- dedicated microprocessor systems
- PC-based controllers

16.3.1 Relay Control

The relay is an electro-mechanical switching device, which can be used to build simple control systems. A small input current through a coil operates, electromagnetically, a set of contacts which can switch a load (motor, heater, pump, etc.) on and off (Fig. 16.6(a)). Thus a low-voltage input circuit controls a high-power load circuit, and relays can be wired together to form a sequential control system.

A relay circuit for controlling a machine tool is illustrated in Fig. 16.6(b). The system is designed to provide push button operation and to prevent the main motor from starting unless the machine guard is closed and the cutting fluid pump is on. There is also a torque overload sensor which disables the machine if the tool jams or the motor is stalled for some other reason. The relays operate in latched mode, and the system will 'fail safe' if the power goes off. Relay

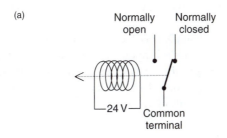

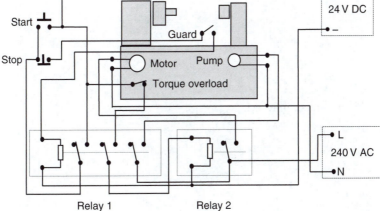

Figure 16.6 Relay control system. (a) Relay operation; (b) Wiring diagram of relay machine control system.

2 (motor) is controlled from Relay 1 (control), operated at 24 V; the motor and pump use a 240 V supply.

Another familiar example of a sequence control system is a domestic washing machine. Traditionally, a motorised rotary switch operates multiple contacts in the required sequence to open valves (filling), switch on motors (washing, spinning and pumping) and heaters. Switched sensors (level, temperature) and safety interlocks (door switch) are also connected. Thus, electro-mechanical devices provide a relatively straightforward solution for simple control applications, or ones where the environment is hostile to delicate electronics. However, switches and relays are inherently unreliable because of the moving parts.

16.3.2   PLC Control

Programmable logic controllers are often used for sequential control in industrial systems. The PLC is a self-contained sequence controller, built around a microprocessor or microcontroller, but with all the electronics and interfaces built in (Fig. 16.7).

The PLC can be programmed to act like a set of relays to give a particular output sequence in response to switched inputs, which can be manual inputs or derived from sensors. It is suitable for controlling systems where motors, heaters, valves and other loads must be switched directly from a power supply. The same machine tool seen in the previous example is now shown under PLC control in Fig. 16.8.

The PLC has inputs labelled X0, X1, X2 and X3. These are detected as 'on' when connected to 24 V via an external switched sensor or control input. The PLC is programmed to operate the outputs, labelled Y0 and Y1, according to the input sequence. The outputs are also simple switched contacts, as in the normally open contact of a relay, which operate a load circuit with an external supply. They are typically designed to handle high-power loads operating with mains voltage, or 3-phase supplies. If necessary, the PLC outputs can control external

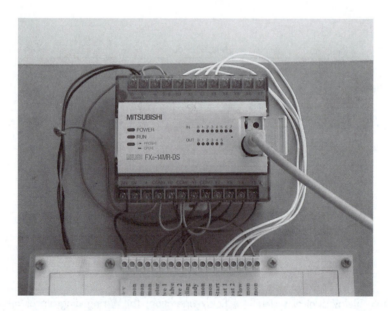

Figure 16.7 Programmable logic controller.

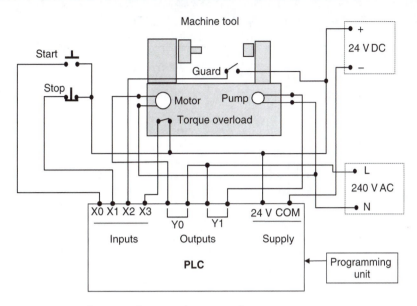

Figure 16.8 Wiring diagram of PLC machine control system.

contactors (load relays) if the load current exceeds the PLC output contact rating. The control and load circuits are electrically isolated from each other, for safety, reliability and ease of use. The PLC inputs use opto-isolators, consisting of an LED and opto-detector. The on/off signal is passed as infra-red light, giving complete electrical isolation between the input and controller internal circuits.

The program for the PLC can be written in 'ladder logic' form (Fig. 16.9) which allows the control program to be defined as if the PLC contained the relay system shown in Fig. 16.6. It is a graphical programming method using a basic set of three symbols: normally open contacts, inverted contacts and output coils. These are labelled to associate them with a physical input (Xn) or output (Yn). The normally open contacts represent external normally open contacts connected to the corresponding input; when the real contact closes, the contact in the program is closed. An inverted contact (not shown in this example) reverses the polarity of the external switch. The PLC will respond as if the ladder program were a wiring circuit. Assuming that the sides of the ladder are supply rails, an output goes on when there is a closed path through the contacts in that rung of the ladder to switch on the coil, which 'operates' the associated output.

In Fig. 16.9, Y1 will come on when the 'Start' input is pressed, if the 'Stop' button is also on (normally closed) and the 'Guard' switch is closed (guard closed). The contact labelled Y1 (Pump) closes as a result, which 'holds on' the output, even when the start button is released. A second Y1 contact then switches on the motor, as long as the overload cut-out is closed (no overload). The machine is then in run mode. If the motor is overloaded, the thermal cut-out operates and switches off the motor, but the pump stays on to maintain coolant feed. If the guard is opened or the stop button pressed, both the motor and the pump go off.

Ladder programming is a relatively easy method for creating this type of sequential control program, compared with assembler or 'C' language. There are other graphical programming methods which are even easier; refer to PLC programming references for more information.

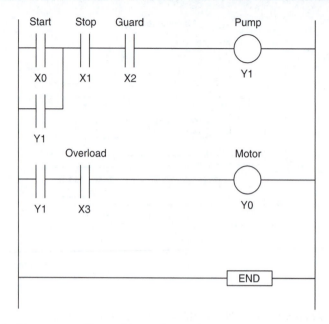

Figure 16.9 Ladder program for machine tool control.

16.3.3 Microcontroller

For comparison with other control technologies, Fig. 16.10 shows the machine tool operated by a microcontroller.

As we know, the microcontroller uses signal levels around 5 V, so the input switches have to be connected with pull-up resistors. The microcontroller can then be programmed (Program 16.2) to operate the output loads via suitable interfaces which allow its outputs to switch the high-power motors. High-current FETs are useful here, as they can operate with 5 V inputs and have no moving parts. The microcontroller has to be programmed in its native language, or 'C', both of which take time to learn, as you know! This is why ladder logic was developed for programming PLCs .

16.3.4 PC Host Control

The PC can be used to control external hardware directly via a suitable interfacing card fitted in an expansion slot on the motherboard (Fig. 16.11). The ISA bus has been used for this since the PC architecture was first devised, but the PCI bus, which is more compact, reliable and faster, is now taking over.

The I/O card connects direct into the peripheral bus of the PC (address, data and control lines) and incorporates port chips which allow external devices to exchange data direct with the PC. A basic card would typically have 16 digital inputs, 16 digital outputs, and eight analogue inputs, and one analogue output. The digital I/O operates at TTL levels, the same as the microcontroller, so the interfacing requirements are similar, except that the internal +5 V supply of the PC can be used to provide the input operating voltages. The analogue inputs required an ADC control sequence equivalent to that required by the analogue inputs in the microcontroller.

Machine tool

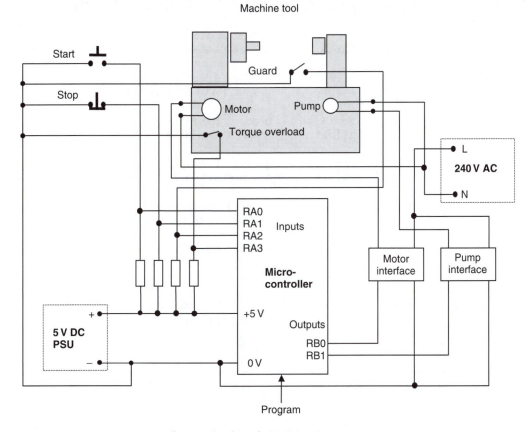

Figure 16.10 Microcontroller-operated machine system.

Program 16.2 PIC machine control program

```
;            MACHINE.ASM
;            M Bates
;            5/12/03
;            Ver 1.0

;            Program to operate a simple machine tool

; Assembler directives ...................................

            PROCESSOR 16F84

PortA       EQU      05
PortB       EQU      06

; Initialise ..........................................

            MOVLW    B'11111100'     ; Initialise outputs
            TRIS     PortB           ; to Motor & Pump

                                        continued...
```

```
; Start main loop .........................................

alloff      CLRW                        ; Switch off
            MOVWF     PortB             ; Motor & Pump

start       BTFSC     PortA,0           ; Check Start button
            GOTO      start             ; & wait is not pressed

stop        BTFSC     PortA,1           ; Check for Stop
            GOTO      alloff            ; & restart if pressed

guard       BTFSC     PortA,2           ; Check for Guard in place
            GOTO      alloff            ; & restart if not safe

            BTFSC     PortA,3           ; Check for Overload
            GOTO      coolit            ; & switch off Motor if true

            BSF       PortB,0           ; Motor ON
            BSF       PortB,1           ; Pump ON
            GOTO      stop              ; and loop

coolit      BCF       PortB,0           ; Switch off Motor, keep Pump on
            GOTO      coolit            ; and wait for reset

            END
```

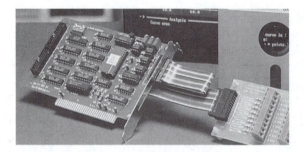

Figure 16.11   I/O card for PC-based controller system.

The PC I/O card could then, for example, control a machine, a process or the temperature control system. The simple machine tool can be operated via a similar interface to the microcontroller (see Fig. 16.12). One of the advantages of using the PC is that the input and display would be provided by the PC keyboard and VDU, so the hardware design required is limited to the interfacing. The PC could be programmed in 'C' to operate the system, and graphical functions are available which would allow a pictorial status display. A simple standard 'C' program which will run an I/O card is shown in Program 16.3, which can be compared with the PLC and microcontroller programs above.

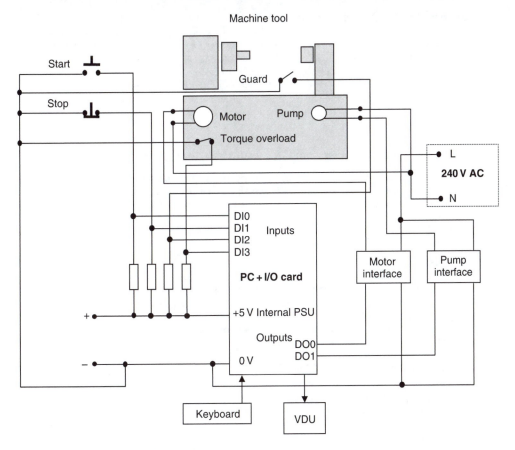

Figure 16.12 PC-controlled machine system.

The most significant features are the function calls to perform input and output. These are:

```
inputnum = inport(inreg);
outport(outreg,xx);
```

For input, the standard function 'inport' needs the address of the input register of the port chip on the I/O card. The 'base' address of the card is the address of the first register, and the other register addresses are calculated by adding an offset to this. The base address is 220h (0x220), and input register has an offset of six locations. The input register address (inreg) is therefore $220 + 6 = 226$h in the PC memory map. When the function is called, the input binary code present at the I/O card input pins is stored in integer variable 'inputnum'; it can then be processed or checked. The input number in the program is determined by the state of the input switches, giving results in the range 0–15 (4 bits).

The output register has an offset of 13 (0x0D), so the output register address (outreg) is 22Dh (0x22D). The decimal equivalent number for the required output binary code follows (xx). The number sent by the function will then appear at the output terminals of the I/O card. The output can be 0, 1, 2 or 3 to set the pump and motor on or off under binary control. The program works by reading the inputs and operating the outputs and displaying messages according to the current state of the switches, using the 'if(condition){block}' syntax to perform

Program 16.3 'C' machine control program

```c
/************************************************************
   MACHINE.C    M Bates    5/12/03    Ver 1.0
   Program to control a simple machine tool
 ************************************************************/

#include <stdio.h>

main()
{
    int     base, inreg, outreg, inputnum, outputnum;

    base   = 0x220 ;           /* I/O ports base address      */
    inreg  = base + 6;         /* Input register address      */
    outreg = base + 13;        /* Output register address     */

    inputnum = 0;
    outputnum = 0;

    clrscr();
    printf("\n\n       Machine Control Program        ");
    printf("\n       **************************        ");
    printf("\n       Ensure guard is closed before     ");
    printf("\n       hitting the start button.....     ");
    printf("\n       Enter to enable controller...     ");
    getch();

/* Control Loop *********************************************/

    do
    { inputnum = inport(inreg);          /* Read the input..... */
    } while (inputnum != 0);

    outport (outreg,3);                  /* Start the machine.. */

    do
    { inputnum = inport(inreg) ;         /* Run ............... */
    } while (inputnum == 1);

    if (inputnum == 3)
    { outport (outreg,0);
        printf("\n\n Stopped .....                    ");
    }

    if (inputnum == 5)
    { outport (outreg,0);
        printf("\n\n\ GUARD OPEN .... close and restart       ");
    }

    if (inputnum == 9)
    { outport (outreg,2);
        printf("\n\n OVERLOAD! Wait for auto reset....       ");
    }
    printf(" \n\n\ Hit space to end, then restart...       ");
    getch();

} /* Finished ***********************************************/
```

a block of code if the condition in the brackets is true. There are many references to help you learn 'C' programming, and some of the syntax has been explained in Chapter 15, so the program will not be analysed any further here.

The program is very basic, and would need development to work with a real system. For example, it does not deal with switch bounce. However, the PC-hosted system has some significant advantages over the other implementations; the keyboard and screen are built in, allowing the system to be controlled from the keyboard and the status displayed in detail using text messages, or even graphics; any calculations required will also be easier to write in 'C'. A more powerful solution is to use a graphical control and instrumentation package such as National Instruments Labview™, which allows the application software to be created without the need for text-based programming. All the necessary control and monitoring function are put together interactively on screen.

As we have seen, the serial (COM) ports on a PC are used to interface with a great variety of systems – process controllers, instrumentation, machines and target systems of various kinds, as well as the PIC programming and debugging systems. Using this interface, the PC can send commands and data, and can receive status information from the target system. This can then be displayed in graphical format, stored on disk or sent on via a network to a higher level supervisory system. The PLCs, for example, are typically programmed and controlled in this way. However, USB and network connections will tend to take over this type of function in future. Networking also allows the PC to be controlled or re-programmed remotely, and to return information to a supervisory system.

16.3.5   Manufacturing and Process Control

Microprocessors and microcontrollers are central to ongoing development of industrial technology. The term 'automation' refers to machines that are specially designed for one production task, but increasingly, flexible manufacturing systems are being introduced; they can produce a range of similar products in smaller quantities at a viable cost. Motor vehicle assembly is a well-established example.

A basic flexible manufacturing system (FMS) is illustrated in Fig. 16.13. It consists of a milling machine, a hydraulic assembly rig and a materials-handling robot. It is designed to machine and assemble a simple product. The mill produces the casing, the robot assembles it and a cover is fitted in the hydraulic press.

A block diagram (Fig. 16.14) shows how the subsystems interconnect. The digital signals in the system operate at 24 V, the higher voltage providing better noise immunity than TTL (5 V) levels. The various controllers signal to each other to control the sequence of operations. For example, when the mill has finished, it asserts (sets active) a 'Mill Ready' signal to the robot controller, which triggers the robot program to pick up the finished workpiece.

The robot slide, the press rig and the mill are all controlled by PLCs, with the main PLC controlling the whole system. In the system illustrated, the robot controller has two microprocessors, because of the complex calculations required for the robot movement. The PLCs are programmed from a PC, via the serial port; this link can be removed for normal running. However, the main system PLC remains connected to its host PC, which then operates as a SCADA (supervisory control and data acquisition) man–machine interface when the system is running. It provides a virtual control panel and graphical status display of the system as it runs, using status bits in the main PLC for its information. The system is stopped and started from the SCADA virtual control panel.

Figure 16.13 Flexible manufacturing system.

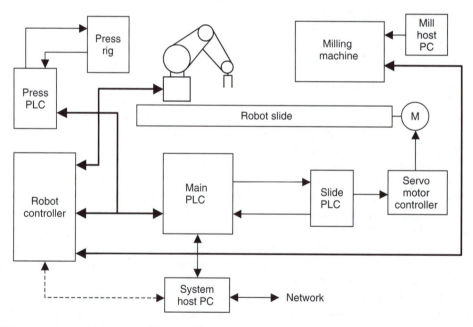

Figure 16.14 Flexible manufacturing system.

16.4 Control System Design

One of the disadvantages of the microcontroller and standard microprocessor board such as the M68k target board is that the hardware is designed to be multipurpose, that is, it has not been designed for any specific application. This means that there are likely to be features of the hardware which are not utilised by a particular application, and the user will be paying for unused facilities. The advantage of a conventional microprocessor is that the hardware can be designed at component level to meet the requirements of the application exactly. For example, the amount of memory, the type and number of I/O ports can be tailored precisely to meet the needs of the application. The decision on which type of hardware to use must therefore be made by balancing the ongoing cost of unused features against the extra development costs of a tailor-made design. Existing expertise will be another factor; it takes time for the design engineer to become familiar with any new developments.

The PLC offers an off-the-shelf hardware package, normally requiring no external electronics to interface it. It is therefore commonly used in industrial control systems, which divide into two main categories: process control and manufacturing systems. It is robust, easy to install and easily programmed.

The PC itself can also be used as a controller, offering a standard operating system, graphics, disk storage, communications and printing. One of the great advantages of the modular design of the PC is that special interface cards can be fitted for control applications. The PC thus can be turned into an oscilloscope, logic analyser, controller or data logger. It is also a universal platform for running design software, for both mechanical and electronic CAD. Thus, you can design hardware, run a control application, write the support documentation and sell your product on the Internet, all using the same machine! Its role as a development platform for PIC microcontroller designs is only one of many uses.

Table 16.1 provides a comparison of the advantages and disadvantages of the different forms of system control outlined above.

We can see that a good working knowledge of all these options is required in order to select the most appropriate technology for any given application. The PIC is one microcontroller among many, and the microcontroller is only one solution to any given problem. Nevertheless, the microcontroller is now meeting an increasing range of needs in current industrial and consumer technology, and the PIC is the best device to start learning with – have fun!

Summary

- Numerous alternative microcontroller families exist, notably the Intel 8051 and derivatives and the Motorola 68HC11.

- The Motorola 68000 (M68k) CISC microprocessor is widely used in education and industry, and has a relatively straightforward architecture, but a complex instruction set.

- Simple sequence control can be implemented using electromechanical relays.

- PLCs have built-in interfacing and user-friendly programming methods to simplify sequence control system design.

- The PC can be used as a universal design, control, monitoring, instrumentation, IT applications and networking hardware platform.

- The control system designer needs to select the most suitable technology for any given application.

Table 16.1 Comparison of control technologies

Technology	Advantages	Disadvantages	Typical applications
Relays	Simple to design No programming needed No electronics skills needed Good electrical isolation	Slow Unreliable High power consumption Not suitable for complex systems	Machine safety interlocks Simple process control High power systems
PLC	Minimal interfacing needed Easy to program Easy to install	Limited processing functionality Limited interfacing flexibility	Machine control Process systems Flexible manufacturing system
Microcontroller	Flexible hardware design Large choice Suitable for small embedded applications	Hardware design skills needed Programming skill needed Limited memory	Smart cards Consumer goods Instrumentation Dedicated controllers
Microprocessor system	Flexible controller design Suitable for larger systems Expandable	Expert hardware design skills needed Good programming skill needed	Automatic machines Specialist control systems Large computers
PC Host	Available off the shelf High cost/performance ratio Built-in data storage and comms Graphical programming and display Basic interfacing skills only needed Standard operating system	High cost of basic unit Large physical size Only suitable for complex applications	Machine tool host SCADA Large systems user interface Instrumentation systems host Networks and distributed systems

1. Outline the differences between the PIC 16F84 and the Intel 8051 microcontrollers.

2. State two advantages and two disadvantages of the conventional processor system over the microcontroller in designing a system to meet a particular specification.

3. Explain briefly the advantages of using a PLC compared to a microprocessor system in control applications.

4. Draw a flowchart for the Program 16.2 to show the control sequence clearly.

5. What are the advantages of using the PC as a controller, compared with the microcontroller?

6. Match up the controller type with the most appropriate programming language or technique:
(i)	Small microcontroller	(a)	'C'
(ii)	CISC microprocessor	(b)	None
(iii)	Relay system	(c)	Graphical
(iv)	PLC	(d)	Assembler
(v)	PC + I/O card	(e)	Ladder logic

1. Log on to the Atmel website and select a microcontroller from the list of available flash devices which is most similar to the 16F84 and compare its features and block diagram. Identify any advantages that the AVR microcontroller has over the PIC.

2. Obtain a copy of the M68000 summary instruction set, and compare it with the PIC 16XXXX instruction set. Identify three instructions which are not available in the PIC.

3. Devise a circuit to switch a motor on and off using push buttons and a single relay (no safety interlocks needed).Why is this safer than using a simple mains switch?

4. Modify the PLC machine tool controller in Fig. 16.8, and its program, to operate an alarm output if the machine stalls or overloads.

5. Devise a block diagram of a domestic washing machine, controlled by a microcontroller. Show interface blocks between the switched actuators and sensors and the microcontroller. Write a description of the operating sequence of the machine, and devise a flowchart for the control sequence, constructed so that it could be implemented in PIC assembly language.

6. By reference to the temperature controller design in Chapter 15, design the hardware for PIC implementation of the system shown in Fig. 16.10. Select a suitable device according to the I/O and memory requirements, test Program 16.2 in the MPLAB simulator and implement the design using the most readily available construction techniques. Devise a target system to simulate the machine tool and confirm correct operation in hardware.

Appendix A
PIC 16F84A Data Sheet

PIC16F84A
Data Sheet

18-pin Enhanced FLASH/EEPROM

8-bit Microcontroller

DS35007B

Note the following details of the code protection feature on PICmicro® MCUs.

- The PICmicro family meets the specifications contained in the Microchip Data Sheet.
- Microchip believes that its family of PICmicro microcontrollers is one of the most secure products of its kind on the market today, when used in the intended manner and under normal conditions.
- There are dishonest and possibly illegal methods used to breach the code protection feature. All of these methods, to our knowledge, require using the PICmicro microcontroller in a manner outside the operating specifications contained in the data sheet. The person doing so may be engaged in theft of intellectual property.
- Microchip is willing to work with the customer who is concerned about the integrity of their code.
- Neither Microchip nor any other semiconductor manufacturer can guarantee the security of their code. Code protection does not mean that we are guaranteeing the product as "unbreakable".
- Code protection is constantly evolving. We at Microchip are committed to continuously improving the code protection features of our product.

If you have any further questions about this matter, please contact the local sales office nearest to you.

Trademarks

The Microchip name and logo, the Microchip logo, PIC, PICmicro, PICMASTER, PICSTART, PRO MATE, KEELOQ, SEEVAL, MPLAB and The Embedded Control Solutions Company are registered trademarks of Microchip Technology Incorporated in the U.S.A. and other countries.

Total Endurance, ICSP, In-Circuit Serial Programming, Filter-Lab, MXDEV, microID, FlexROM, fuzzyLAB, MPASM, MPLINK, MPLIB, PICC, PICDEM, PICDEM.net, ICEPIC, Migratable Memory, FanSense, ECONOMONITOR, Select Mode and microPort are trademarks of Microchip Technology Incorporated in the U.S.A.

Serialized Quick Term Programming (SQTP) is a service mark of Microchip Technology Incorporated in the U.S.A.

All other trademarks mentioned herein are property of their respective companies.

Microchip received QS-9000 quality system certification for its worldwide headquarters, design and wafer fabrication facilities in Chandler and Tempe, Arizona in July 1999. The Company's quality system processes and procedures are QS-9000 compliant for its PICmicro® 8-bit MCUs, KEELOQ® code hopping devices, Serial EEPROMs and microperipheral products. In addition, Microchip's quality system for the design and manufacture of development systems is ISO 9001 certified.

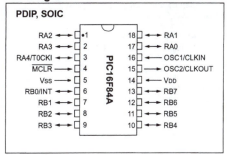

PIC16F84A

18-pin *Enhanced* FLASH/EEPROM 8-Bit Microcontroller

High Performance RISC CPU Features:

- Only 35 single word instructions to learn
- All instructions single-cycle except for program branches which are two-cycle
- Operating speed: DC - 20 MHz clock input
  DC - 200 ns instruction cycle
- 1024 words of program memory
- 68 bytes of Data RAM
- 64 bytes of Data EEPROM
- 14-bit wide instruction words
- 8-bit wide data bytes
- 15 Special Function Hardware registers
- Eight-level deep hardware stack
- Direct, indirect and relative addressing modes
- Four interrupt sources:
  - External RB0/INT pin
  - TMR0 timer overflow
  - PORTB<7:4> interrupt-on-change
  - Data EEPROM write complete

Peripheral Features:

- 13 I/O pins with individual direction control
- High current sink/source for direct LED drive
  - 25 mA sink max. per pin
  - 25 mA source max. per pin
- TMR0: 8-bit timer/counter with 8-bit programmable prescaler

Special Microcontroller Features:

- 10,000 erase/write cycles *Enhanced* FLASH Program memory typical
- 10,000,000 typical erase/write cycles EEPROM Data memory typical
- EEPROM Data Retention > 40 years
- In-Circuit Serial Programming™ (ICSP™) - via two pins
- Power-on Reset (POR), Power-up Timer (PWRT), Oscillator Start-up Timer (OST)
- Watchdog Timer (WDT) with its own On-Chip RC Oscillator for reliable operation
- Code protection
- Power saving SLEEP mode
- Selectable oscillator options

Pin Diagrams

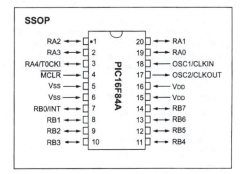

CMOS Enhanced FLASH/EEPROM Technology:

- Low power, high speed technology
- Fully static design
- Wide operating voltage range:
  - Commercial: 2.0V to 5.5V
  - Industrial: 2.0V to 5.5V
- Low power consumption:
  - < 2 mA typical @ 5V, 4 MHz
  - 15 µA typical @ 2V, 32 kHz
  - < 0.5 µA typical standby current @ 2V

PIC16F84A

Table of Contents

TO OUR VALUED CUSTOMERS

It is our intention to provide our valued customers with the best documentation possible to ensure successful use of your Microchip products. To this end, we will continue to improve our publications to better suit your needs. Our publications will be refined and enhanced as new volumes and updates are introduced.

If you have any questions or comments regarding this publication, please contact the Marketing Communications Department via E-mail at **docerrors@mail.microchip.com** or fax the **Reader Response Form** in the back of this data sheet to (480) 792-4150. We welcome your feedback.

Most Current Data Sheet

To obtain the most up-to-date version of this data sheet, please register at our Worldwide Web site at:

　　　http://www.microchip.com

You can determine the version of a data sheet by examining its literature number found on the bottom outside corner of any page. The last character of the literature number is the version number, (e.g., DS30000A is version A of document DS30000).

Errata

An errata sheet, describing minor operational differences from the data sheet and recommended workarounds, may exist for current devices. As device/documentation issues become known to us, we will publish an errata sheet. The errata will specify the revision of silicon and revision of document to which it applies.

To determine if an errata sheet exists for a particular device, please check with one of the following:

- Microchip's Worldwide Web site; http://www.microchip.com
- Your local Microchip sales office (see last page)
- The Microchip Corporate Literature Center; U.S. FAX: (480) 792-7277

When contacting a sales office or the literature center, please specify which device, revision of silicon and data sheet (include literature number) you are using.

Customer Notification System

Register on our web site at **www.microchip.com/cn** to receive the most current information on all of our products.

PIC16F84A

1.0   DEVICE OVERVIEW

This document contains device specific information for the operation of the PIC16F84A device. Additional information may be found in the PICmicro™ Mid-Range Reference Manual, (DS33023), which may be downloaded from the Microchip website. The Reference Manual should be considered a complementary document to this data sheet, and is highly recommended reading for a better understanding of the device architecture and operation of the peripheral modules.

The PIC16F84A belongs to the mid-range family of the PICmicro® microcontroller devices. A block diagram of the device is shown in Figure 1-1.

The program memory contains 1K words, which translates to 1024 instructions, since each 14-bit program memory word is the same width as each device instruction. The data memory (RAM) contains 68 bytes. Data EEPROM is 64 bytes.

There are also 13 I/O pins that are user-configured on a pin-to-pin basis. Some pins are multiplexed with other device functions. These functions include:

- External interrupt
- Change on PORTB interrupt
- Timer0 clock input

Table 1-1 details the pinout of the device with descriptions and details for each pin.

FIGURE 1-1:   PIC16F84A BLOCK DIAGRAM

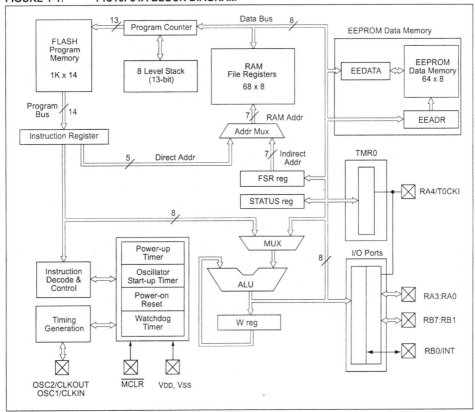

PIC16F84A

TABLE 1-1:   PIC16F84A PINOUT DESCRIPTION

Pin Name	PDIP No.	SOIC No.	SSOP No.	I/O/P Type	Buffer Type	Description
OSC1/CLKIN	16	16	18	I	ST/CMOS[3]	Oscillator crystal input/external clock source input.
OSC2/CLKOUT	15	15	19	O	—	Oscillator crystal output. Connects to crystal or resonator in Crystal Oscillator mode. In RC mode, OSC2 pin outputs CLKOUT, which has 1/4 the frequency of OSC1 and denotes the instruction cycle rate.
MCLR	4	4	4	I/P	ST	Master Clear (Reset) input/programming voltage input. This pin is an active low RESET to the device.
RA0	17	17	19	I/O	TTL	PORTA is a bi-directional I/O port.
RA1	18	18	20	I/O	TTL	
RA2	1	1	1	I/O	TTL	
RA3	2	2	2	I/O	TTL	
RA4/T0CKI	3	3	3	I/O	ST	Can also be selected to be the clock input to the TMR0 timer/counter. Output is open drain type.
RB0/INT	6	6	7	I/O	TTL/ST[1]	PORTB is a bi-directional I/O port. PORTB can be software programmed for internal weak pull-up on all inputs. RB0/INT can also be selected as an external interrupt pin.
RB1	7	7	8	I/O	TTL	
RB2	8	8	9	I/O	TTL	
RB3	9	9	10	I/O	TTL	
RB4	10	10	11	I/O	TTL	Interrupt-on-change pin.
RB5	11	11	12	I/O	TTL	Interrupt-on-change pin.
RB6	12	12	13	I/O	TTL/ST[2]	Interrupt-on-change pin. Serial programming clock.
RB7	13	13	14	I/O	TTL/ST[2]	Interrupt-on-change pin. Serial programming data.
Vss	5	5	5,6	P	—	Ground reference for logic and I/O pins.
Vdd	14	14	15,16	P	—	Positive supply for logic and I/O pins.

Legend: I= input   O = Output   I/O = Input/Output   P = Power
              — = Not used   TTL = TTL input   ST = Schmitt Trigger input

Note 1: This buffer is a Schmitt Trigger input when configured as the external interrupt.
      2: This buffer is a Schmitt Trigger input when used in Serial Programming mode.
      3: This buffer is a Schmitt Trigger input when configured in RC oscillator mode and a CMOS input otherwise.

PIC16F84A

2.0 MEMORY ORGANIZATION

There are two memory blocks in the PIC16F84A. These are the program memory and the data memory. Each block has its own bus, so that access to each block can occur during the same oscillator cycle.

The data memory can further be broken down into the general purpose RAM and the Special Function Registers (SFRs). The operation of the SFRs that control the "core" are described here. The SFRs used to control the peripheral modules are described in the section discussing each individual peripheral module.

The data memory area also contains the data EEPROM memory. This memory is not directly mapped into the data memory, but is indirectly mapped. That is, an indirect address pointer specifies the address of the data EEPROM memory to read/write. The 64 bytes of data EEPROM memory have the address range 0h-3Fh. More details on the EEPROM memory can be found in Section 3.0.

Additional information on device memory may be found in the PICmicro™ Mid-Range Reference Manual, (DS33023).

2.1 Program Memory Organization

The PIC16FXX has a 13-bit program counter capable of addressing an 8K x 14 program memory space. For the PIC16F84A, the first 1K x 14 (0000h-03FFh) are physically implemented (Figure 2-1). Accessing a location above the physically implemented address will cause a wraparound. For example, for locations 20h, 420h, 820h, C20h, 1020h, 1420h, 1820h, and 1C20h, the instruction will be the same.

The RESET vector is at 0000h and the interrupt vector is at 0004h.

FIGURE 2-1: PROGRAM MEMORY MAP AND STACK - PIC16F84A

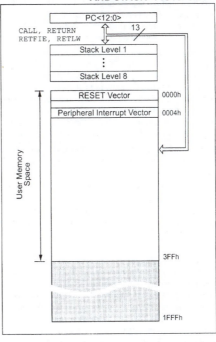

PIC16F84A

2.2   Data Memory Organization

The data memory is partitioned into two areas. The first is the Special Function Registers (SFR) area, while the second is the General Purpose Registers (GPR) area. The SFRs control the operation of the device.

Portions of data memory are banked. This is for both the SFR area and the GPR area. The GPR area is banked to allow greater than 116 bytes of general purpose RAM. The banked areas of the SFR are for the registers that control the peripheral functions. Banking requires the use of control bits for bank selection. These control bits are located in the STATUS Register. Figure 2-2 shows the data memory map organization.

Instructions MOVWF and MOVF can move values from the W register to any location in the register file ("F"), and vice-versa.

The entire data memory can be accessed either directly using the absolute address of each register file or indirectly through the File Select Register (FSR) (Section 2.5). Indirect addressing uses the present value of the RP0 bit for access into the banked areas of data memory.

Data memory is partitioned into two banks which contain the general purpose registers and the special function registers. Bank 0 is selected by clearing the RP0 bit (STATUS<5>). Setting the RP0 bit selects Bank 1. Each Bank extends up to 7Fh (128 bytes). The first twelve locations of each Bank are reserved for the Special Function Registers. The remainder are General Purpose Registers, implemented as static RAM.

2.2.1   GENERAL PURPOSE REGISTER FILE

Each General Purpose Register (GPR) is 8-bits wide and is accessed either directly or indirectly through the FSR (Section 2.5).

The GPR addresses in Bank 1 are mapped to addresses in Bank 0. As an example, addressing location 0Ch or 8Ch will access the same GPR.

FIGURE 2-2:   REGISTER FILE MAP - PIC16F84A

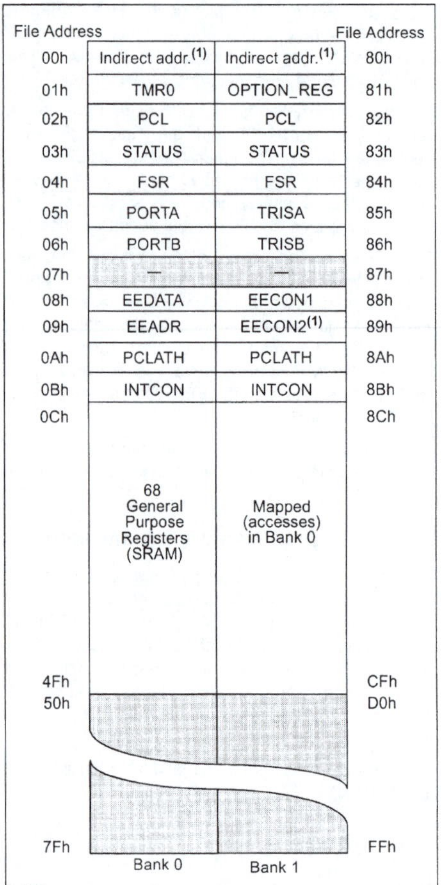

File Address	Bank 0	Bank 1	File Address
00h	Indirect addr.[1]	Indirect addr.[1]	80h
01h	TMR0	OPTION_REG	81h
02h	PCL	PCL	82h
03h	STATUS	STATUS	83h
04h	FSR	FSR	84h
05h	PORTA	TRISA	85h
06h	PORTB	TRISB	86h
07h	—	—	87h
08h	EEDATA	EECON1	88h
09h	EEADR	EECON2[1]	89h
0Ah	PCLATH	PCLATH	8Ah
0Bh	INTCON	INTCON	8Bh
0Ch			8Ch
	68 General Purpose Registers (SRAM)	Mapped (accesses) in Bank 0	
4Fh			CFh
50h			D0h
7Fh			FFh

☐ Unimplemented data memory location, read as '0'.

Note 1: Not a physical register.

PIC16F84A

2.3 Special Function Registers

The Special Function Registers (Figure 2-2 and Table 2-1) are used by the CPU and Peripheral functions to control the device operation. These registers are static RAM.

The special function registers can be classified into two sets, core and peripheral. Those associated with the core functions are described in this section. Those related to the operation of the peripheral features are described in the section for that specific feature.

TABLE 2-1: SPECIAL FUNCTION REGISTER FILE SUMMARY

Addr	Name	Bit 7	Bit 6	Bit 5	Bit 4	Bit 3	Bit 2	Bit 1	Bit 0	Value on Power-on RESET	Details on page
Bank 0											
00h	INDF	Uses contents of FSR to address Data Memory (not a physical register)								---- ----	11
01h	TMR0	8-bit Real-Time Clock/Counter								xxxx xxxx	20
02h	PCL	Low Order 8 bits of the Program Counter (PC)								0000 0000	11
03h	STATUS[2]	IRP	RP1	RP0	$\overline{TO}$	$\overline{PD}$	Z	DC	C	0001 1xxx	8
04h	FSR	Indirect Data Memory Address Pointer 0								xxxx xxxx	11
05h	PORTA[4]	—	—	—	RA4/T0CKI	RA3	RA2	RA1	RA0	---x xxxx	16
06h	PORTB[5]	RB7	RB6	RB5	RB4	RB3	RB2	RB1	RB0/INT	xxxx xxxx	18
07h	—	Unimplemented location, read as '0'								—	—
08h	EEDATA	EEPROM Data Register								xxxx xxxx	13,14
09h	EEADR	EEPROM Address Register								xxxx xxxx	13,14
0Ah	PCLATH	—	—	—	Write Buffer for upper 5 bits of the PC[1]					---0 0000	11
0Bh	INTCON	GIE	EEIE	T0IE	INTE	RBIE	T0IF	INTF	RBIF	0000 000x	10
Bank 1											
80h	INDF	Uses Contents of FSR to address Data Memory (not a physical register)								---- ----	11
81h	OPTION_REG	RBPU	INTEDG	T0CS	T0SE	PSA	PS2	PS1	PS0	1111 1111	9
82h	PCL	Low order 8 bits of Program Counter (PC)								0000 0000	11
83h	STATUS [2]	IRP	RP1	RP0	$\overline{TO}$	$\overline{PD}$	Z	DC	C	0001 1xxx	8
84h	FSR	Indirect data memory address pointer 0								xxxx xxxx	11
85h	TRISA	—	—	—	PORTA Data Direction Register					---1 1111	16
86h	TRISB	PORTB Data Direction Register								1111 1111	18
87h	—	Unimplemented location, read as '0'								—	—
88h	EECON1	—	—	—	—	EEIF	WRERR	WREN	WR	RD	---0 x000	13
89h	EECON2	EEPROM Control Register 2 (not a physical register)								---- ----	14
0Ah	PCLATH	—	—	—	Write buffer for upper 5 bits of the PC[1]					---0 0000	11
0Bh	INTCON	GIE	EEIE	T0IE	INTE	RBIE	T0IF	INTF	RBIF	0000 000x	10

Legend: x = unknown, u = unchanged. - = unimplemented, read as '0', q = value depends on condition

Note   1:   The upper byte of the program counter is not directly accessible. PCLATH is a slave register for PC<12:8>. The contents of PCLATH can be transferred to the upper byte of the program counter, but the contents of PC<12:8> are never transferred to PCLATH.
   2:   The $\overline{TO}$ and $\overline{PD}$ status bits in the STATUS register are not affected by a $\overline{MCLR}$ Reset.
   3:   Other (non power-up) RESETS include: external RESET through $\overline{MCLR}$ and the Watchdog Timer Reset.
   4:   On any device RESET, these pins are configured as inputs.
   5:   This is the value that will be in the port output latch.

PIC16F84A

2.3.1   STATUS REGISTER

The STATUS register contains the arithmetic status of the ALU, the RESET status and the bank select bit for data memory.

As with any register, the STATUS register can be the destination for any instruction. If the STATUS register is the destination for an instruction that affects the Z, DC or C bits, then the write to these three bits is disabled. These bits are set or cleared according to device logic. Furthermore, the $\overline{TO}$ and $\overline{PD}$ bits are not writable. Therefore, the result of an instruction with the STATUS register as destination may be different than intended.

For example, CLRF STATUS will clear the upper three bits and set the Z bit. This leaves the STATUS register as 000u u1uu (where u = unchanged).

Only the BCF, BSF, SWAPF and MOVWF instructions should be used to alter the STATUS register (Table 7-2), because these instructions do not affect any status bit.

Note 1: The IRP and RP1 bits (STATUS<7:6>) are not used by the PIC16F84A and should be programmed as cleared. Use of these bits as general purpose R/W bits is NOT recommended, since this may affect upward compatibility with future products.

2: The C and DC bits operate as a borrow and digit borrow out bit, respectively, in subtraction. See the SUBLW and SUBWF instructions for examples.

3: When the STATUS register is the destination for an instruction that affects the Z, DC or C bits, then the write to these three bits is disabled. The specified bit(s) will be updated according to device logic

REGISTER 2-1:   STATUS REGISTER (ADDRESS 03h, 83h)

R/W-0	R/W-0	R/W-0	R-1	R-1	R/W-x	R/W-x	R/W-x
IRP	RP1	RP0	$\overline{TO}$	$\overline{PD}$	Z	DC	C

bit 7   bit 0

bit 7-6   **Unimplemented:** Maintain as '0'

bit 5   **RP0**: Register Bank Select bits (used for direct addressing)
01 = Bank 1 (80h - FFh)
00 = Bank 0 (00h - 7Fh)

bit 4 $\overline{TO}$: Time-out bit
1 = After power-up, CLRWDT instruction, or SLEEP instruction
0 = A WDT time-out occurred

bit 3 $\overline{PD}$: Power-down bit
1 = After power-up or by the CLRWDT instruction
0 = By execution of the SLEEP instruction

bit 2   **Z**: Zero bit
1 = The result of an arithmetic or logic operation is zero
0 = The result of an arithmetic or logic operation is not zero

bit 1   **DC**: Digit carry/borrow bit (ADDWF, ADDLW, SUBLW, SUBWF instructions) (for $\overline{borrow}$, the polarity is reversed)
1 = A carry-out from the 4th low order bit of the result occurred
0 = No carry-out from the 4th low order bit of the result

bit 0   **C**: Carry/borrow bit (ADDWF, ADDLW, SUBLW, SUBWF instructions) (for $\overline{borrow}$, the polarity is reversed)
1 = A carry-out from the Most Significant bit of the result occurred
0 = No carry-out from the Most Significant bit of the result occurred

Note: A subtraction is executed by adding the two's complement of the second operand. For rotate (RRF, RLF) instructions, this bit is loaded with either the high or low order bit of the source register.

Legend:		
R = Readable bit	W = Writable bit	U = Unimplemented bit, read as '0'
- n = Value at POR	'1' = Bit is set	'0' = Bit is cleared x = Bit is unknown

PIC16F84A

2.3.2 OPTION REGISTER

The OPTION register is a readable and writable register which contains various control bits to configure the TMR0/WDT prescaler, the external INT interrupt, TMR0, and the weak pull-ups on PORTB.

Note:	When the prescaler is assigned to the WDT (PSA = '1'), TMR0 has a 1:1 prescaler assignment.

REGISTER 2-2: **OPTION REGISTER (ADDRESS 81h)**

R/W-1	R/W-1	R/W-1	R/W-1	R/W-1	R/W-1	R/W-1	R/W-1
RBPU	INTEDG	T0CS	T0SE	PSA	PS2	PS1	PS0

bit 7   bit 0

bit 7   **RBPU:** PORTB Pull-up Enable bit

1 = PORTB pull-ups are disabled
0 = PORTB pull-ups are enabled by individual port latch values

bit 6   **INTEDG**: Interrupt Edge Select bit

1 = Interrupt on rising edge of RB0/INT pin
0 = Interrupt on falling edge of RB0/INT pin

bit 5   **T0CS**: TMR0 Clock Source Select bit

1 = Transition on RA4/T0CKI pin
0 = Internal instruction cycle clock (CLKOUT)

bit 4   **T0SE**: TMR0 Source Edge Select bit

1 = Increment on high-to-low transition on RA4/T0CKI pin
0 = Increment on low-to-high transition on RA4/T0CKI pin

bit 3   **PSA**: Prescaler Assignment bit

1 = Prescaler is assigned to the WDT
0 = Prescaler is assigned to the Timer0 module

bit 2-0   **PS2:PS0**: Prescaler Rate Select bits

Bit Value	TMR0 Rate	WDT Rate
000	1 : 2	1 : 1
001	1 : 4	1 : 2
010	1 : 8	1 : 4
011	1 : 16	1 : 8
100	1 : 32	1 : 16
101	1 : 64	1 : 32
110	1 : 128	1 : 64
111	1 : 256	1 : 128

Legend:			
R = Readable bit	W = Writable bit	U = Unimplemented bit, read as '0'	
- n = Value at POR	'1' = Bit is set	'0' = Bit is cleared	x = Bit is unknown

PIC16F84A

2.3.3   INTCON REGISTER

The INTCON register is a readable and writable register that contains the various enable bits for all interrupt sources.

> **Note:** Interrupt flag bits are set when an interrupt condition occurs, regardless of the state of its corresponding enable bit or the global enable bit, GIE (INTCON<7>).

REGISTER 2-3:   INTCON REGISTER (ADDRESS 0Bh, 8Bh)

R/W-0	R/W-0	R/W-0	R/W-0	R/W-0	R/W-0	R/W-0	R/W-x
GIE	EEIE	T0IE	INTE	RBIE	T0IF	INTF	RBIF

bit 7   bit 0

bit 7   **GIE:** Global Interrupt Enable bit
1 = Enables all unmasked interrupts
0 = Disables all interrupts

bit 6   **EEIE**: EE Write Complete Interrupt Enable bit
1 = Enables the EE Write Complete interrupts
0 = Disables the EE Write Complete interrupt

bit 5   **T0IE**: TMR0 Overflow Interrupt Enable bit
1 = Enables the TMR0 interrupt
0 = Disables the TMR0 interrupt

bit 4   **INTE**: RB0/INT External Interrupt Enable bit
1 = Enables the RB0/INT external interrupt
0 = Disables the RB0/INT external interrupt

bit 3   **RBIE**: RB Port Change Interrupt Enable bit
1 = Enables the RB port change interrupt
0 = Disables the RB port change interrupt

bit 2   **T0IF**: TMR0 Overflow Interrupt Flag bit
1 = TMR0 register has overflowed (must be cleared in software)
0 = TMR0 register did not overflow

bit 1   **INTF**: RB0/INT External Interrupt Flag bit
1 = The RB0/INT external interrupt occurred (must be cleared in software)
0 = The RB0/INT external interrupt did not occur

bit 0   **RBIF**: RB Port Change Interrupt Flag bit
1 = At least one of the RB7:RB4 pins changed state (must be cleared in software)
0 = None of the RB7:RB4 pins have changed state

Legend:			
R = Readable bit	W = Writable bit	U = Unimplemented bit, read as '0'	
- n = Value at POR	'1' = Bit is set	'0' = Bit is cleared	x = Bit is unknown

PIC16F84A

2.4   PCL and PCLATH

The program counter (PC) specifies the address of the instruction to fetch for execution. The PC is 13 bits wide. The low byte is called the PCL register. This register is readable and writable. The high byte is called the PCH register. This register contains the PC<12:8> bits and is not directly readable or writable. If the program counter (PC) is modified or a conditional test is true, the instruction requires two cycles. The second cycle is executed as a NOP. All updates to the PCH register go through the PCLATH register.

2.4.1   STACK

The stack allows a combination of up to 8 program calls and interrupts to occur. The stack contains the return address from this branch in program execution.

Mid-range devices have an 8 level deep x 13-bit wide hardware stack. The stack space is not part of either program or data space and the stack pointer is not readable or writable. The PC is PUSHed onto the stack when a CALL instruction is executed or an interrupt causes a branch. The stack is POPed in the event of a RETURN, RETLW or a RETFIE instruction execution. PCLATH is not modified when the stack is PUSHed or POPed.

After the stack has been PUSHed eight times, the ninth push overwrites the value that was stored from the first push. The tenth push overwrites the second push (and so on).

2.5   Indirect Addressing; INDF and FSR Registers

The INDF register is not a physical register. Addressing INDF actually addresses the register whose address is contained in the FSR register (FSR is a *pointer*). This is indirect addressing.

EXAMPLE 2-1:   INDIRECT ADDRESSING

- Register file 05 contains the value 10h
- Register file 06 contains the value 0Ah
- Load the value 05 into the FSR register
- A read of the INDF register will return the value of 10h
- Increment the value of the FSR register by one (FSR = 06)
- A read of the INDF register now will return the value of 0Ah.

Reading INDF itself indirectly (FSR = 0) will produce 00h. Writing to the INDF register indirectly results in a no-operation (although STATUS bits may be affected).

A simple program to clear RAM locations 20h-2Fh using indirect addressing is shown in Example 2-2.

EXAMPLE 2-2:   HOW TO CLEAR RAM USING INDIRECT ADDRESSING

```
          movlw   0x20    ;initialize pointer
          movwf   FSR     ;to RAM
NEXT      clrf    INDF    ;clear INDF register
          incf    FSR     ;inc pointer
          btfss   FSR,4   ;all done?
          goto    NEXT    ;NO, clear next
CONTINUE
          :               ;YES, continue
```

An effective 9-bit address is obtained by concatenating the 8-bit FSR register and the IRP bit (STATUS<7>), as shown in Figure 2-3. However, IRP is not used in the PIC16F84A.

PIC16F84A

FIGURE 2-3:   DIRECT/INDIRECT ADDRESSING

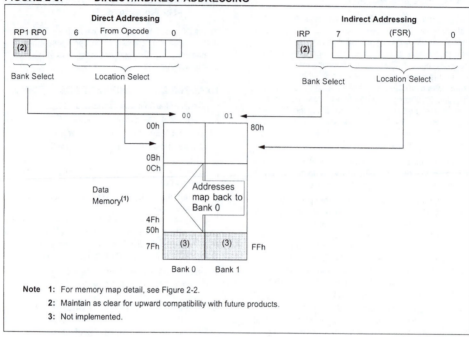

Note 1: For memory map detail, see Figure 2-2.

2: Maintain as clear for upward compatibility with future products.

3: Not implemented.

PIC16F84A

3.0   DATA EEPROM MEMORY

The EEPROM data memory is readable and writable during normal operation (full VDD range). This memory is not directly mapped in the register file space. Instead it is indirectly addressed through the Special Function Registers. There are four SFRs used to read and write this memory. These registers are:

- EECON1
- EECON2 (not a physically implemented register)
- EEDATA
- EEADR

EEDATA holds the 8-bit data for read/write, and EEADR holds the address of the EEPROM location being accessed. PIC16F84A devices have 64 bytes of data EEPROM with an address range from 0h to 3Fh.

The EEPROM data memory allows byte read and write. A byte write automatically erases the location and writes the new data (erase before write). The EEPROM data memory is rated for high erase/write cycles. The write time is controlled by an on-chip timer. The write-time will vary with voltage and temperature as well as from chip to chip. Please refer to AC specifications for exact limits.

When the device is code protected, the CPU may continue to read and write the data EEPROM memory. The device programmer can no longer access this memory.

Additional information on the Data EEPROM is available in the PICmicro™ Mid-Range Reference Manual (DS33023).

REGISTER 3-1:   EECON1 REGISTER (ADDRESS 88h)

U-0	U-0	U-0	R/W-0	R/W-x	R/W-0	R/S-0	R/S-0
—	—	—	EEIF	WRERR	WREN	WR	RD

bit 7   bit 0

bit 7-5   **Unimplemented:** Read as '0'

bit 4   **EEIF:** EEPROM Write Operation Interrupt Flag bit
1 = The write operation completed (must be cleared in software)
0 = The write operation is not complete or has not been started

bit 3   **WRERR:** EEPROM Error Flag bit
1 = A write operation is prematurely terminated
    (any $\overline{\text{MCLR}}$ Reset or any WDT Reset during normal operation)
0 = The write operation completed

bit 2   **WREN:** EEPROM Write Enable bit
1 = Allows write cycles
0 = Inhibits write to the EEPROM

bit 1   **WR:** Write Control bit
1 = Initiates a write cycle. The bit is cleared by hardware once write is complete. The WR bit
    can only be set (not cleared) in software.
0 = Write cycle to the EEPROM is complete

bit 0   **RD:** Read Control bit
1 = Initiates an EEPROM read RD is cleared in hardware. The RD bit can only be set (not
    cleared) in software.
0 = Does not initiate an EEPROM read

Legend:			
R = Readable bit	W = Writable bit	U = Unimplemented bit, read as '0'	
- n = Value at POR	'1' = Bit is set	'0' = Bit is cleared	x = Bit is unknown

PIC16F84A

3.1 Reading the EEPROM Data Memory

To read a data memory location, the user must write the address to the EEADR register and then set control bit RD (EECON1<0>). The data is available, in the very next cycle, in the EEDATA register; therefore, it can be read in the next instruction. EEDATA will hold this value until another read or until it is written to by the user (during a write operation).

EXAMPLE 3-1:   DATA EEPROM READ

```
BCF     STATUS, RP0  ; Bank 0
MOVLW   CONFIG_ADDR  ;
MOVWF   EEADR        ; Address to read
BSF     STATUS, RP0  ; Bank 1
BSF     EECON1, RD   ; EE Read
BCF     STATUS, RP0  ; Bank 0
MOVF    EEDATA, W    ; W = EEDATA
```

3.2 Writing to the EEPROM Data Memory

To write an EEPROM data location, the user must first write the address to the EEADR register and the data to the EEDATA register. Then the user must follow a specific sequence to initiate the write for each byte.

EXAMPLE 3-2:   DATA EEPROM WRITE

```
        BSF     STATUS, RP0  ; Bank 1
        BCF     INTCON, GIE  ; Disable INTs.
        BSF     EECON1, WREN ; Enable Write
        MOVLW   55h          ;
        MOVWF   EECON2       ; Write 55h
        MOVLW   AAh          ;
        MOVWF   EECON2       ; Write AAh
        BSF     EECON1, WR   ; Set WR bit
                             ;  begin write
        BSF     INTCON, GIE  ; Enable INTs.
```
Required Sequence

The write will not initiate if the above sequence is not exactly followed (write 55h to EECON2, write AAh to EECON2, then set WR bit) for each byte. We strongly recommend that interrupts be disabled during this code segment.

Additionally, the WREN bit in EECON1 must be set to enable write. This mechanism prevents accidental writes to data EEPROM due to errant (unexpected) code execution (i.e., lost programs). The user should keep the WREN bit clear at all times, except when updating EEPROM. The WREN bit is not cleared by hardware.

After a write sequence has been initiated, clearing the WREN bit will not affect this write cycle. The WR bit will be inhibited from being set unless the WREN bit is set.

At the completion of the write cycle, the WR bit is cleared in hardware and the EE Write Complete Interrupt Flag bit (EEIF) is set. The user can either enable this interrupt or poll this bit. EEIF must be cleared by software.

3.3 Write Verify

Depending on the application, good programming practice may dictate that the value written to the Data EEPROM should be verified (Example 3-3) to the desired value to be written. This should be used in applications where an EEPROM bit will be stressed near the specification limit.

Generally, the EEPROM write failure will be a bit which was written as a '0', but reads back as a '1' (due to leakage off the bit).

EXAMPLE 3-3:   WRITE VERIFY

```
        BCF     STATUS, RP0  ; Bank 0
        :                    ; Any code
        :                    ; can go here
        MOVF    EEDATA, W    ; Must be in Bank 0
        BSF     STATUS, RP0  ; Bank 1
READ
        BSF     EECON1, RD   ; YES, Read the
                             ; value written
        BCF     STATUS, RP0  ; Bank 0
                             ;
                             ; Is the value written
                             ; (in W reg) and
                             ; read (in EEDATA)
                             ; the same?
                             ;
        SUBWF   EEDATA, W    ;
        BTFSS   STATUS, Z    ; Is difference 0?
        GOTO    WRITE_ERR    ; NO, Write error
```

TABLE 3-1:   REGISTERS/BITS ASSOCIATED WITH DATA EEPROM

Address	Name	Bit 7	Bit 6	Bit 5	Bit 4	Bit 3	Bit 2	Bit 1	Bit 0	Value on Power-on Reset	Value on all other RESETS
08h	EEDATA	EEPROM Data Register								xxxx xxxx	uuuu uuuu
09h	EEADR	EEPROM Address Register								xxxx xxxx	uuuu uuuu
88h	EECON1	—	—	—	EEIF	WRERR	WREN	WR	RD	---0 x000	---0 q000
89h	EECON2	EEPROM Control Register 2								---- ----	---- ----

Legend: x = unknown, u = unchanged, - = unimplemented, read as '0', q = value depends upon condition.
          Shaded cells are not used by data EEPROM.

© 2001 Microchip Technology Inc.

PIC16F84A

4.0 I/O PORTS

Some pins for these I/O ports are multiplexed with an alternate function for the peripheral features on the device. In general, when a peripheral is enabled, that pin may not be used as a general purpose I/O pin.

Additional information on I/O ports may be found in the PICmicro™ Mid-Range Reference Manual (DS33023).

4.1 PORTA and TRISA Registers

PORTA is a 5-bit wide, bi-directional port. The corresponding data direction register is TRISA. Setting a TRISA bit (= 1) will make the corresponding PORTA pin an input (i.e., put the corresponding output driver in a Hi-Impedance mode). Clearing a TRISA bit (= 0) will make the corresponding PORTA pin an output (i.e., put the contents of the output latch on the selected pin).

> **Note:** On a Power-on Reset, these pins are configured as inputs and read as '0'.

Reading the PORTA register reads the status of the pins, whereas writing to it will write to the port latch. All write operations are read-modify-write operations. Therefore, a write to a port implies that the port pins are read. This value is modified and then written to the port data latch.

Pin RA4 is multiplexed with the Timer0 module clock input to become the RA4/T0CKI pin. The RA4/T0CKI pin is a Schmitt Trigger input and an open drain output. All other RA port pins have TTL input levels and full CMOS output drivers.

EXAMPLE 4-1: INITIALIZING PORTA

```
BCF     STATUS, RP0 ;
CLRF    PORTA       ; Initialize PORTA by
                    ; clearing output
                    ; data latches
BSF     STATUS, RP0 ; Select Bank 1
MOVLW   0x0F        ; Value used to
                    ; initialize data
                    ; direction
MOVWF   TRISA       ; Set RA<3:0> as inputs
                    ; RA4 as output
                    ; TRISA<7:5> are always
                    ; read as '0'.
```

FIGURE 4-1: BLOCK DIAGRAM OF PINS RA3:RA0

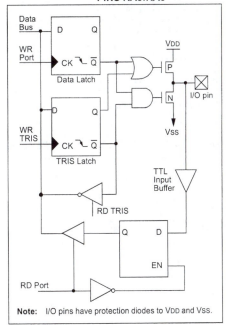

Note: I/O pins have protection diodes to VDD and VSS.

FIGURE 4-2: BLOCK DIAGRAM OF PIN RA4

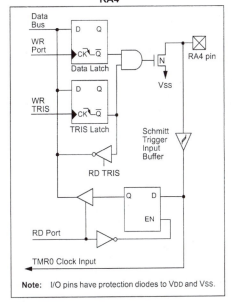

Note: I/O pins have protection diodes to VDD and VSS.

PIC16F84A

TABLE 4-1:   PORTA FUNCTIONS

Name	Bit0	Buffer Type	Function
RA0	bit0	TTL	Input/output
RA1	bit1	TTL	Input/output
RA2	bit2	TTL	Input/output
RA3	bit3	TTL	Input/output
RA4/T0CKI	bit4	ST	Input/output or external clock input for TMR0. Output is open drain type.

Legend: TTL = TTL input, ST = Schmitt Trigger input

TABLE 4-2:   SUMMARY OF REGISTERS ASSOCIATED WITH PORTA

Address	Name	Bit 7	Bit 6	Bit 5	Bit 4	Bit 3	Bit 2	Bit 1	Bit 0	Value on Power-on Reset	Value on all other RESETS
05h	PORTA	—	—	—	RA4/T0CKI	RA3	RA2	RA1	RA0	---x xxxx	---u uuuu
85h	TRISA	—	—	—	TRISA4	TRISA3	TRISA2	TRISA1	TRISA0	---1 1111	---1 1111

Legend: x = unknown, u = unchanged, - = unimplemented, read as '0'. Shaded cells are unimplemented, read as '0'.

PIC16F84A

4.2 PORTB and TRISB Registers

PORTB is an 8-bit wide, bi-directional port. The corresponding data direction register is TRISB. Setting a TRISB bit (= 1) will make the corresponding PORTB pin an input (i.e., put the corresponding output driver in a Hi-Impedance mode). Clearing a TRISB bit (= 0) will make the corresponding PORTB pin an output (i.e., put the contents of the output latch on the selected pin).

EXAMPLE 4-2: INITIALIZING PORTB

```
BCF     STATUS, RP0 ;
CLRF    PORTB       ; Initialize PORTB by
                    ; clearing output
                    ; data latches
BSF     STATUS, RP0 ; Select Bank 1
MOVLW   0xCF        ; Value used to
                    ; initialize data
                    ; direction
MOVWF   TRISB       ; Set RB<3:0> as inputs
                    ; RB<5:4> as outputs
                    ; RB<7:6> as inputs
```

Each of the PORTB pins has a weak internal pull-up. A single control bit can turn on all the pull-ups. This is performed by clearing bit RBPU (OPTION<7>). The weak pull-up is automatically turned off when the port pin is configured as an output. The pull-ups are disabled on a Power-on Reset.

Four of PORTB's pins, RB7:RB4, have an interrupt-on-change feature. Only pins configured as inputs can cause this interrupt to occur (i.e., any RB7:RB4 pin configured as an output is excluded from the interrupt-on-change comparison). The input pins (of RB7:RB4) are compared with the old value latched on the last read of PORTB. The "mismatch" outputs of RB7:RB4 are OR'ed together to generate the RB Port Change Interrupt with flag bit RBIF (INTCON<0>).

This interrupt can wake the device from SLEEP. The user, in the Interrupt Service Routine, can clear the interrupt in the following manner:

a) Any read or write of PORTB. This will end the mismatch condition.
b) Clear flag bit RBIF.

A mismatch condition will continue to set flag bit RBIF. Reading PORTB will end the mismatch condition and allow flag bit RBIF to be cleared.

The interrupt-on-change feature is recommended for wake-up on key depression operation and operations where PORTB is only used for the interrupt-on-change feature. Polling of PORTB is not recommended while using the interrupt-on-change feature.

FIGURE 4-3: BLOCK DIAGRAM OF PINS RB7:RB4

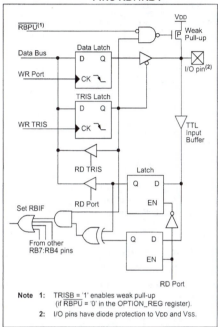

Note 1: TRISB = '1' enables weak pull-up (if RBPU = '0' in the OPTION_REG register).
2: I/O pins have diode protection to VDD and VSS.

FIGURE 4-4: BLOCK DIAGRAM OF PINS RB3:RB0

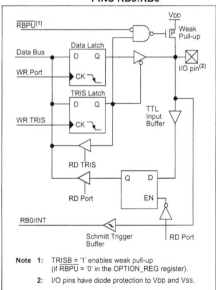

Note 1: TRISB = '1' enables weak pull-up (if RBPU = '0' in the OPTION_REG register).
2: I/O pins have diode protection to VDD and VSS.

PIC16F84A

TABLE 4-3:   PORTB FUNCTIONS

Name	Bit	Buffer Type	I/O Consistency Function
RB0/INT	bit0	TTL/ST[1]	Input/output pin or external interrupt input. Internal software programmable weak pull-up.
RB1	bit1	TTL	Input/output pin. Internal software programmable weak pull-up.
RB2	bit2	TTL	Input/output pin. Internal software programmable weak pull-up.
RB3	bit3	TTL	Input/output pin. Internal software programmable weak pull-up.
RB4	bit4	TTL	Input/output pin (with interrupt-on-change). Internal software programmable weak pull-up.
RB5	bit5	TTL	Input/output pin (with interrupt-on-change). Internal software programmable weak pull-up.
RB6	bit6	TTL/ST[2]	Input/output pin (with interrupt-on-change). Internal software programmable weak pull-up. Serial programming clock.
RB7	bit7	TTL/ST[2]	Input/output pin (with interrupt-on-change). Internal software programmable weak pull-up. Serial programming data.

Legend: TTL = TTL input, ST = Schmitt Trigger.
Note 1: This buffer is a Schmitt Trigger input when configured as the external interrupt.
2: This buffer is a Schmitt Trigger input when used in Serial Programming mode.

TABLE 4-4:   SUMMARY OF REGISTERS ASSOCIATED WITH PORTB

Address	Name	Bit 7	Bit 6	Bit 5	Bit 4	Bit 3	Bit 2	Bit 1	Bit 0	Value on Power-on Reset	Value on all other RESETS
06h	PORTB	RB7	RB6	RB5	RB4	RB3	RB2	RB1	RB0/INT	xxxx xxxx	uuuu uuuu
86h	TRISB	TRISB7	TRISB6	TRISB5	TRISB4	TRISB3	TRISB2	TRISB1	TRISB0	1111 1111	1111 1111
81h	OPTION_REG	RBPU	INTEDG	T0CS	T0SE	PSA	PS2	PS1	PS0	1111 1111	1111 1111
0Bh,8Bh	INTCON	GIE	EEIE	T0IE	INTE	RBIE	T0IF	INTF	RBIF	0000 000x	0000 000u

Legend: x = unknown, u = unchanged. Shaded cells are not used by PORTB.

© 2001 Microchip Technology Inc.

PIC16F84A

5.0 TIMER0 MODULE

The Timer0 module timer/counter has the following features:

- 8-bit timer/counter
- Readable and writable
- Internal or external clock select
- Edge select for external clock
- 8-bit software programmable prescaler
- Interrupt-on-overflow from FFh to 00h

Figure 5-1 is a simplified block diagram of the Timer0 module.

Additional information on timer modules is available in the PICmicro™ Mid-Range Reference Manual (DS33023).

5.1 Timer0 Operation

Timer0 can operate as a timer or as a counter.

Timer mode is selected by clearing bit T0CS (OPTION_REG<5>). In Timer mode, the Timer0 module will increment every instruction cycle (without prescaler). If the TMR0 register is written, the increment is inhibited for the following two instruction cycles. The user can work around this by writing an adjusted value to the TMR0 register.

Counter mode is selected by setting bit T0CS (OPTION_REG<5>). In Counter mode, Timer0 will increment, either on every rising or falling edge of pin RA4/T0CKI. The incrementing edge is determined by the Timer0 Source Edge Select bit, T0SE (OPTION_REG<4>). Clearing bit T0SE selects the rising edge. Restrictions on the external clock input are discussed below.

When an external clock input is used for Timer0, it must meet certain requirements. The requirements ensure the external clock can be synchronized with the internal phase clock (T$_{OSC}$). Also, there is a delay in the actual incrementing of Timer0 after synchronization.

Additional information on external clock requirements is available in the PICmicro™ Mid-Range Reference Manual, (DS33023).

5.2 Prescaler

An 8-bit counter is available as a prescaler for the Timer0 module, or as a postscaler for the Watchdog Timer, respectively (Figure 5-2). For simplicity, this counter is being referred to as "prescaler" throughout this data sheet. Note that there is only one prescaler available which is mutually exclusively shared between the Timer0 module and the Watchdog Timer. Thus, a prescaler assignment for the Timer0 module means that there is no prescaler for the Watchdog Timer, and vice-versa.

The prescaler is not readable or writable.

The PSA and PS2:PS0 bits (OPTION_REG<3:0>) determine the prescaler assignment and prescale ratio.

Clearing bit PSA will assign the prescaler to the Timer0 module. When the prescaler is assigned to the Timer0 module, prescale values of 1:2, 1:4, ..., 1:256 are selectable.

Setting bit PSA will assign the prescaler to the Watchdog Timer (WDT). When the prescaler is assigned to the WDT, prescale values of 1:1, 1:2, ..., 1:128 are selectable.

When assigned to the Timer0 module, all instructions writing to the TMR0 register (e.g., `CLRF 1`, `MOVWF 1`, `BSF 1`, etc.) will clear the prescaler. When assigned to WDT, a `CLRWDT` instruction will clear the prescaler along with the WDT.

> **Note:** Writing to TMR0 when the prescaler is assigned to Timer0 will clear the prescaler count, but will not change the prescaler assignment.

FIGURE 5-1: **TIMER0 BLOCK DIAGRAM**

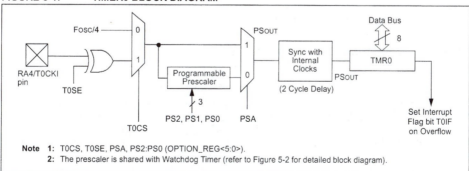

Note 1: T0CS, T0SE, PSA, PS2:PS0 (OPTION_REG<5:0>).
   2: The prescaler is shared with Watchdog Timer (refer to Figure 5-2 for detailed block diagram).

PIC16F84A

5.2.1 SWITCHING PRESCALER ASSIGNMENT

The prescaler assignment is fully under software control (i.e., it can be changed "on the fly" during program execution).

Note:	To avoid an unintended device RESET, a specific instruction sequence (shown in the PICmicro™ Mid-Range Reference Manual, DS33023) must be executed when changing the prescaler assignment from Timer0 to the WDT. This sequence must be followed even if the WDT is disabled.

5.3 Timer0 Interrupt

The TMR0 interrupt is generated when the TMR0 register overflows from FFh to 00h. This overflow sets bit T0IF (INTCON<2>). The interrupt can be masked by clearing bit T0IE (INTCON<5>). Bit T0IF must be cleared in software by the Timer0 module Interrupt Service Routine before re-enabling this interrupt. The TMR0 interrupt cannot awaken the processor from SLEEP since the timer is shut-off during SLEEP.

FIGURE 5-2: BLOCK DIAGRAM OF THE TIMER0/WDT PRESCALER

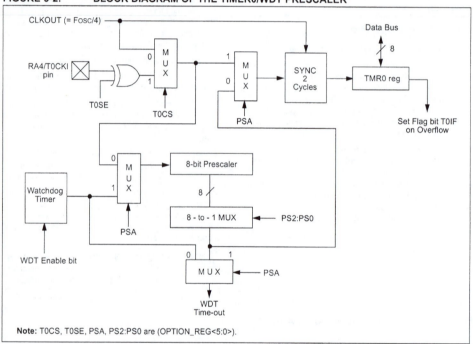

Note: T0CS, T0SE, PSA, PS2:PS0 are (OPTION_REG<5:0>).

TABLE 5-1: REGISTERS ASSOCIATED WITH TIMER0

Address	Name	Bit 7	Bit 6	Bit 5	Bit 4	Bit 3	Bit 2	Bit 1	Bit 0	Value on POR, BOR	Value on all other RESETS
01h	TMR0	Timer0 Module Register								xxxx xxxx	uuuu uuuu
0Bh,8Bh	INTCON	GIE	EEIE	T0IE	INTE	RBIE	T0IF	INTF	RBIF	0000 000x	0000 000u
81h	OPTION_REG	RBPU	INTEDG	T0CS	T0SE	PSA	PS2	PS1	PS0	1111 1111	1111 1111
85h	TRISA	—	—	—	PORTA Data Direction Register					---1 1111	---1 1111

Legend: x = unknown, u = unchanged, - = unimplemented locations read as '0'. Shaded cells are not used by Timer0.

PIC16F84A

6.0 SPECIAL FEATURES OF THE CPU

What sets a microcontroller apart from other processors are special circuits to deal with the needs of real time applications. The PIC16F84A has a host of such features intended to maximize system reliability, minimize cost through elimination of external components, provide power saving operating modes and offer code protection. These features are:

- OSC Selection
- RESET
  - Power-on Reset (POR)
  - Power-up Timer (PWRT)
  - Oscillator Start-up Timer (OST)
- Interrupts
- Watchdog Timer (WDT)
- SLEEP
- Code Protection
- ID Locations
- In-Circuit Serial Programming™ (ICSP™)

The PIC16F84A has a Watchdog Timer which can be shut-off only through configuration bits. It runs off its own RC oscillator for added reliability. There are two timers that offer necessary delays on power-up. One is the Oscillator Start-up Timer (OST), intended to keep the chip in RESET until the crystal oscillator is stable. The other is the Power-up Timer (PWRT), which provides a fixed delay of 72 ms (nominal) on power-up only. This design keeps the device in RESET while the power supply stabilizes. With these two timers on-chip, most applications need no external RESET circuitry.

SLEEP mode offers a very low current power-down mode. The user can wake-up from SLEEP through external RESET, Watchdog Timer Time-out or through an interrupt. Several oscillator options are provided to allow the part to fit the application. The RC oscillator option saves system cost while the LP crystal option saves power. A set of configuration bits are used to select the various options.

Additional information on special features is available in the PICmicro™ Mid-Range Reference Manual (DS33023).

6.1 Configuration Bits

The configuration bits can be programmed (read as '0'), or left unprogrammed (read as '1'), to select various device configurations. These bits are mapped in program memory location 2007h.

Address 2007h is beyond the user program memory space and it belongs to the special test/configuration memory space (2000h - 3FFFh). This space can only be accessed during programming.

REGISTER 6-1: PIC16F84A CONFIGURATION WORD

R/P-u	R/P-u	R/P-u	R/P-u	R/P-u	R/P-u	R/P-u	R/P-u	R/P-u	R/P-u	R/P-u	R/P-u	R/P-u	R/P-u
CP	CP	CP	CP	CP	CP	CP	CP	CP	CP	PWRTE	WDTE	F0SC1	F0SC0

bit13   bit0

bit 13-4 **CP:** Code Protection bit
1 = Code protection disabled
0 = All program memory is code protected

bit 3 **PWRTE**: Power-up Timer Enable bit
1 = Power-up Timer is disabled
0 = Power-up Timer is enabled

bit 2 **WDTE**: Watchdog Timer Enable bit
1 = WDT enabled
0 = WDT disabled

bit 1-0 **FOSC1:FOSC0**: Oscillator Selection bits
11 = RC oscillator
10 = HS oscillator
01 = XT oscillator
00 = LP oscillator

PIC16F84A

6.2 Oscillator Configurations

6.2.1 OSCILLATOR TYPES

The PIC16F84A can be operated in four different oscillator modes. The user can program two configuration bits (FOSC1 and FOSC0) to select one of these four modes:

- LP   Low Power Crystal
- XT   Crystal/Resonator
- HS   High Speed Crystal/Resonator
- RC   Resistor/Capacitor

6.2.2 CRYSTAL OSCILLATOR/CERAMIC RESONATORS

In XT, LP, or HS modes, a crystal or ceramic resonator is connected to the OSC1/CLKIN and OSC2/CLKOUT pins to establish oscillation (Figure 6-1).

FIGURE 6-1:   **CRYSTAL/CERAMIC RESONATOR OPERATION (HS, XT OR LP OSC CONFIGURATION)**

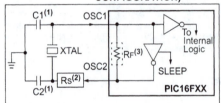

Note 1: See Table 6-1 for recommended values of C1 and C2.

      2: A series resistor (Rs) may be required for AT strip cut crystals.

The PIC16F84A oscillator design requires the use of a parallel cut crystal. Use of a series cut crystal may give a frequency out of the crystal manufacturers specifications. When in XT, LP, or HS modes, the device can have an external clock source to drive the OSC1/CLKIN pin (Figure 6-2).

FIGURE 6-2:   **EXTERNAL CLOCK INPUT OPERATION (HS, XT OR LP OSC CONFIGURATION)**

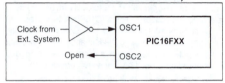

TABLE 6-1:   **CAPACITOR SELECTION FOR CERAMIC RESONATORS**

Ranges Tested:			
Mode	**Freq**	**OSC1/C1**	**OSC2/C2**
XT	455 kHz	47 - 100 pF	47 - 100 pF
	2.0 MHz	15 - 33 pF	15 - 33 pF
	4.0 MHz	15 - 33 pF	15 - 33 pF
HS	8.0 MHz	15 - 33 pF	15 - 33 pF
	10.0 MHz	15 - 33 pF	15 - 33 pF

Note: Recommended values of C1 and C2 are identical to the ranges tested in this table. Higher capacitance increases the stability of the oscillator, but also increases the start-up time. These values are for design guidance only. Since each resonator has its own characteristics, the user should consult the resonator manufacturer for the appropriate values of external components.

Note: When using resonators with frequencies above 3.5 MHz, the use of HS mode rather than XT mode, is recommended. HS mode may be used at any VDD for which the controller is rated.

TABLE 6-2: **CAPACITOR SELECTION FOR CRYSTAL OSCILLATOR**

Mode	Freq	OSC1/C1	OSC2/C2
LP	32 kHz	68 - 100 pF	68 - 100 pF
	200 kHz	15 - 33 pF	15 - 33 pF
XT	100 kHz	100 - 150 pF	100 - 150 pF
	2 MHz	15 - 33 pF	15 - 33 pF
	4 MHz	15 - 33 pF	15 - 33 pF
HS	4 MHz	15 - 33 pF	15 - 33 pF
	20 MHz	15 - 33 pF	15 - 33 pF

Note: Higher capacitance increases the stability of the oscillator, but also increases the start-up time. These values are for design guidance only. Rs may be required in HS mode, as well as XT mode, to avoid over-driving crystals with low drive level specification. Since each crystal has its own characteristics, the user should consult the crystal manufacturer for appropriate values of external components.
For V_{DD} > 4.5V, C1 = C2 ≈ 30 pF is recommended.

6.2.3 RC OSCILLATOR

For timing insensitive applications, the RC device option offers additional cost savings. The RC oscillator frequency is a function of the supply voltage, the resistor (R_{EXT}) values, capacitor (C_{EXT}) values, and the operating temperature. In addition to this, the oscillator frequency will vary from unit to unit due to normal process parameter variation. Furthermore, the difference in lead frame capacitance between package types also affects the oscillation frequency, especially for low C_{EXT} values. The user needs to take into account variation, due to tolerance of the external R and C components. Figure 6-3 shows how an R/C combination is connected to the PIC16F84A.

FIGURE 6-3: **RC OSCILLATOR MODE**

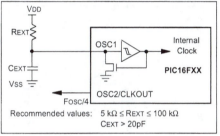

Recommended values: 5 kΩ ≤ R_{EXT} ≤ 100 kΩ
C_{EXT} > 20pF

PIC16F84A

6.3   RESET

The PIC16F84A differentiates between various kinds of RESET:

- Power-on Reset (POR)
- $\overline{\text{MCLR}}$ during normal operation
- $\overline{\text{MCLR}}$ during SLEEP
- WDT Reset (during normal operation)
- WDT Wake-up (during SLEEP)

Figure 6-4 shows a simplified block diagram of the On-Chip RESET Circuit. The $\overline{\text{MCLR}}$ Reset path has a noise filter to ignore small pulses. The electrical specifications state the pulse width requirements for the $\overline{\text{MCLR}}$ pin.

Some registers are not affected in any RESET condition; their status is unknown on a POR and unchanged in any other RESET. Most other registers are reset to a "RESET state" on POR, $\overline{\text{MCLR}}$ or WDT Reset during normal operation and on $\overline{\text{MCLR}}$ during SLEEP. They are not affected by a WDT Reset during SLEEP, since this RESET is viewed as the resumption of normal operation.

Table 6-3 gives a description of RESET conditions for the program counter (PC) and the STATUS register. Table 6-4 gives a full description of RESET states for all registers.

The $\overline{\text{TO}}$ and $\overline{\text{PD}}$ bits are set or cleared differently in different RESET situations (Section 6.7). These bits are used in software to determine the nature of the RESET.

FIGURE 6-4:   SIMPLIFIED BLOCK DIAGRAM OF ON-CHIP RESET CIRCUIT

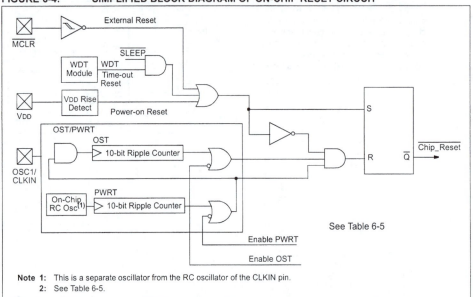

Note 1:   This is a separate oscillator from the RC oscillator of the CLKIN pin.
    2:    See Table 6-5.

TABLE 6-3:   RESET CONDITION FOR PROGRAM COUNTER AND THE STATUS REGISTER

Condition	Program Counter	STATUS Register
Power-on Reset	000h	0001 1xxx
$\overline{\text{MCLR}}$ during normal operation	000h	000u uuuu
$\overline{\text{MCLR}}$ during SLEEP	000h	0001 0uuu
WDT Reset (during normal operation)	000h	0000 1uuu
WDT Wake-up	PC + 1	uuu0 0uuu
Interrupt wake-up from SLEEP	PC + 1[1]	uuu1 0uuu

Legend: u = unchanged, x = unknown
Note 1:   When the wake-up is due to an interrupt and the GIE bit is set, the PC is loaded with the interrupt vector (0004h).

PIC16F84A

TABLE 6-4: RESET CONDITIONS FOR ALL REGISTERS

Register	Address	Power-on Reset	MCLR during: – normal operation – SLEEP WDT Reset during normal operation	Wake-up from SLEEP: – through interrupt – through WDT Time-out
W	—	xxxx xxxx	uuuu uuuu	uuuu uuuu
INDF	00h	---- ----	---- ----	---- ----
TMR0	01h	xxxx xxxx	uuuu uuuu	uuuu uuuu
PCL	02h	0000 0000	0000 0000	PC + 1(2)
STATUS	03h	0001 1xxx	000q quuu(3)	uuuq quuu(3)
FSR	04h	xxxx xxxx	uuuu uuuu	uuuu uuuu
PORTA(4)	05h	---x xxxx	---u uuuu	---u uuuu
PORTB(5)	06h	xxxx xxxx	uuuu uuuu	uuuu uuuu
EEDATA	08h	xxxx xxxx	uuuu uuuu	uuuu uuuu
EEADR	09h	xxxx xxxx	uuuu uuuu	uuuu uuuu
PCLATH	0Ah	---0 0000	---0 0000	---u uuuu
INTCON	0Bh	0000 000x	0000 000u	uuuu uuuu(1)
INDF	80h	---- ----	---- ----	---- ----
OPTION_REG	81h	1111 1111	1111 1111	uuuu uuuu
PCL	82h	0000 0000	0000 0000	PC + 1(2)
STATUS	83h	0001 1xxx	000q quuu(3)	uuuq quuu(3)
FSR	84h	xxxx xxxx	uuuu uuuu	uuuu uuuu
TRISA	85h	---1 1111	---1 1111	---u uuuu
TRISB	86h	1111 1111	1111 1111	uuuu uuuu
EECON1	88h	---0 x000	---0 q000	---0 uuuu
EECON2	89h	---- ----	---- ----	---- ----
PCLATH	8Ah	---0 0000	---0 0000	---u uuuu
INTCON	8Bh	0000 000x	0000 000u	uuuu uuuu(1)

Legend: u = unchanged, x = unknown, - = unimplemented bit, read as '0', q = value depends on condition

Note 1: One or more bits in INTCON will be affected (to cause wake-up).
    2: When the wake-up is due to an interrupt and the GIE bit is set, the PC is loaded with the interrupt vector (0004h).
    3: Table 6-3 lists the RESET value for each specific condition.
    4: On any device RESET, these pins are configured as inputs.
    5: This is the value that will be in the port output latch.

PIC16F84A

6.4   Power-on Reset (POR)

A Power-on Reset pulse is generated on-chip when VDD rise is detected (in the range of 1.2V - 1.7V). To take advantage of the POR, just tie the $\overline{\text{MCLR}}$ pin directly (or through a resistor) to VDD. This will eliminate external RC components usually needed to create Power-on Reset. A minimum rise time for VDD must be met for this to operate properly. See Electrical Specifications for details.

When the device starts normal operation (exits the RESET condition), device operating parameters (voltage, frequency, temperature, etc.) must be met to ensure operation. If these conditions are not met, the device must be held in RESET until the operating conditions are met.

For additional information, refer to Application Note AN607, "*Power-up Trouble Shooting.*"

The POR circuit does not produce an internal RESET when VDD declines.

6.5   Power-up Timer (PWRT)

The Power-up Timer (PWRT) provides a fixed 72 ms nominal time-out (TPWRT) from POR (Figures 6-6 through 6-9). The Power-up Timer operates on an internal RC oscillator. The chip is kept in RESET as long as the PWRT is active. The PWRT delay allows the VDD to rise to an acceptable level (possible exception shown in Figure 6-9).

A configuration bit, $\overline{\text{PWRTE}}$, can enable/disable the PWRT. See Register 6-1 for the operation of the $\overline{\text{PWRTE}}$ bit for a particular device.

The power-up time delay TPWRT will vary from chip to chip due to VDD, temperature, and process variation. See DC parameters for details.

6.6   Oscillator Start-up Timer (OST)

The Oscillator Start-up Timer (OST) provides a 1024 oscillator cycle delay (from OSC1 input) after the PWRT delay ends (Figure 6-6, Figure 6-7, Figure 6-8 and Figure 6-9). This ensures the crystal oscillator or resonator has started and stabilized.

The OST time-out (TOST) is invoked only for XT, LP and HS modes and only on Power-on Reset or wake-up from SLEEP.

When VDD rises very slowly, it is possible that the TPWRT time-out and TOST time-out will expire before VDD has reached its final value. In this case (Figure 6-9), an external Power-on Reset circuit may be necessary (Figure 6-5).

FIGURE 6-5:   EXTERNAL POWER-ON RESET CIRCUIT (FOR SLOW VDD POWER-UP)

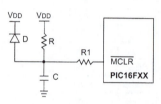

Note   1:   External Power-on Reset circuit is required only if VDD power-up rate is too slow. The diode D helps discharge the capacitor quickly when VDD powers down.

2:   R < 40 kΩ is recommended to make sure that voltage drop across R does not exceed 0.2V (max leakage current spec on $\overline{\text{MCLR}}$ pin is 5 µA). A larger voltage drop will degrade VIH level on the $\overline{\text{MCLR}}$ pin.

3:   R1 = 100Ω to 1 kΩ will limit any current flowing into $\overline{\text{MCLR}}$ from external capacitor C, in the event of a $\overline{\text{MCLR}}$ pin breakdown due to ESD or EOS.

PIC16F84A

FIGURE 6-6:　TIME-OUT SEQUENCE ON POWER-UP ($\overline{\text{MCLR}}$ NOT TIED TO V$_{DD}$): CASE 1

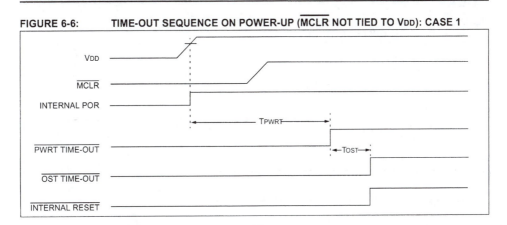

FIGURE 6-7:　TIME-OUT SEQUENCE ON POWER-UP ($\overline{\text{MCLR}}$ NOT TIED TO V$_{DD}$): CASE 2

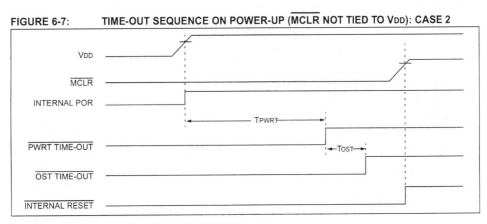

FIGURE 6-8:　TIME-OUT SEQUENCE ON POWER-UP ($\overline{\text{MCLR}}$ TIED TO V$_{DD}$): FAST V$_{DD}$ RISE TIME

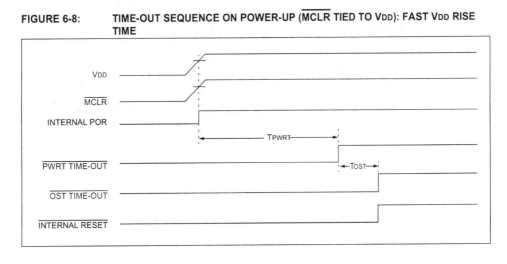

PIC16F84A

FIGURE 6-9:   TIME-OUT SEQUENCE ON POWER-UP ($\overline{\text{MCLR}}$ TIED TO V_{DD}): SLOW V_{DD} RISE TIME

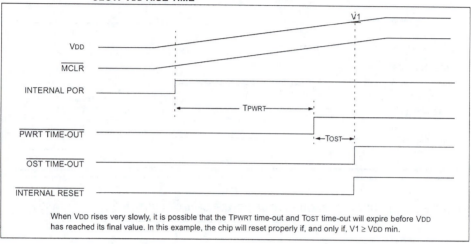

When V_{DD} rises very slowly, it is possible that the T_{PWRT} time-out and T_{OST} time-out will expire before V_{DD} has reached its final value. In this example, the chip will reset properly if, and only if, V1 ≥ V_{DD} min.

6.7 Time-out Sequence and Power-down Status Bits ($\overline{\text{TO}}$/$\overline{\text{PD}}$)

On power-up (Figures 6-6 through 6-9), the time-out sequence is as follows:

1. PWRT time-out is invoked after a POR has expired.
2. Then, the OST is activated.

The total time-out will vary based on oscillator configuration and PWRTE configuration bit status. For example, in RC mode with the PWRT disabled, there will be no time-out at all.

TABLE 6-5:   TIME-OUT IN VARIOUS SITUATIONS

Oscillator Configuration	Power-up		Wake-up from SLEEP
	PWRT Enabled	**PWRT Disabled**	
XT, HS, LP	72 ms + 1024T_{OSC}	1024T_{OSC}	1024T_{OSC}
RC	72 ms	—	—

Since the time-outs occur from the POR pulse, if $\overline{\text{MCLR}}$ is kept low long enough, the time-outs will expire. Then bringing $\overline{\text{MCLR}}$ high, execution will begin immediately (Figure 6-6). This is useful for testing purposes or to synchronize more than one PIC16F84A device when operating in parallel.

Table 6-6 shows the significance of the $\overline{\text{TO}}$ and $\overline{\text{PD}}$ bits. Table 6-3 lists the RESET conditions for some special registers, while Table 6-4 lists the RESET conditions for all the registers.

TABLE 6-6:   STATUS BITS AND THEIR SIGNIFICANCE

$\overline{\text{TO}}$	$\overline{\text{PD}}$	Condition
1	1	Power-on Reset
0	x	Illegal, $\overline{\text{TO}}$ is set on $\overline{\text{POR}}$
x	0	Illegal, $\overline{\text{PD}}$ is set on $\overline{\text{POR}}$
0	1	WDT Reset (during normal operation)
0	0	WDT Wake-up
1	1	$\overline{\text{MCLR}}$ during normal operation
1	0	$\overline{\text{MCLR}}$ during SLEEP or interrupt wake-up from SLEEP

PIC16F84A

6.8 Interrupts

The PIC16F84A has 4 sources of interrupt:

- External interrupt RB0/INT pin
- TMR0 overflow interrupt
- PORTB change interrupts (pins RB7:RB4)
- Data EEPROM write complete interrupt

The interrupt control register (INTCON) records individual interrupt requests in flag bits. It also contains the individual and global interrupt enable bits.

The global interrupt enable bit, GIE (INTCON<7>), enables (if set) all unmasked interrupts or disables (if cleared) all interrupts. Individual interrupts can be disabled through their corresponding enable bits in INTCON register. Bit GIE is cleared on RESET.

The "return from interrupt" instruction, RETFIE, exits interrupt routine as well as sets the GIE bit, which re-enables interrupts.

The RB0/INT pin interrupt, the RB port change interrupt and the TMR0 overflow interrupt flags are contained in the INTCON register.

When an interrupt is responded to, the GIE bit is cleared to disable any further interrupt, the return address is pushed onto the stack and the PC is loaded with 0004h. For external interrupt events, such as the RB0/INT pin or PORTB change interrupt, the interrupt latency will be three to four instruction cycles. The exact latency depends when the interrupt event occurs. The latency is the same for both one and two cycle instructions. Once in the Interrupt Service Routine, the source(s) of the interrupt can be determined by polling the interrupt flag bits. The interrupt flag bit(s) must be cleared in software before re-enabling interrupts to avoid infinite interrupt requests.

> **Note:** Individual interrupt flag bits are set regardless of the status of their corresponding mask bit or the GIE bit.

6.8.1 INT INTERRUPT

External interrupt on RB0/INT pin is edge triggered: either rising if INTEDG bit (OPTION_REG<6>) is set, or falling if INTEDG bit is clear. When a valid edge appears on the RB0/INT pin, the INTF bit (INTCON<1>) is set. This interrupt can be disabled by clearing control bit INTE (INTCON<4>). Flag bit INTF must be cleared in software via the Interrupt Service Routine before re-enabling this interrupt. The INT interrupt can wake the processor from SLEEP (Section 6.11) only if the INTE bit was set prior to going into SLEEP. The status of the GIE bit decides whether the processor branches to the interrupt vector following wake-up.

6.8.2 TMR0 INTERRUPT

An overflow (FFh → 00h) in TMR0 will set flag bit T0IF (INTCON<2>). The interrupt can be enabled/disabled by setting/clearing enable bit T0IE (INTCON<5>) (Section 5.0).

6.8.3 PORTB INTERRUPT

An input change on PORTB<7:4> sets flag bit RBIF (INTCON<0>). The interrupt can be enabled/disabled by setting/clearing enable bit RBIE (INTCON<3>) (Section 4.2).

> **Note:** For a change on the I/O pin to be recognized, the pulse width must be at least T$_{CY}$ wide.

6.8.4 DATA EEPROM INTERRUPT

At the completion of a data EEPROM write cycle, flag bit EEIF (EECON1<4>) will be set. The interrupt can be enabled/disabled by setting/clearing enable bit EEIE (INTCON<6>) (Section 3.0).

FIGURE 6-10:   INTERRUPT LOGIC

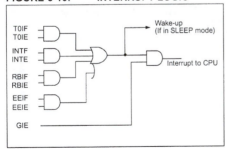

PIC16F84A

6.9 Context Saving During Interrupts

During an interrupt, only the return PC value is saved on the stack. Typically, users wish to save key register values during an interrupt (e.g., W register and STATUS register). This is implemented in software.

The code in Example 6-1 stores and restores the STATUS and W register's values. The user defined registers, W_TEMP and STATUS_TEMP are the temporary storage locations for the W and STATUS registers values.

Example 6-1 does the following:

a) Stores the W register.
b) Stores the STATUS register in STATUS_TEMP.
c) Executes the Interrupt Service Routine code.
d) Restores the STATUS (and bank select bit) register.
e) Restores the W register.

EXAMPLE 6-1: SAVING STATUS AND W REGISTERS IN RAM

```
PUSH    MOVWF   W_TEMP          ; Copy W to TEMP register,
        SWAPF   STATUS,    W    ; Swap status to be saved into W
        MOVWF   STATUS_TEMP     ; Save status to STATUS_TEMP register
ISR     :                       :
        :                       ; Interrupt Service Routine
        :                       ; should configure Bank as required
        :                       ;
POP     SWAPF   STATUS_TEMP,W   ; Swap nibbles in STATUS_TEMP register
                                ; and place result into W
        MOVWF   STATUS          ; Move W into STATUS register
                                ; (sets bank to original state)
        SWAPF   W_TEMP,    F    ; Swap nibbles in W_TEMP and place result in W_TEMP
        SWAPF   W_TEMP,    W    ; Swap nibbles in W_TEMP and place result into W
```

6.10 Watchdog Timer (WDT)

The Watchdog Timer is a free running On-Chip RC Oscillator which does not require any external components. This RC oscillator is separate from the RC oscillator of the OSC1/CLKIN pin. That means that the WDT will run even if the clock on the OSC1/CLKIN and OSC2/CLKOUT pins of the device has been stopped, for example, by execution of a SLEEP instruction. During normal operation, a WDT time-out generates a device RESET. If the device is in SLEEP mode, a WDT wake-up causes the device to wake-up and continue with normal operation. The WDT can be permanently disabled by programming configuration bit WDTE as a '0' (Section 6.1).

6.10.1 WDT PERIOD

The WDT has a nominal time-out period of 18 ms, (with no prescaler). The time-out periods vary with temperature, VDD and process variations from part to part (see DC specs). If longer time-out periods are desired, a prescaler with a division ratio of up to 1:128 can be assigned to the WDT under software control by writing to the OPTION_REG register. Thus, time-out periods up to 2.3 seconds can be realized.

The CLRWDT and SLEEP instructions clear the WDT and the postscaler (if assigned to the WDT) and prevent it from timing out and generating a device RESET condition.

The $\overline{TO}$ bit in the STATUS register will be cleared upon a WDT time-out.

PIC16F84A

6.10.2   WDT PROGRAMMING CONSIDERATIONS

It should also be taken into account that under worst case conditions (VDD = Min., Temperature = Max., Max. WDT Prescaler), it may take several seconds before a WDT time-out occurs.

FIGURE 6-11:   **WATCHDOG TIMER BLOCK DIAGRAM**

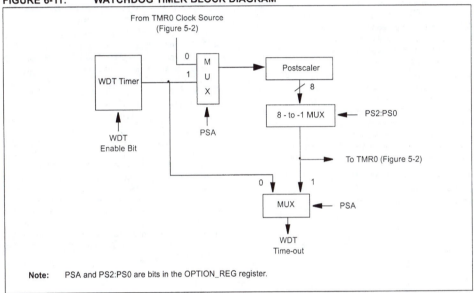

Note: PSA and PS2:PS0 are bits in the OPTION_REG register.

TABLE 6-7:   **SUMMARY OF REGISTERS ASSOCIATED WITH THE WATCHDOG TIMER**

Addr	Name	Bit 7	Bit 6	Bit 5	Bit 4	Bit 3	Bit 2	Bit 1	Bit 0	Value on Power-on Reset	Value on all other RESETS
2007h	Config. bits	(2)	(2)	(2)	(2)	PWRTE[1]	WDTE	FOSC1	FOSC0	(2)	
81h	OPTION_REG	RBPU	INTEDG	T0CS	T0SE	PSA	PS2	PS1	PS0	1111 1111	1111 1111

Legend: x = unknown. Shaded cells are not used by the WDT.
Note   1:   See Register 6-1 for operation of the PWRTE bit.
      2:   See Register 6-1 and Section 6.12 for operation of the code and data protection bits.

PIC16F84A

6.11 Power-down Mode (SLEEP)

A device may be powered down (SLEEP) and later powered up (wake-up from SLEEP).

6.11.1 SLEEP

The Power-down mode is entered by executing the SLEEP instruction.

If enabled, the Watchdog Timer is cleared (but keeps running), the $\overline{PD}$ bit (STATUS<3>) is cleared, the $\overline{TO}$ bit (STATUS<4>) is set, and the oscillator driver is turned off. The I/O ports maintain the status they had before the SLEEP instruction was executed (driving high, low, or hi-impedance).

For the lowest current consumption in SLEEP mode, place all I/O pins at either VDD or VSS, with no external circuitry drawing current from the I/O pins, and disable external clocks. I/O pins that are hi-impedance inputs should be pulled high or low externally to avoid switching currents caused by floating inputs. The T0CKI input should also be at VDD or VSS. The contribution from on-chip pull-ups on PORTB should be considered.

The $\overline{MCLR}$ pin must be at a logic high level (VIHMC).

It should be noted that a RESET generated by a WDT time-out does not drive the $\overline{MCLR}$ pin low.

6.11.2 WAKE-UP FROM SLEEP

The device can wake-up from SLEEP through one of the following events:

1. External RESET input on $\overline{MCLR}$ pin.
2. WDT wake-up (if WDT was enabled).
3. Interrupt from RB0/INT pin, RB port change, or data EEPROM write complete.

Peripherals cannot generate interrupts during SLEEP, since no on-chip Q clocks are present.

The first event ($\overline{MCLR}$ Reset) will cause a device RESET. The two latter events are considered a continuation of program execution. The $\overline{TO}$ and $\overline{PD}$ bits can be used to determine the cause of a device RESET. The $\overline{PD}$ bit, which is set on power-up, is cleared when SLEEP is invoked. The $\overline{TO}$ bit is cleared if a WDT time-out occurred (and caused wake-up).

While the SLEEP instruction is being executed, the next instruction (PC + 1) is pre-fetched. For the device to wake-up through an interrupt event, the corresponding interrupt enable bit must be set (enabled). Wake-up occurs regardless of the state of the GIE bit. If the GIE bit is clear (disabled), the device continues execution at the instruction after the SLEEP instruction. If the GIE bit is set (enabled), the device executes the instruction after the SLEEP instruction and then branches to the interrupt address (0004h). In cases where the execution of the instruction following SLEEP is not desirable, the user should have a NOP after the SLEEP instruction.

FIGURE 6-12:   WAKE-UP FROM SLEEP THROUGH INTERRUPT

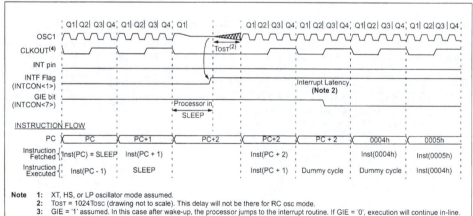

Note   1:   XT, HS, or LP oscillator mode assumed.
        2:   TOST = 1024TOSC (drawing not to scale). This delay will not be there for RC osc mode.
        3:   GIE = '1' assumed. In this case after wake-up, the processor jumps to the interrupt routine. If GIE = '0', execution will continue in-line.
        4:   CLKOUT is not available in these osc modes, but shown here for timing reference.

PIC16F84A

6.11.3   WAKE-UP USING INTERRUPTS

When global interrupts are disabled (GIE cleared) and any interrupt source has both its interrupt enable bit and interrupt flag bit set, one of the following will occur:

- If the interrupt occurs **before** the execution of a SLEEP instruction, the SLEEP instruction will complete as a NOP. Therefore, the WDT and WDT postscaler will not be cleared, the $\overline{TO}$ bit will not be set and $\overline{PD}$ bits will not be cleared.

- If the interrupt occurs **during or after** the execution of a SLEEP instruction, the device will immediately wake-up from SLEEP. The SLEEP instruction will be completely executed before the wake-up. Therefore, the WDT and WDT postscaler will be cleared, the $\overline{TO}$ bit will be set and the $\overline{PD}$ bit will be cleared.

Even if the flag bits were checked before executing a SLEEP instruction, it may be possible for flag bits to become set before the SLEEP instruction completes. To determine whether a SLEEP instruction executed, test the $\overline{PD}$ bit. If the $\overline{PD}$ bit is set, the SLEEP instruction was executed as a NOP.

To ensure that the WDT is cleared, a CLRWDT instruction should be executed before a SLEEP instruction.

6.12   Program Verification/Code Protection

If the code protection bit(s) have not been programmed, the on-chip program memory can be read out for verification purposes.

6.13   ID Locations

Four memory locations (2000h - 2004h) are designated as ID locations to store checksum or other code identification numbers. These locations are not accessible during normal execution but are readable and writable only during program/verify. Only the four Least Significant bits of ID location are usable.

6.14   In-Circuit Serial Programming

PIC16F84A microcontrollers can be serially programmed while in the end application circuit. This is simply done with two lines for clock and data, and three other lines for power, ground, and the programming voltage. Customers can manufacture boards with unprogrammed devices, and then program the microcontroller just before shipping the product, allowing the most recent firmware or custom firmware to be programmed.

For complete details of Serial Programming, please refer to the In-Circuit Serial Programming™ (ICSP™) Guide, (DS30277).

PIC16F84A

NOTES:

PIC16F84A

7.0 INSTRUCTION SET SUMMARY

Each PIC16CXX instruction is a 14-bit word, divided into an OPCODE which specifies the instruction type and one or more operands which further specify the operation of the instruction. The PIC16CXX instruction set summary in Table 7-2 lists **byte-oriented**, **bit-oriented**, and **literal and control** operations. Table 7-1 shows the opcode field descriptions.

For **byte-oriented** instructions, 'f' represents a file register designator and 'd' represents a destination designator. The file register designator specifies which file register is to be used by the instruction.

The destination designator specifies where the result of the operation is to be placed. If 'd' is zero, the result is placed in the W register. If 'd' is one, the result is placed in the file register specified in the instruction.

For **bit-oriented** instructions, 'b' represents a bit field designator which selects the number of the bit affected by the operation, while 'f' represents the address of the file in which the bit is located.

For **literal and control** operations, 'k' represents an eight or eleven bit constant or literal value.

TABLE 7-1: OPCODE FIELD DESCRIPTIONS

Field	Description
f	Register file address (0x00 to 0x7F)
W	Working register (accumulator)
b	Bit address within an 8-bit file register
k	Literal field, constant data or label
x	Don't care location (= 0 or 1) The assembler will generate code with x = 0. It is the recommended form of use for compatibility with all Microchip software tools.
d	Destination select; d = 0: store result in W, d = 1: store result in file register f. Default is d = 1
PC	Program Counter
TO	Time-out bit
PD	Power-down bit

The instruction set is highly orthogonal and is grouped into three basic categories:

- **Byte-oriented** operations
- **Bit-oriented** operations
- **Literal and control** operations

All instructions are executed within one single instruction cycle, unless a conditional test is true or the program counter is changed as a result of an instruction. In this case, the execution takes two instruction cycles with the second cycle executed as a NOP. One instruction cycle consists of four oscillator periods. Thus, for an oscillator frequency of 4 MHz, the normal instruction execution time is 1 μs. If a conditional test is true or the program counter is changed as a result of an instruction, the instruction execution time is 2 μs.

Table 7-2 lists the instructions recognized by the MPASM™ Assembler.

Figure 7-1 shows the general formats that the instructions can have.

> **Note:** To maintain upward compatibility with future PIC16CXX products, <u>do not use</u> the OPTION and TRIS instructions.

All examples use the following format to represent a hexadecimal number:

0xhh

where h signifies a hexadecimal digit.

FIGURE 7-1: GENERAL FORMAT FOR INSTRUCTIONS

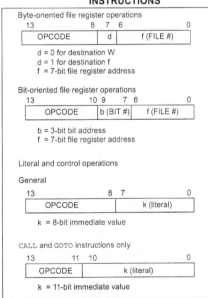

A description of each instruction is available in the PICmicro™ Mid-Range Reference Manual (DS33023).

PIC16F84A

TABLE 7-2:   PIC16CXXX INSTRUCTION SET

Mnemonic, Operands		Description	Cycles	14-Bit Opcode		Status Affected	Notes
				MSb	LSb		
BYTE-ORIENTED FILE REGISTER OPERATIONS							
ADDWF	f, d	Add W and f	1	00 0111	dfff ffff	C,DC,Z	1,2
ANDWF	f, d	AND W with f	1	00 0101	dfff ffff	Z	1,2
CLRF	f	Clear f	1	00 0001	1fff ffff	Z	2
CLRW	-	Clear W	1	00 0001	0xxx xxxx	Z	
COMF	f, d	Complement f	1	00 1001	dfff ffff	Z	1,2
DECF	f, d	Decrement f	1	00 0011	dfff ffff	Z	1,2
DECFSZ	f, d	Decrement f, Skip if 0	1 (2)	00 1011	dfff ffff		1,2,3
INCF	f, d	Increment f	1	00 1010	dfff ffff	Z	1,2
INCFSZ	f, d	Increment f, Skip if 0	1 (2)	00 1111	dfff ffff		1,2,3
IORWF	f, d	Inclusive OR W with f	1	00 0100	dfff ffff	Z	1,2
MOVF	f, d	Move f	1	00 1000	dfff ffff	Z	1,2
MOVWF	f	Move W to f	1	00 0000	1fff ffff		
NOP	-	No Operation	1	00 0000	0xx0 0000		
RLF	f, d	Rotate Left f through Carry	1	00 1101	dfff ffff	C	1,2
RRF	f, d	Rotate Right f through Carry	1	00 1100	dfff ffff	C	1,2
SUBWF	f, d	Subtract W from f	1	00 0010	dfff ffff	C,DC,Z	1,2
SWAPF	f, d	Swap nibbles in f	1	00 1110	dfff ffff		1,2
XORWF	f, d	Exclusive OR W with f	1	00 0110	dfff ffff	Z	1,2
BIT-ORIENTED FILE REGISTER OPERATIONS							
BCF	f, b	Bit Clear f	1	01 00bb	bfff ffff		1,2
BSF	f, b	Bit Set f	1	01 01bb	bfff ffff		1,2
BTFSC	f, b	Bit Test f, Skip if Clear	1 (2)	01 10bb	bfff ffff		3
BTFSS	f, b	Bit Test f, Skip if Set	1 (2)	01 11bb	bfff ffff		3
LITERAL AND CONTROL OPERATIONS							
ADDLW	k	Add literal and W	1	11 111x	kkkk kkkk	C,DC,Z	
ANDLW	k	AND literal with W	1	11 1001	kkkk kkkk	Z	
CALL	k	Call subroutine	2	10 0kkk	kkkk kkkk		
CLRWDT	-	Clear Watchdog Timer	1	00 0000	0110 0100	TO,PD	
GOTO	k	Go to address	2	10 1kkk	kkkk kkkk		
IORLW	k	Inclusive OR literal with W	1	11 1000	kkkk kkkk	Z	
MOVLW	k	Move literal to W	1	11 00xx	kkkk kkkk		
RETFIE	-	Return from interrupt	2	00 0000	0000 1001		
RETLW	k	Return with literal in W	2	11 01xx	kkkk kkkk		
RETURN	-	Return from Subroutine	2	00 0000	0000 1000		
SLEEP	-	Go into standby mode	1	00 0000	0110 0011	TO,PD	
SUBLW	k	Subtract W from literal	1	11 110x	kkkk kkkk	C,DC,Z	
XORLW	k	Exclusive OR literal with W	1	11 1010	kkkk kkkk	Z	

Note   1:   When an I/O register is modified as a function of itself (e.g., `MOVF PORTB, 1`), the value used will be that value present on the pins themselves. For example, if the data latch is '1' for a pin configured as input and is driven low by an external device, the data will be written back with a '0'.

   2:   If this instruction is executed on the TMR0 register (and, where applicable, d = 1), the prescaler will be cleared if assigned to the Timer0 Module.

   3:   If Program Counter (PC) is modified or a conditional test is true, the instruction requires two cycles. The second cycle is executed as a `NOP`.

Note: Additional information on the mid-range instruction set is available in the PICmicro™ Mid-Range MCU Family Reference Manual (DS33023).

PIC16F84A

7.1 Instruction Descriptions

ADDLW — **Add Literal and W**

Syntax:	[*label*] ADDLW k
Operands:	$0 \le k \le 255$
Operation:	$(W) + k \rightarrow (W)$
Status Affected:	C, DC, Z
Description:	The contents of the W register are added to the eight-bit literal 'k' and the result is placed in the W register.

ADDWF — **Add W and f**

Syntax:	[*label*] ADDWF f,d
Operands:	$0 \le f \le 127$ $d \in [0,1]$
Operation:	$(W) + (f) \rightarrow (destination)$
Status Affected:	C, DC, Z
Description:	Add the contents of the W register with register 'f'. If 'd' is 0, the result is stored in the W register. If 'd' is 1, the result is stored back in register 'f'.

ANDLW — **AND Literal with W**

Syntax:	[*label*] ANDLW k
Operands:	$0 \le k \le 255$
Operation:	$(W) .AND. (k) \rightarrow (W)$
Status Affected:	Z
Description:	The contents of W register are AND'ed with the eight-bit literal 'k'. The result is placed in the W register.

ANDWF — **AND W with f**

Syntax:	[*label*] ANDWF f,d
Operands:	$0 \le f \le 127$ $d \in [0,1]$
Operation:	$(W) .AND. (f) \rightarrow (destination)$
Status Affected:	Z
Description:	AND the W register with register 'f'. If 'd' is 0, the result is stored in the W register. If 'd' is 1, the result is stored back in register 'f'.

BCF — **Bit Clear f**

Syntax:	[*label*] BCF f,b
Operands:	$0 \le f \le 127$ $0 \le b \le 7$
Operation:	$0 \rightarrow (f<b>)$
Status Affected:	None
Description:	Bit 'b' in register 'f' is cleared.

BSF — **Bit Set f**

Syntax:	[*label*] BSF f,b
Operands:	$0 \le f \le 127$ $0 \le b \le 7$
Operation:	$1 \rightarrow (f<b>)$
Status Affected:	None
Description:	Bit 'b' in register 'f' is set.

BTFSS — **Bit Test f, Skip if Set**

Syntax:	[*label*] BTFSS f,b
Operands:	$0 \le f \le 127$ $0 \le b < 7$
Operation:	skip if $(f<b>) = 1$
Status Affected:	None
Description:	If bit 'b' in register 'f' is '0', the next instruction is executed. If bit 'b' is '1', then the next instruction is discarded and a NOP is executed instead, making this a 2TCY instruction.

PIC16F84A

BTFSC   **Bit Test, Skip if Clear**

Syntax:	[*label*] BTFSC f,b
Operands:	$0 \leq f \leq 127$ $0 \leq b \leq 7$
Operation:	skip if (f<b>) = 0
Status Affected:	None
Description:	If bit 'b' in register 'f' is '1', the next instruction is executed. If bit 'b' in register 'f' is '0', the next instruction is discarded, and a NOP is executed instead, making this a 2TCY instruction.

CALL   **Call Subroutine**

Syntax:	[*label*] CALL k
Operands:	$0 \leq k \leq 2047$
Operation:	(PC)+ 1 → TOS, k → PC<10:0>, (PCLATH<4:3>) → PC<12:11>
Status Affected:	None
Description:	Call Subroutine. First, return address (PC+1) is pushed onto the stack. The eleven-bit immediate address is loaded into PC bits <10:0>. The upper bits of the PC are loaded from PCLATH. CALL is a two-cycle instruction.

CLRF   **Clear f**

Syntax:	[*label*] CLRF f
Operands:	$0 \leq f \leq 127$
Operation:	00h → (f) 1 → Z
Status Affected:	Z
Description:	The contents of register 'f' are cleared and the Z bit is set.

CLRW   **Clear W**

Syntax:	[*label*] CLRW
Operands:	None
Operation:	00h → (W) 1 → Z
Status Affected:	Z
Description:	W register is cleared. Zero bit (Z) is set.

CLRWDT   **Clear Watchdog Timer**

Syntax:	[*label*] CLRWDT
Operands:	None
Operation:	00h → WDT 0 → WDT prescaler, $1 \to \overline{TO}$ $1 \to \overline{PD}$
Status Affected:	$\overline{TO}$, $\overline{PD}$
Description:	CLRWDT instruction resets the Watchdog Timer. It also resets the prescaler of the WDT. Status bits $\overline{TO}$ and $\overline{PD}$ are set.

COMF   **Complement f**

Syntax:	[*label*] COMF f,d
Operands:	$0 \leq f \leq 127$ $d \in [0,1]$
Operation:	$(\bar{f}) \to$ (destination)
Status Affected:	Z
Description:	The contents of register 'f' are complemented. If 'd' is 0, the result is stored in W. If 'd' is 1, the result is stored back in register 'f'.

DECF   **Decrement f**

Syntax:	[*label*] DECF f,d
Operands:	$0 \leq f \leq 127$ $d \in [0,1]$
Operation:	(f) - 1 → (destination)
Status Affected:	Z
Description:	Decrement register 'f'. If 'd' is 0, the result is stored in the W register. If 'd' is 1, the result is stored back in register 'f'.

PIC16F84A

DECFSZ	**Decrement f, Skip if 0**
Syntax:	[*label*] DECFSZ f,d
Operands:	$0 \le f \le 127$ $d \in [0,1]$
Operation:	(f) - 1 → (destination); skip if result = 0
Status Affected:	None
Description:	The contents of register 'f' are decremented. If 'd' is 0, the result is placed in the W register. If 'd' is 1, the result is placed back in register 'f'. If the result is 1, the next instruction is executed. If the result is 0, then a NOP is executed instead, making it a 2TCY instruction.

GOTO	**Unconditional Branch**
Syntax:	[*label*] GOTO k
Operands:	$0 \le k \le 2047$
Operation:	k → PC<10:0> PCLATH<4:3> → PC<12:11>
Status Affected:	None
Description:	GOTO is an unconditional branch. The eleven-bit immediate value is loaded into PC bits <10:0>. The upper bits of PC are loaded from PCLATH<4:3>. GOTO is a two-cycle instruction.

INCF	**Increment f**
Syntax:	[*label*] INCF f,d
Operands:	$0 \le f \le 127$ $d \in [0,1]$
Operation:	(f) + 1 → (destination)
Status Affected:	Z
Description:	The contents of register 'f' are incremented. If 'd' is 0, the result is placed in the W register. If 'd' is 1, the result is placed back in register 'f'.

INCFSZ	**Increment f, Skip if 0**
Syntax:	[*label*] INCFSZ f,d
Operands:	$0 \le f \le 127$ $d \in [0,1]$
Operation:	(f) + 1 → (destination), skip if result = 0
Status Affected:	None
Description:	The contents of register 'f' are incremented. If 'd' is 0, the result is placed in the W register. If 'd' is 1, the result is placed back in register 'f'. If the result is 1, the next instruction is executed. If the result is 0, a NOP is executed instead, making it a 2TCY instruction.

IORLW	**Inclusive OR Literal with W**
Syntax:	[*label*] IORLW k
Operands:	$0 \le k \le 255$
Operation:	(W) .OR. k → (W)
Status Affected:	Z
Description:	The contents of the W register are OR'ed with the eight-bit literal 'k'. The result is placed in the W register.

IORWF	**Inclusive OR W with f**
Syntax:	[*label*] IORWF f,d
Operands:	$0 \le f \le 127$ $d \in [0,1]$
Operation:	(W) .OR. (f) → (destination)
Status Affected:	Z
Description:	Inclusive OR the W register with register 'f'. If 'd' is 0, the result is placed in the W register. If 'd' is 1, the result is placed back in register 'f'.

PIC16F84A

MOVF	**Move f**
Syntax:	[*label*] MOVF f,d
Operands:	0 ≤ f ≤ 127 d ∈ [0,1]
Operation:	(f) → (destination)
Status Affected:	Z
Description:	The contents of register f are moved to a destination dependant upon the status of d. If d = 0, destination is W register. If d = 1, the destination is file register f itself. d = 1 is useful to test a file register, since status flag Z is affected.

MOVLW	**Move Literal to W**
Syntax:	[*label*] MOVLW k
Operands:	0 ≤ k ≤ 255
Operation:	k → (W)
Status Affected:	None
Description:	The eight-bit literal 'k' is loaded into W register. The don't cares will assemble as 0's.

MOVWF	**Move W to f**
Syntax:	[*label*] MOVWF f
Operands:	0 ≤ f ≤ 127
Operation:	(W) → (f)
Status Affected:	None
Description:	Move data from W register to register 'f'.

NOP	**No Operation**
Syntax:	[*label*] NOP
Operands:	None
Operation:	No operation
Status Affected:	None
Description:	No operation.

RETFIE	**Return from Interrupt**
Syntax:	[*label*] RETFIE
Operands:	None
Operation:	TOS → PC, 1 → GIE
Status Affected:	None

RETLW	**Return with Literal in W**
Syntax:	[*label*] RETLW k
Operands:	0 ≤ k ≤ 255
Operation:	k → (W); TOS → PC
Status Affected:	None
Description:	The W register is loaded with the eight-bit literal 'k'. The program counter is loaded from the top of the stack (the return address). This is a two-cycle instruction.

RETURN	**Return from Subroutine**
Syntax:	[*label*] RETURN
Operands:	None
Operation:	TOS → PC
Status Affected:	None
Description:	Return from subroutine. The stack is POPed and the top of the stack (TOS) is loaded into the program counter. This is a two-cycle instruction.

PIC16F84A

RLF	Rotate Left f through Carry
Syntax:	[*label*] RLF f,d
Operands:	0 ≤ f ≤ 127 d ∈ [0,1]
Operation:	See description below
Status Affected:	C
Description:	The contents of register 'f' are rotated one bit to the left through the Carry Flag. If 'd' is 0, the result is placed in the W register. If 'd' is 1, the result is stored back in register 'f'.

```
 ┌──────────────────────────────┐
 └─►│C│◄─┤ Register f ├◄─────────┘
```

RRF	Rotate Right f through Carry
Syntax:	[*label*] RRF f,d
Operands:	0 ≤ f ≤ 127 d ∈ [0,1]
Operation:	See description below
Status Affected:	C
Description:	The contents of register 'f' are rotated one bit to the right through the Carry Flag. If 'd' is 0, the result is placed in the W register. If 'd' is 1, the result is placed back in register 'f'.

```
 ┌──────────────────────────────┐
 └─►│C│─►┤ Register f ├─►─────────┘
```

SLEEP	
Syntax:	[*label*] SLEEP
Operands:	None
Operation:	00h → WDT, 0 → WDT prescaler, 1 → $\overline{TO}$, 0 → $\overline{PD}$
Status Affected:	$\overline{TO}$, $\overline{PD}$
Description:	The power-down status bit, $\overline{PD}$ is cleared. Time-out status bit, $\overline{TO}$ is set. Watchdog Timer and its prescaler are cleared. The processor is put into SLEEP mode with the oscillator stopped.

SUBLW	Subtract W from Literal
Syntax:	[*label*] SUBLW k
Operands:	0 ≤ k ≤ 255
Operation:	k - (W) → (W)
Status Affected:	C, DC, Z
Description:	The W register is subtracted (2's complement method) from the eight-bit literal 'k'. The result is placed in the W register.

SUBWF	Subtract W from f
Syntax:	[*label*] SUBWF f,d
Operands:	0 ≤ f ≤ 127 d ∈ [0,1]
Operation:	(f) - (W) → (destination)
Status Affected:	C, DC, Z
Description:	Subtract (2's complement method) W register from register 'f'. If 'd' is 0, the result is stored in the W register. If 'd' is 1, the result is stored back in register 'f'.

SWAPF	Swap Nibbles in f
Syntax:	[*label*] SWAPF f,d
Operands:	0 ≤ f ≤ 127 d ∈ [0,1]
Operation:	(f<3:0>) → (destination<7:4>), (f<7:4>) → (destination<3:0>)
Status Affected:	None
Description:	The upper and lower nibbles of register 'f' are exchanged. If 'd' is 0, the result is placed in W register. If 'd' is 1, the result is placed in register 'f'.

PIC16F84A

XORLW	**Exclusive OR Literal with W**
Syntax:	[*label*] XORLW k
Operands:	0 ≤ k ≤ 255
Operation:	(W) .XOR. k → (W)
Status Affected:	Z
Description:	The contents of the W register are XOR'ed with the eight-bit literal 'k'. The result is placed in the W register.

XORWF	**Exclusive OR W with f**
Syntax:	[*label*] XORWF f,d
Operands:	0 ≤ f ≤ 127 d ∈ [0,1]
Operation:	(W) .XOR. (f) → (destination)
Status Affected:	Z
Description:	Exclusive OR the contents of the W register with register 'f'. If 'd' is 0, the result is stored in the W register. If 'd' is 1, the result is stored back in register 'f'.

Appendix B
DIZI-2 Board and Lock Application

DIZI-2 Demonstration Board

A circuit was required to demonstrate a range of basic microcontroller programming techniques via a set of simple applications for the PIC 16F84. The DIZI (DIsplay, buZZer and Interrupt) board was designed to allow the special hardware features of the PIC chip to be exercised, including interrupts, timer and EEPROM memory. In-circuit programming was not incorporated, in order to emphasise the stand-alone operation of the microcontroller. The chip would be programmed separately and then physically transferred to the target system. The enhanced DIZI-2 board described in this appendix incorporates an on-board battery supply, a finger pot to provide an analogue input and hardware switch debouncing to improve the reliability of the push button operation.

The circuit is built on a 100×100 mm piece of stripboard which has copper tracks for making the component connections on a standard 0.1" grid. The design includes a 2×1.5 V battery pack on the board. The power is switched on via a non-latching push button so that it cannot be left on accidentally, and thereby exhaust the batteries; it must be held on manually while the circuit is in operation.

The basic demonstration programs in Part B of this book can be run on the DIZI-2 board, while the motor programs must be run on the MOTA demo board. The simple introductory circuits in Part B and the motor board can be constructed using the same techniques as will be described for the DIZI board. The reader who is inexperienced in prototype construction is encouraged to attempt these tasks. The binary output counts from the BINx programs will be seen on the corresponding LED segments of the DIZI display, although the binary number is not displayed so clearly as it would be on a set of eight discrete LEDs, or an LED bar graph module.

DIZI-2 Board Design

A seven-segment display allows decimal and hexadecimal digits to be shown. A range of applications with a numerical output can thus be demonstrated, for example, the electronic DICE in Chapter 12 and the LOCK application detailed below. Port B has eight I/O bits; seven are used for the LED display, leaving RB0 free for use as both an audio output and a push button (interrupt) input. A small audio transducer, a peizo electric buzzer, provides a simple and effective way of monitoring audio output frequencies or generating status signals.

Port A has five pins. RA4 can be used as an input to the TMR0 counter, so this was allocated as another push button input. A 4-bit switch bank is useful for setting coded inputs (for example,

BCD inputs for the LOCK application), so RA0–RA3 were allocated for this purpose. The switch and push button inputs have 100k pull-up resistors, and the push buttons have $22\,\mu F$ debouncing capacitors. These are fitted to such inputs because, when a switch closes, the metal contacts can bounce open again several times before finally closing. The CR network prevents this from causing multiple transitions on the logic signal input to which the switch is connected, because after the capacitor has quickly discharged on the first contact, it must recharge via the 100k. This takes a relatively long time, preventing the voltage from jumping back to a high level when the contacts re-open. By the same process, the CR network also ensures a smooth transition from low to high logic levels when the push button is released.

In addition, the CR network on RA4 was modified with a potentiometer (pot) connected as a variable resistance in series with the 100k, so that it could also be used to demonstrate the analogue input process in the LOCK program.

Circuit Diagram

Developing circuits for microcontrollers is not always too difficult, because the same circuit elements can often be re-used in different designs. Thus, in the DIZI circuit (Fig. B.1), the switch inputs, display outputs and the clock circuit are standard arrangements of components. Remember, however, that, unlike the PIC, some microcontrollers cannot drive LEDs directly, but need a current driver stage inserted between each output and the LED.

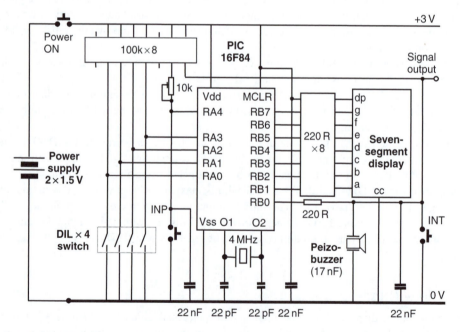

Figure B.1   DIZI-2 board circuit diagram.

DIZI-2 Board Layout

The layout of a PCB or prototype circuit is derived from the circuit diagram. The pins on DIL (Dual In-Line) chips are spaced 0.1″ apart, so the circuits must be laid out on a 0.1″ grid. When the pin out of each component has been established by reference to the data sheet or catalogue information, the connections can be mapped out on a square grid on paper. Alternatively, it is not too difficult to use the basic drawing tools in a wordprocessor to do the same job. Examples of such connection diagrams are given in the text. A more refined layout drawing for the DIZI-2 board is shown in Fig. B.2.

The board is viewed from the front (component) side, with the tracks on the back shown vertically. The components are numbered for reference to the parts listed below. The chips are all placed in the same orientation, so that pin 1 is bottom left. The seven-segment display has the decimal point bottom right. The ICs must obviously be fitted across the tracks, so that their pin connections are separated. The PIC chip should be fitted in a socket so that it can be removed for programming.

Horizontal links of tinned copper wire (TCW) complete the connections required. A solder joint is shown as a solid black dot. The broader solid lines indicate a continuous link across the tracks on the back of the board, where a set of adjacent tracks must be connected. Where required, the tracks are cut with a hand drill; these positions are shown as small white circles. The tracks must also be cut between the opposite pairs of pins on each DIL component, and, in this case, under the clock circuit capacitors (11).

A computer-drawing method allows component positioning to be easily adjusted so that the minimum area of stripboard is used. However, with experience, the circuit may be built directly onto the board without necessarily drawing the layout, perhaps with some wastage of board area.

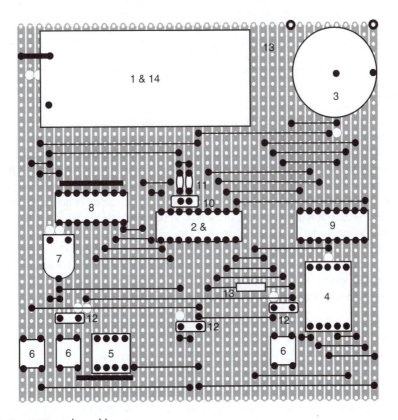

Figure B.2 DIZI stripboard layout.

Parts List

A parts list is required to specify the exact component when ordering from a suitable supplier. The availability of components varies over time, so updating is sometimes necessary. For example, the finger preset pot originally used in the design is no longer listed by the UK supplier, and had to be replaced with an equivalent part. The layout then had to be amended because the pin arrangement of the new component was different.

Layout	Description
1	Battery box, 2 x AA cells, PCB mounting
2	Microcontroller, PIC 16LF84-04
3	Peizo electric sounder, PCB mounting
4	Seven-segment LED display, 0.5", Common Cathode
5	Piano DIL switch, 4-way
6	Tactile switch, PCB mounting (3 off)
	Caps for above: Red
	Blue
	Yellow
7	Preset potentiometer, 10k, H-mount
8	DIL isolated resistor network, 100k x 8
9	DIL isolated resistor network, 220 R x 8
10	Quartz crystal, general purpose, 4 MHz
11	Capacitor, 22 pF, ceramic (2 off)
12	Capacitor, 22 nF, polyester (3 off)
13	Stripboard, SRBP 3939 100 x 100 mm
14	Batteries, 1.5 V, size AA, Duracell (2 off)
15	18-pin DIL IC socket

Construction

When the layout has been checked against the circuit diagram, the main components can be inserted in the board and retained by, if necessary, slightly bending the corner pins outwards. All the pins should then be soldered to the tracks using the minimum amount of solder necessary, whilst ensuring that the joint is covered evenly with no cavities. At the same time, the soldering iron should be in contact with the joint for the shortest possible time, to avoid component overheating. The TCW links can also be retained before soldering by bending the ends towards each other. If a very neat job is required, one end can be soldered and the link stretched slightly before fixing the other end, to ensure that the link has no kinks in it, and that adjacent links do not touch; insulated TCW may be used on longer links if necessary. The tracks should then be cut where necessary, and the track side brushed with a small stiff brush to clear any debris. Rake between the tracks with a small screwdriver or knife to ensure that there are no short circuits left between adjacent tracks and solder joints.

Static Testing

Thoroughly re-inspect the board for correct connections, and that there are no debris, solder splashes or whiskers or dry joints. With the batteries not yet fitted, check with a multimeter that there is no short circuit between the power supplies. Fit the batteries, but not the PIC chip, and hold down the power button. The display decimal point should light. Check the supply voltages on the supply tracks and PIC socket pins: Pin 5 = 0 V and Pin 14 = +3 V. Check that the voltages at the PIC inputs change correctly as the switches are toggled. A DMM (Digital Multimeter) or oscilloscope is required for this test, because of the high impedance of the pull-up resistors. Connect a temporary link between Pin 14 (+3 V) on the PIC IC socket and each PIC output in turn, RB0–RB7. The peizo-buzzer should produce an audible 'tick' and the LED segments should light.

Test Program

To complete testing of the DIZI-2 board, a program should be blown into the PIC which exercises all the hardware, while remaining as simple as possible so that there is no question

Program B.1   DIZI board test program

```
; diz1.asm

; Test DIZI hardware ....................

        GOTO     inter   ; jump over delay

; Delay Subroutine .....................

delay   MOVLW    0FF      ; Load FF
        MOVWF    0C       ; into counter
down    DECFSZ   0C       ; and decrement
        GOTO     down     ; until zero
        RETURN

; Check Interrupt Button ................

inter   BTFSC    06,0     ; Test Button RB0
        GOTO     inter    ; until pressed

; Check Display ........................

        MOVLW    00       ; Set PortB bits
        TRIS     06       ; as outputs
        MOVLW    0FF      ; Switch on all
        MOVWF    06       ; display segments

; Check Input Button ...................

input   BTFSC    05,4     ; Test Button RA4
        GOTO     input    ; until pressed

; Check DIP Switches and Buzzer ..........

again   MOVF     05,W     ; get DIL input &
        MOVWF    06       ; send to display
        RLF      06       ; rotate bits left

                             continued...
```

```
BSF    06,0    ; set buzzer high
CALL   delay   ; delay about 1ms
BCF    06,0    ; reset buzzer low
CALL   delay   ; delay about 1ms

GOTO   again   ; and keep going..
END            ; End of code
```

of the software being faulty. A suitable program is listed in Program B.1; it does not test the analogue input operation, which will be covered later. When this program has been loaded, the following test procedure will confirm correct hardware operation.

Step	Test	Result
1	Power Button On	Decimal point ON
2	Button B pressed and released	All display segments ON
3	Button A pressed and released	Buzzer sounds
4	Operate DIL switches	Segments a, b, c, d change

If faults are found, it is quite possible that there are still short circuits on the board. Check also that all the tracks have been cut as required, and that all connections are correct.

Analogue to Digital Conversion

One feature of the DIZI board not described in the main text is the analogue input. A similar input is available on the MOTA board, so the method for using it will now be explained. Some PIC chips, and other microcontrollers, have built-in ADCs, which allow analogue voltages to be converted to binary form for input to the processor. An ADC would be needed, for example, if a temperature is to be measured by the controller in a process system. The general block diagram for a counting ADC is shown in Fig. B.3.

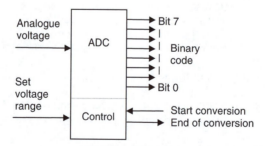

Figure B.3 General analogue to digital converter.

The 8-bit ADC shown in Fig. B.3 converts the analogue voltage present at its input to an 8-bit binary number, which means it can detect 256 different voltage levels. If the input range is set to, say, 0–2.55 V, then 255 steps of 10 mV can be detected. When the 'Start Conversion' (GO bit) signal goes active, the DAC converts the binary number into a corresponding voltage. The 'End of Conversion' signal to the processor is to indicate completion of the conversion process.

CR-ADC

The hardware for the standard ADC above is fairly complex, while the control process is relatively simple. An alternative is to use simple external hardware with a software conversion procedure, if there is no hardware ADC available. The CR-ADC is based on the measurement, using a counter register, of the rise time in a CR network connected to the processor system input. The CR converter will generally be slower and less accurate than a hardware-based ADC, but may be quite adequate in simple applications.

The components connected to RA4 in the DIZI circuit are shown in Fig. B.4. The capacitor-charging curve in Fig. B.5 shows the time constant of the circuit as 2.3 ms, assuming that the pot is set midway. This is the time taken to reach 63% of the final value (3 V) as the C charges via R. The PIC chip is a CMOS device, so the voltage level at which an input changes from logic 0 to 1, the threshold voltage, is around half the supply voltage, 1.5 V. Therefore the time taken to reach this level, here called the charging time, is estimated at 1.5 ms. This could be calculated more accurately from the formula for the charging of a capacitor, but as long as the circuit operation is consistent, it is not necessary for this application.

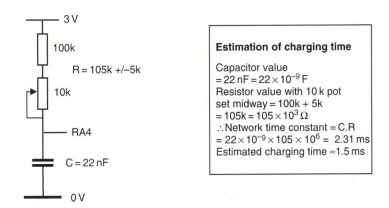

Estimation of charging time

Capacitor value
$= 22\,nF = 22 \times 10^{-9}\,F$
Resistor value with 10 k pot
set midway $= 100k + 5k$
$= 105k = 105 \times 10^3\,\Omega$
$\therefore$ Network time constant $= C.R$
$= 22 \times 10^{-9} \times 105 \times 10^6 = 2.31\,ms$
Estimated charging time $\approx 1.5\,ms$

Figure B.4 CR conversion network.

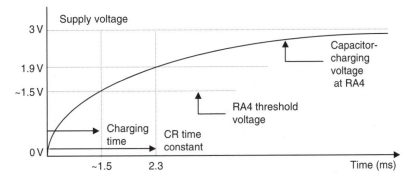

Figure B.5 CR network characteristic.

The resistance, R, varies between 100k and 110k, depending on the position of the pot. The variation in the pot value will produce a corresponding variation in the rise time of the circuit. The rise time can be measured by discharging the capacitor and then counting while the voltage rises back towards the threshold. The capacitor is discharged by setting RA4 as an output and then setting the port data bit to zero. RA4 is then reconfigured as an input and checked at fixed time intervals while a register is incremented. The count is stopped when RA4 goes high.

The waveform which will be seen at RA4 is illustrated in Fig. B.6 (not to scale). A register labelled PotVal is incremented, and RA4 checked, within a loop taking, in the LOCK program, $20\,\mu s$ to execute (see Fig. B.7). An adjustable delay routine allows the timing to be modified to suit the application and CR component values.

The result of the process is that a count is obtained which represents the setting of the pot. This could be converted to a resistance value if required, but in the LOCK program all we need is a variation in the displayed digit between 0 and 9, to allow the user to input a decimal combination. Therefore, the delay associated with the count was simply adjusted to give one decade on the display with one turn of the pot. Only the low four bits of the count were required, so any decade of values could be used. The upper end of the 4-bit range, hex numbers A–F, are displayed as '-'. These can be used as 'hidden' digits for extra security, if required.

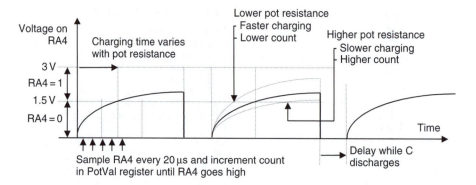

Figure B.6   Conversion waveform at RA4.

```
Set RA4 as Output
Clear RA4 to 0V to discharge C
Clear Counter Register
Set RA4 as Input
Test RA4 while C charges through R:
        Increment Counter Register
        Delay 20us
        Until RA4 = 1
Convert Count to Resistance or Pot Position
```

Figure B.7   CR-ADC algorithm.

A similar method can be used to operate the analogue input on the MOTA board (Fig. 13.2). The motor speed could then be controlled by a voltage source or sensor, or analogue feedback employed to implement closed loop control.

EEPROM Memory

Non-volatile read and write memory is very useful because data input by the user or acquired by the processor during its operation can be retained while the power is off. One important application area is data security and encryption. PIC devices are used, for example, in smart cards for controlling access to satellite television broadcast channels. The LOCK application illustrates this feature of the PIC 16F84 by using the EEPROM memory to store a 4-digit security code.

The PIC 16F84 has 64 bytes of EEPROM, with addresses 00–3F. The memory is accessed via EEDATA and EEADR in the SFR set. The EEPROM address is loaded into EEADR, and the data byte to be stored in EEDATA. An artificially complex write initialisation sequence is then executed to actually write to the EEPROM memory, using EECON1 and EECON2 page 1 SFRs. The sequence is designed to reduce the possibility of an accidental write to the EEPROM, because a high level of reliability is required for security applications. This code sequence is given in the data sheet and LOCK program listing.

The read sequence, for retrieving the data, is more straightforward. Using EECON1, the data in the address pointed to by EEADR is returned in EEDATA. For accessing sequences of locations, EEADR can be incremented directly.

LOCK Application

In this demonstration application, a sequence of four decimal digits is stored in the PIC EEPROM memory from the DIL switch inputs. This sets the combination for the lock. To 'open' the lock, the pot is rotated, and the input decimal digits are displayed and entered. This simulates the rotary action of mechanical combination locks. If the sequence of four input digits matches those previously stored in EEPROM, a siren sound is made to indicate the opening of the lock.

In the actual application, a solenoid-operated lock mechanism would be activated from this output, by replacing the siren sequence with an instruction to set an output bit. A suitable current driver interface for the solenoid would be required. Only the Power button, Enter button, Digit Select pot and display would be accessible to the user in the final design. The hardware would need to be reconfigured so that the unit would appear as shown in Fig. B.8 to the user. The DIL switch bank and its button for setting the entry code would be concealed.

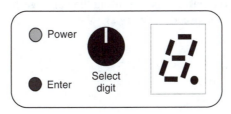

Figure B.8 Lock user interface.

Program Structure

The program contains the following blocks:

1. declaration of register and bit Labels

2. initialisation of registers

3. sequence 1 – Store combination

4. sequence 2 – Check combination

5. end 1 – Continuous siren output

6. end 2 – Sleep

7. Subroutine 1 – Display code table

8. Subroutine 2 – Variable delay

9. Subroutine 3 – Output one tone cycle

10. Subroutine 4 – Get digit from pot

The program blocks should be ordered in such a way that labels referenced have already been defined when the program is assembled. Thus, the subroutines should be placed *before* the main sequences in the source code. However, when actually developing the code, if you are working 'top down', the subroutines may actually be written after the main sequences. To place them correctly in the source code, they can then be inserted when editing, or cut and pasted later.

The program has two main sequences, for inputting and checking a combination, and two alternate endings. The processor goes to sleep after completion of the input sequence, or an incorrect digit match. The DIZI board must be re-powered to try again, as there are no other interrupts enabled to restart it. If the combination checks out correctly, the siren ending is used, which continues until the power goes off.

Pseudocode

The program is outlined below using 'pseudocode', which is a text method of designing the program, which may be used instead of a flowchart. The pseudocode is developed in a word processor or the program source code editor until the statements are detailed enough to be converted into assembly code statements. In this case, it must be written in a form which allows it to be easily converted to PIC assembly language. The program structure and logic can thus be worked out before attempting to write the source code itself. To use pseudocode effectively, the programmer must be reasonably expert in using the language syntax.

The conventions used in the pseudocode are as follows:

- block structure applied
- target hardware specified
- Register and bit labels defined
- user inputs included in the sequence
- GOTO [deslab]
        – Jump to destination address label

- CALL [subnam]
    - – Call subroutine at address label
    - – Values passed to and received from subroutine defined
- GOTO [addlab] UNTIL (condition)
    - – Implemented using Bit Test, Skip & GoTo operation
- (regname)
    - = Contents of register labelled 'regname'
- program block type defined:
    - INIT = Initialise
    - MAIN = Main program
    - SEQn = Sequence ending with GOTO
    - ENDn = End operation
    - SUBn = Subroutine, optionally receiving and/or returning values

LOCK List File

The list file for the LOCK program contains the source code and machine code. If the reader wishes to test the program, the machine code (column 2 of the list file) can be entered directly into the program memory buffer in the programming software. This avoids the need to type in the source code, if the hex file itself is not available.

The source code file uses the following conventions:

- full details of hardware and operation of application in source code;
- SFR, user and bit labels defined in separate blocks for clarity;
- block and line comments in source code;
- lower case for address labels;
- upper case for instruction mnemonics and SFR labels;
- capitalisation of user register labels;
- identification and separation of block types.

The pseudocode and list files are reproduced in Programs B.2 and B.3.

Program B.2 Lock program pseudocode

```
LOCK PROGRAM PSEUDOCODE          MPB           29/8/99
****************************************************
Hardware: DIZI PIC 16F84 Demo Board
****************************************************
General Purpose Register Labels:
    0C = Period = Delay Period Preload Value
    0D = Count  = Delay Counter
    0E = PotVal = Count from ADC conversion
    0F = DigVal = Low 4 bits of PotVal
User Bit Labels:
    butA (RA4 input) – Normally 1
    butB (RB0 input) – Normally 1
    buzO (RB0 output)
See Data Sheet for SFR Labels and addresses


                                        continued...
```

```
{Power Button On}
INIT: Initialise Port B *************************

    Port A defaults to inputs
    RA0 - RA3 = DIL Switches = 4-bit input
    RA4 = Input = butA = INP Button
    RB0 = Input = butB = INT Button
    RB1 - RB7 = Output = 7 Seg Display


MAIN: Select Set or Check Combination ***********

select              {Press Button A or B}
                    If (butA)=0, GOTO [stocom]
                    If (butB)=0, GOTO [checom]
                    GOTO [select]


SEQ1: Store 4 digits in EEPROM, beep after each *

stocom              {Release Button A}
                    CALL [delay] with (W)=FF
                    GOTO [stocom] UNTIL (butA)=1

                    Clear (EEADR)
getdil              {Set DIL Switches or Press A}
                    Read (PORTA) into (W)
                    Calc (W) AND 0F
                    Store (W) in (EEDATA)
                    CALL [codtab] with (W)=00-0F
                    {Returns with '7SegCode' in (W)}
                    Output (W) to (PORTB)
                    GOTO [getdil] UNTIL (butA)=0

waita               {Release Button A}
                    GOTO [waita] UNTIL (butA)=1
                    Store (EEDATA) in (EEADR)
                    CALL [beep]
                    Increment (EEADR) from 00 to 04
                    GOTO [getdil] UNTIL (EEADR)=4
                    CALL [beep]
                    CALL [beep]
                    GOTO [done]


SEQ2: Check 4 digits from pot for match *************

checom              {Release Button B}
                    CALL [delay] with (W)=FF
                    GOTO [checom] UNTIL (butB)=1

                    Clear (EEADR)
potin               {Adjust Pot or Press Button B}

                    CALL [getpot] for (DigVal)
                    {Returns with (DigVal)=00-0F}
                    GOTO [potin] UNTIL (butB)=0
```

```
                        Read (EEDATA) at (EEADR)
                        Compare (EEDATA) with (DigVal)
                        If (Z)=0 GOTO [done]

waitb                   {Release Button B}
                        GOTO [waitb] UNTIL (butB)=1
                        CALL [beep]
                        Increment (EEADR)
                        GOTO [potin] UNTIL EEADR=4
                        GOTO [siren]
```

END1: Sequences matches, sound siren ****************

```
    siren               CALL [beep]
                        GOTO [siren]
```

END2: Digit compare failed, finish ****************

```
    done                Clear (PORTB)
                        Sleep
```

SUBROUTINES *

SUB1: Get Display Code
    Receives: Table Offset in W
    Returns: 7-Segment Display Code in W

```
codtab                  Add (W) to (PCL)
                        RETURN with '7SegCode' in (W)
```

SUB2: Variable Delay
    Receives: (Count) in W

```
delay                   Load (Count) from (W)
                        Decrement (Count) UNTIL (Count)=0
                        RETURN
```

SUB3: Outputs one cycle of sound output
    Receives: (Period)

```
beep                    Load (Period) with FF
                        Set RB0 as Output

cycle                   Set (BuzO)=1
                        CALL [delay] with (Period) in W
                        Set BuzO=0
                        CALL [delay] with (Period) in W
                        Decrement (Period) from FF to 00
                        GOTO [cycle] UNTIL (Period)=0
                        Reset RB0 as Input
                        RETURN
```

continued...

```
SUB4: Get Pot Value using CR ADC method

       Returns: (DigVal)=00-0F

getpot            Set RA4 as Output
                  Clear (RA4)
                  CALL [delay] with (W)=FF
                  Reset RA4 as Input

                  Clear (PotVal)
check             Increment (PotVal) from 00 to XX
                  CALL [delay] with (W)=3
                  GOTO [check] UNTIL (RA4)=1

                  (DigVal) = (PotVal) AND 0F
                  CALL [codtab] with (DigVal)=00-0F
                  RETURN


END OF LOCK PROGRAM ********************************
```

Program B.3   Lock program list file

```
00001 ;****************************************************************
00002 ; LOCK.ASM                                         MPB    17/8/99
00003 ;****************************************************************
00004 ;
00005 ; Four digit combination lock simulation demonstrates the hardware
00006 ; features of the DIZI demo board and the PIC 16F84.
00007 ;
00008 ; Hardware:    DIZI Demo Board with PIC 16F84 (4MHz)
00009 ; Setup:       RA0-RA3      DIL Switch Inputs
00010 ;              RA4          Push Button Input / Analogue Input
00011 ;              RB0          Push Button Input / Audio Output
00012 ;              RB1-RB7      7-Segment Display Output
00013 ; Fuses:       WDT off, PuT on, CP off
00014 ;
00015 ; Operation ---------------------------------------------------
00016 ;
00017 ; To set the combination, a sequence of 4 digits is input on the DIL
00018 ; piano switches; this is retained in the EEPROM when power is off.
00019 ; To 'open' the lock, a sequence of 4 digits is input via
00020 ; the potentiometer. These are compared with the stored data,
00021 ; and an audio output generated to indicate the correct sequence.
00022 ; The processor halts if any digit fails to match, and the
00023 ; program must be restarted.
00024 ;
00025 ;            To set a combination:
```

```
      00026 ;              1.  Hold Power On Button
      00027 ;              2.  Press Button A
      00028 ;              3.  Set a digit on DIL switches and Press A - beeps
      00029 ;              4.  Repeat step 3 for 3 more digits
      00030 ;              5.  Release Power Button
      00031 ;
      00032 ;              To check a combination:
      00033 ;              1.  Hold Power On Button
      00034 ;              2.  Press Button B
      00035 ;              3.  Set a digit on pot and Press B - beeps if matched
      00036 ;              4.  Repeat step 3 for 3 more digits
      00037 ;                  - if digits all match, siren is sounded
      00038 ;                  - if any digit fails to match, the processor
            ;                    halts
      00039 ;              5.  Release Power Button
      00040 ;
      00041 ; ********************************************************
      00042                 PROCESSOR 16F84        ; Processor Type Directive
      00043 ; ********************************************************
      00044
      00045 ; EQU: Special Function Register Equates...................
      00046
0002  00047 PCL    EQU    02    ; Program Counter Low
0005  00048 PORTA  EQU    05    ; Port A Data
0006  00049 PORTB  EQU    06    ; Port B Data
0003  00050 STATUS EQU    03    ; Flags
0008  00051 EEDATA EQU    08    ; EEPROM Memory Data
0009  00052 EEADR  EQU    09    ; EEPROM Memory Address
0008  00053 EECON1 EQU    08    ; EEPROM Control Register 1
0009  00054 EECON2 EQU    09    ; EEPROM Control Register 2
      00055
      00056 ; EQU: User Register Equates..............................
      00057
000C  00058 Period EQU    0C    ; Period of Output Sound
000D  00059 Count  EQU    0D    ; Delay Down Counter
000E  00060 PotVal EQU    0E    ; Analogue Input Value
000F  00061 DigVal EQU    0F    ; Current Digit Value 00 to 09
      00062
      00063 ; EQU: SFR Bit Equates....................................
      00064
0005  00065 RP0    EQU    5     ; STATUS - Register Page Select
0000  00066 RD     EQU    0     ; EECON1 - EEPROM Memory Read Byte Initiate
0001  00067 WR     EQU    1     ; EECON1 - EEPROM Memory Write Byte Initiate
0002  00068 WREN   EQU    2     ; EECON1 - EEPROM Memory Write Enable
0002  00069 Z      EQU    2     ; STATUS - Zero Flag
      00070
      00071 ; EQU: User Bit Equates...................................
      00072
0004  00073 butA   EQU    4     ; PORTA - RA4 Input Button
0000  00074 butB   EQU    0     ; PORTB - RB0 Input Button
0000  00075 buzO   EQU    0     ; PORTB - RB0 Output Buzzer
      00076
      00077 ; ********************************************************
```

continued ...

```
            00078
            00079 ; INIT: Initialise Port B (Port A defaults to inputs)
            00080
0000 3001 00081 start  MOVLW  001          ; RB0 = Input, RB1-RB7 = Outputs
0001 0066 00082        TRIS   PORTB         ; Set Data Direction
0002 0086 00083        MOVWF  PORTB         ; Clear Data
0003 286D 00084        GOTO   select        ; Select Combination Read or
                                              Write
            00085
            00086 ; SUBROUTINES ****************************************
            00087
            00088 ; SUB1: 7-Segment Code Table using PCL + offset in W
                        Returns
            00089 ;       digit display codes, with '-' for numbers A to F
            00090
0004 0782 00091 codtab ADDWF  PCL          ; Add offset to Program Counter
0005 347E 00092        RETLW  B'01111110'  ; Return with display code for '0'
0006 340C 00093        RETLW  B'00001100'  ; Return with display code for '1'
0007 34B6 00094        RETLW  B'10110110'  ; Return with display code for '2'
0008 349E 00095        RETLW  B'10011110'  ; Return with display code for '3'
0009 34CC 00096        RETLW  B'11001100'  ; Return with display code for '4'
000A 34DA 00097        RETLW  B'11011010'  ; Return with display code for '5'
000B 34FA 00098        RETLW  B'11111010'  ; Return with display code for '6'
000C 340E 00099        RETLW  B'00001110'  ; Return with display code for '7'
000D 34FE 00100        RETLW  B'11111110'  ; Return with display code for '8'
000E 34DE 00101        RETLW  B'11011110'  ; Return with display code for '9'
000F 3480 00102        RETLW  B'10000000'  ; Return with display code for '-'
0010 3480 00103        RETLW  B'10000000'  ; Return with display code for '-'
0011 3480 00104        RETLW  B'10000000'  ; Return with display code for '-'
0012 3480 00105        RETLW  B'10000000'  ; Return with display code for '-'
0013 3480 00106        RETLW  B'10000000'  ; Return with display code for '-'
0014 3480 00107        RETLW  B'10000000'  ; Return with display code for '-'
            00108
            00109 ; ---------------------------------------------
            00110 ; SUB2:   Delay routine
            00111 ;         Receives delay count in W
            00112
0015 008D 00113 delay  MOVWF     Count      ; Load counter from W
0016 0B8D 00114 loop   DECFSZ    Count      ; and decrement
0017 2816 00115        GOTO      loop       ; until zero
0018 0008 00116        RETURN               ; and return
            00117
            00118 ; ---------------------------------------------
            00119 ; SUB3: Output One Beep Cycle to BuzO
            00120
0019 30FF 00121 beep   MOVLW     0FF        ; Load FF into
001A 008C 00122        MOVWF     Period     ; Period counter
            00123
001B 3000 00124        MOVLW     B'00000000' ; Set RB0
001C 0066 00125        TRIS      PORTB      ; as output
            00126
            00127 ; Do one cycle of rising tone....
            00128
001D 1406 00129 cycle  BSF       PORTB,buzO ; Output High
001E 080C 00130        MOVF      Period,W   ; Load W with Period value
001F 2015 00131        CALL      delay      ; and delay for Period
            00132
```

```
0020 1006  00133          BCF      PORTB,buzO    ; Output Low
0021 2015  00134          CALL     delay         ; and delay for same Period
0022 0B8C  00135          DECFSZ   Period        ; Decrement Period
0023 281D  00136          GOTO     cycle         ; and do next cycle until 0
           00137
           00138 ; Set RB0 to input again...............................
           00139
0024 3001  00140          MOVLW    B'00000001'   ; Reset RB0
0025 0066  00141          TRIS     PORTB         ; as input
0026 0008  00142          RETURN                 ; from tone cycle
           00143
           00144 ; -------------------------------------------------
           00145 ; SUB4: Get pot value (Rv) using rise time due to C and R
                 ;       on RA4
           00146 ;        Returns with digit value (0-F) in DigVal
           00147
           00148 ; Discharge external capacitor on RA4
           00149
0027 300F  00150 getpot   MOVLW    B'00001111'   ; Set RA4
0028 0065  00151          TRIS     PORTA         ; as output
0029 1205  00152          BCF      PORTA,4       ; and discharge C setting
                                                    output low
002A 30FF  00153          MOVLW    0FF           ; Delay for about 1ms
002B 2015  00154          CALL     delay         ; to ensure C is discharged
002C 301F  00155          MOVLW    B'00011111'   ; Reset RA4
002D 0065  00156          TRIS     PORTA         ; as input
           00157
           00158 ; Increment a counter until RA4 goes high due to
                 ; charging of C
           00159
002E 018E  00160          CLRF     PotVal        ; Clear input value counter
002F 0A8E  00161 check    INCF     PotVal        ; increment counter
0030 3003  00162          MOVLW    03            ; Set delay count to 3
0031 2015  00163          CALL     delay         ; and delay between
                                                    input checks
0032 1E05  00164          BTFSS    PORTA,4       ; Check input bit RA4
0033 282F  00165          GOTO     check         ; and repeat if not yet high
           00166
           00167 ; Mask out high bits of count value, and store & display
           00168 ; 4-bit digit value, 0-F
           00169
0034 080E  00170          MOVF     PotVal,W      ; Put count value in W
0035 390F  00171          ANDLW    00F           ; and set high 4 bits to 0
0036 008F  00172          MOVWF    DigVal        ; Store 4-bit value
0037 2004  00173          CALL     codtab        ; Get 7-segment code, 0-9
0038 0086  00174          MOVWF    PORTB         ; and display
           00175
0039 0008  00176          RETURN                 ; with DigVal from setting
                                                    of pot
           00177
           00178 ; MAIN SEQUENCES ************************************
           00179
           00180 ;SEQ1: Store 4 Digits in non volatile EEPROM
           00181 ; Beep after each digit, and twice when 4 done
           00182
```

continued ...

```
                  00183 ; Complete Button A input operation
                  00184
003A 30FF 00185 stocom MOVLW 0FF            ; Delay for about 1ms
003B 2015 00186        CALL  delay          ; to avoid Button A switch bounce
003C 1E05 00187        BTFSS PORTA,butA     ; Wait for Button A
003D 283A 00188        GOTO  stocom         ; to be released
                  00189
                  00190 ; Read 4-bit binary number from DIL switches into EEDATA
                        and display
                  00191
003E 0189 00192        CLRF  EEADR          ; Zero EEPROM address register
003F 0805 00193 getdil MOVF  PORTA,W        ; Read DIL switches
0040 390F 00194        ANDLW 0F             ; and set high 4 bits to 0
0041 0088 00195        MOVWF EEDATA         ; Put DIL value in EEPROM data
                  00196
0042 2004 00197        CALL  codtab         ; Display DIL input as decimal
0043 0086 00198        MOVWF PORTB          ;
                  00199
0044 1A05 00200        BTFSC PORTA,butA     ; Check if Button A pressed
0045 283F 00201        GOTO  getdil         ; If not, keep reading DIL input
                  00202
                  00203 ; Store the current DIL input in EEPROM at current address
                  00204
0046 1683 00205 store  BSF   STATUS,RP0     ; Select Register Bank 1
0047 1508 00206        BSF   EECON1,WREN    ; Enable EEPROM write
0048 3055 00207        MOVLW 055            ; Write initialisation sequence
0049 0089 00208        MOVWF EECON2         ;
004A 30AA 00209        MOVLW 0AA            ;
004B 0089 00210        MOVWF EECON2         ;
004C 1488 00211        BSF   EECON1,WR      ; Write data into current address
004D 1283 00212        BCF   STATUS,RP0     ; Re-select Register Bank 0
                  00213
004E 1E05 00214 waita  BTFSS PORTA,butA     ; Wait for Button A to be released
004F 284E 00215        GOTO  waita          ;
0050 2019 00216        CALL  beep           ; Beep to indicate digit write
                                              done
                  00217
                  00218 ; Checkif 4 digits have been stored yet, if not, get next
0051 0A89 00220        INCF  EEADR          ; Select next EEPROM address
0052 1D09 00221        BTFSS EEADR,2        ; Is the address now = 4?
0053 283F 00222        GOTO  getdil         ; If not, get next digit
0054 2019 00224        CALL  beep           ; Beep twice when 4 digits stored
0055 2019 00225        CALL  beep           ;
0056 2874 00226        GOTO  done           ; Go to sleep when done
                  00228 ; ------------------------------------------------
                  00230 ; SEQ2:Check PotVal v EEPROM
                  00231
0057 30FF 00232 checom MOVLW 0FF            ; Delay for about 1ms
0058 2015 00233        CALL  delay          ; to avoid Button B switch bounce
0059 1C06 00234        BTFSS PORTB,butB     ; Wait for Button B to be released
005A 2857 00235        GOTO  checom         ;
                  00236
                  00237 ; Read the value set on the input pot
                  00238
005B 0189 00239        CLRF  EEADR          ; Zero EEPROM address
005C 2027 00240 potin  CALL  getpot         ; Get a digit value set on pot (Rv)
```

```
005D 1806 00241         BTFSC PORTB,butB ; Check in Button pressed again
005E 285C 00242         GOTO  potin      ; If not, keep reading the pot
          00243
          00244 ; Get a digit value from EEPROM and compare with the pot
                  input
          00245
005F 1683 00246         BSF   STATUS,RP0 ; Select Register Bank 1
0060 1408 00247         BSF   EECON1,RD  ; Read selected EEPROM location
0061 1283 00248         BCF   STATUS,RP0 ; Re-select Register Bank 0
0062 0808 00249         MOVF  EEDATA,W   ; Copy EEPROM data to W
          00250
0063 068F 00251         XORWF DigVal     ; Compare the input with EEPROM
                                           data
0064 1D03 00252         BTFSS STATUS,Z   ; If it does not match, go to sleep
0065 2874 00253         GOTO  done       ;
          00254
          00255 ; If digit match obtained, check if 4 done and do next if not
          00256
0066 1C06 00257 waitb   BTFSS PORTB,butB ; Wait for Button B to be released
0067 2866 00258         GOTO  waitb      ;
0068 2019 00259         CALL  beep       ; Beep to confirm successful match
          00260
0069 0A89 00261         INCF  EEADR      ; Select next EEPROM location
006A 1D09 00262         BTFSS EEADR,2    ; 4 digits checked yet?
006B 285C 00263         GOTO  potin      ; If not, do the next
006C 2872 00264         GOTO  siren      ; When 4 digits done, run siren
          00265
          00266 ; *********************************************************
          00267
          00268 ; MAIN: Select Set or Check Combination
          00269
006D 1E05 00270 select  BTFSS PORTA,butA ; Button A pressed?
006E 283A 00271         GOTO  stocom     ; If so, store a combination
006F 1C06 00272         BTFSS PORTB,butB ; Button B pressed?
0070 2857 00273         GOTO  checom     ; If so, check a combination
0071 286D 00274         GOTO  select     ; repeat endlessly
          00275
          00276 ; *********************************************************
          00277
          00278 ; END1: When combination successfully matched, make siren
                         sound
          00279
0072 2019 00280 siren   CALL  beep       ; Do a tone cycle
0073 2872 00281         GOTO  siren      ; and repeat endlessly
          00282
          00283 ; -----------------------------------------------------
          00284
          00285 ; END2: When a digit check fails, go to sleep, and try again
          00286
0074 0186 00287 done    CLRF  PORTB      ; Switch off display
0075 0063 00288         SLEEP            ; Processor halts
          00289
          00290 ; *********************************************************
          00291         END              ; of program source code
```

continued ...

```
SYMBOL TABLE
   LABEL          VALUE

Count           0000000D
DigVal          0000000F
EEADR           00000009
EECON1          00000008
EECON2          00000009
EEDATA          00000008
PCL             00000002
PORTA           00000005
PORTB           00000006
Period          0000000C
PotVal          0000000E
RD              00000000
RP0             00000005
STATUS          00000003
WR              00000001
WREN            00000002
Z               00000002
__16C84         00000001
beep            00000019
butA            00000004
butB            00000000
buzO            00000000
check           0000002F
checom          00000057
codtab          00000004
cycle           0000001D
delay           00000015
done            00000074
getdil          0000003F
getpot          00000027
loop            00000016
potin           0000005C
select          0000006D
siren           00000072
start           00000000
stocom          0000003A
store           00000046
waita           0000004E
waitb           00000066

MEMORY USAGE MAP ('X' = Used, '-' = Unused)
0000 : XXXXXXXXXXXXXXXX XXXXXXXXXXXXXXXX XXXXXXXXXXXXXXXX XXXXXXXXXXXX
0040 : XXXXXXXXXXXXXXXX XXXXXXXXXXXXXXXX XXXXXXXXXXXXXXXX XXXXXX -----

All other memory blocks unused.
```

Index